CW00351869

The Geographies of Canada

P.I.E. Peter Lang

Bruxelles · Bern · Berlin · Frankfurt am Main · New York · Oxford · Wien

Canadian Studies

The series Canadian Studies examines the many facets of Canadian reality from a multidisciplinary perspective. Contributions from both the humanities and the social sciences are invited. The editor welcomes manuscripts whose primary object is "Canada" in the widest possible sense of the term. The series therefore covers a variety of fields such as literature, history, sociology, politics, economics, geography, law, media, museology, etc. as well as comparative studies.

One of the most innovative features of this series is its focus on the latest research conducted outside Canada. It therefore illuminates various aspects of the country in a new and significant manner and encourages a constant and innovative dialogue between Canadian scholars and the community of Canadian Studies specialists worldwide.

***Series Editor*: Serge Jaumain**
Centre d'études canadiennes
Université Libre de Bruxelles (Belgium)

Rémy TREMBLAY & Hugues CHICOINE (eds.)

The Geographies of Canada

"Canadian Studies"
No.24

1 avenue Maurice, B-1050 Bruxelles, Belgique
info@peterlang.com ; www.peterlang.com

ISSN 1781-3867
ISBN 978-2-87574-017-5
D/2013/5678/04

Printed in Germany

Library of Congress Cataloging-in-Publication Data
The geographies of Canada / Rémy Tremblay & Hughes Chicoine (eds.). pages cm. -- ("Canadian studies" series ; no. 24) Includes bibliographical references and index. ISBN 978-2-87574-017-5 1. Canada--Geography--Textbooks. 2. Regionalism--Canada--Textbooks. 3. Human geography--Canada--Textbooks. I. Tremblay, Rémy. II. Chicoine, Hughes.
F1011.3.G36 2013 917.1--dc23 2012046641

CIP available from the British Library, GB and the Library of Congress, USA.

Bibliographic information published by "Die Deutsche Nationalbibliothek"
"Die Deutsche Nationalbibliothek" lists this publication in the "Deutsche National-bibliografie"; detailed bibliographic data is available in the Internet at http://dnb.d-nb.de.

Contents

PART II
REGIONAL PERSPECTIVES

Introduction

The present book outlines the territorial evolution of Canada as both a political and cultural space where 83% of the population lives on small patches of the territory. The book offers a wealth of information on an array of particulars where differences are noticeable throughout the territory, from distribution of immigrants to mother tongues, education level, regional mobility, etc. Readers are invited to discover how Canada's population is aging and how waves of immigration continue to contribute to growth. Territorial awareness appears to be a necessary background when studying Canada. Side by side in the seemingly unending description of such a vast country, one will need to integrate true statements such as: fish was the first major export commodity from the present Canada territory, and public sector employment varies greatly between the provinces and territories. Concerning the Canadian economy it may be argued there are as many economies as there are provinces and territories, yet financial differences between provinces and territories are not larger than they are, in part due to federal government transfer payments to the provinces, a long tradition in Canada (Hecht). It was not until 1949 that the final piece of the puzzle, Newfoundland, was added to Canada's political territory, yet the foundation of the new country was achieved in less than a century, creating a territorial foundation which amounted to six times the size of the original colonies (Nicol). Strategically important territory that extends over lands and waters almost to the North Pole is a defining element of Canadian territory and sovereignty in the Arctic. These vast lands constitute 40% of Canadian territory; however, they concentrate only 0.3% of Canada's total population. Most of these areas belong to the Arctic biome (Laidler and Petrov). Further, right up to the eve of BC's joining the Canadian federation in 1871 and then beyond, Native peoples (particularly those on the coast and in the south and central portion of the colony) were marginalized within their own territory, set up on tiny reserves laid out by representatives of settler society, and generally considered to be inferior members of a newly emerging society that was based upon British social, political, economic, and cultural systems (Rossiter). In sum, the geology, soils, forests, water bodies and movement, and climate vary greatly over this vast territory (Cameron). This book also offers an analysis of the occupation of the territory: geographical and historical distribution of the urban population on one hand, demographic portrait, including territorial inoccupation (Maret, Lewis and Breux).

For our purposes, as editors and authors of a textbook on Canada, the book is a statement of considerable interest. It is a tacit admission that geography and distance matter and have always mattered. In this introduction we suggest that geography and distance matter in three specific ways. First, geography and distance impact everyday life in Canada and contribute to the character of the Canadian identity. Secondly, geography and distance have influenced, or perhaps determined (depending on who you speak to) the territorial evolution, politics and culture of the country our contributors write about in the chapters to come. Finally, geography and distance matter because the where, how and why of the distribution of people and activities across Canada are important topics of study in and of themselves. In other words, the study of the geography of Canada is a legitimate pursuit for its own sake.

This introduction is divided into three sections. In the first section we briefly outline how and why distance and geography have influenced the territorial evolution of 'Canada'. The political, social and cultural development of Canada as an independent country and the contemporary reality of Canada's internal and international politics then follow. The reason for beginning in this way is to provide some background to the descriptions of Canada's environment, cities, people and economy provided in later chapters. History and geography are intimately intertwined and so, while this textbook provides an introduction to the *geographies* of Canada, in many ways it also provides an introduction to the *history* of Canada. This introduction is designed to build upon and reinforce this relationship. We conclude this section with an examination of regionalism or regionally based politics, one of the most powerful set of forces influencing Canada's development and Canadian politics both now and in the past. The second section details the rationale for and possible uses of this book. We conclude with a final section describing the structure of this book.

Geography and distance matter in a number of ways to Canadians and Canada. Canadians identify themselves as Northerns. This suggests that just as Canada's identity has been shaped by past exploration and conquest of the northern and arctic regions, the country's future also depends on the continued exploitation of resources found in these regions, and now as described by Andrey Petrov and Gita J. Laidler later in this volume, Canada's ability to preserve and protect her northern borders. Likewise, Canadian scholars' increased interest in and adaptation of American historian Frederick Jackson Turner's frontier thesis during the 1920s and 1930s helped historians establish a separate Canadian history and identity by articulating ideas of unexplored northern (and western) lands.

In popular culture the fascination with geography, distance and the north are captured in familiar ideas about the 'wilderness', the 'frontier', the 'frozen north' and even 'cottage country' (a rural yet civilised area where city dwellers escape during the summer to enjoy the simple pleasures of 'the wilderness') shared by many Canadians. This fascination with geography, distance and the north is also reflected in Canadian popular literature. Best-selling works by Ian and Sally Wilson about their journey by canoe and dog team across the Northwest Territories to Hudson Bay (1992), their trek to the Klondike (1997) and finally, their 2000 trip replicating the route taken by Canada's voyageurs along the Winnipeg and Saskatchewan rivers from Grand Portage to Cumberland House exemplify the types of stories that continually reaffirm Canadians close connection to the wilderness.

More pertinently in the present context, this book suggests that geography and distance have had important (yet perhaps under-appreciated) impacts on Canada's discovery, territorial evolution, history and politics. A review of how the territory that became Canada was discovered might lead to the conclusion that Canada was discovered as Europeans searched for somewhere else, the fabled Northwest Passage and the treasures of the Orient that lay beyond. 'Geography', more precisely Canada's distance from Europe, inhospitable climate and barren landscape meant that the European nations, predominately England and France, who sought dominance over northern North America struggled to find any reason to remain until the late 1500s. Canada was simply too cold and too empty!

Canada's resources and the distribution of those resources also shaped exploration. The French and English explored the interior of what is now Ontario and Québec in search of resources that would make them rich. Instead the English and French found nothing much of value except the poor hapless beaver. While distance from Europe protected Canada from the worst ravages of many of the wars fought in Europe during the 1600s and 1700s; the fur trade in beaver pelts intensified and changed the nature of the rivalry between England and France. From the late 1600s the English and French fought each other not only in Europe but in the lands around Hudson Bay for control of the fur trade. Conflict over the fur trade quickly escalated into all-out war (King William's war). That Canada-based conflict then mingled with European conflicts, specifically the War of the League of Augsburg (1688-1697). The eventual French victory in King William's War was an isolated but important one. The French victory meant that the settlement patterns in the early 1700s endure to this day in what are modern Ontario and

Québec; the French are concentrated in Québec and the English are not. Equally important, the failure of the British to control the fur trade forced them, under the auspices of first the Northwest Company and then the Hudson Bay Company to explore further west toward the Rockies and into what is now Manitoba, Alberta and Saskatchewan.

France's control of northern North America was short-lived. Subsequent defeats at the hands of British in 'Canada' resulted in the transfer of a vast territory from France to England in 1763. New France was absorbed into the British Empire and the economic and social exchanges that would shape Canada until World War II were established. Geographic distance, Canada's climate and the types of resources available all helped shape the new 'country'. As Alfred Hecht in Chapter 4 and Heather Nicol in Chapter 5 show in their discussions of Staples Theory the dependent relationship of colony and mother country established and fostered the development of resource-based economy that still exists to some degree today.

The cession of New France to Britain created fundamental social, cultural and political divisions which generated not only a seemingly never ending stream of problems for the territory's administrators but impacted the foundation and constitutional framework of Canada, the independent country. The administration of British North America during the late 1700s and 1800s, whether in the form of Upper and Lower Canada or Ontario and Québec was based on a simple geographic reality: the concentration of a Roman Catholic French-speaking population in New France and the concentration of a much larger English-speaking protestant population elsewhere. For the future country of Canada the first consequence of this geographic reality was the Québec Act (1774). The Québec Act codified Québec's difference from the rest of the country for the first time, and thus established the French – English dualism (the two solitudes) that would be enshrined in the Constitutional Act (1791), the Act of Union (1840) and the British North America Act. The ambiguity concerning Québec's boundaries produced by the changes made to the Proclamation of 1763 by the Québec Act would become important in the 1960s and 1970s as debate raged about Québec separation and land claims to ensue.

Distance and the vastness of British North America also impacted Confederation, or the creation of the Dominion of Canada. One of the most pressing questions facing Sir John A. MacDonald (Canada's first Prime Minister) was how to overcome the vast distances that separated the various component provinces and build a country. Railway construction in general and specifically the construction of a trans-Canada or Canada-Pacific railway were viewed by Prime Minister MacDonald and others as the practical means by which the new country of Canada would be built. The entry of Prince Edward Island and British Columbia into Canada was facilitated in one fashion or another by the conquest of distance through the construction of railways. In the case of British Columbia it was the promise of a trans-Canada railway within ten years, whereas Prince Edward Island was lured into Confederation by the promise of help in financing its already existing but beleaguered railways.

As we have briefly discussed geography and distance have directly influenced the development of Canada. Geography and distance have also influenced the development of Canada through the creation and existence of regions. For our purposes a 'region' is both a classificatory concept that delineates an area based on shared characteristics (physical, human and/or functional) and a political, social and cultural entity formed and reformed. For non-geographers regions and the related ideas of regionalism and regional identity are comparatively obscure concepts, the preserve of academics and perhaps politicians at election time. Regionalism and regional identity are however potent forces in Canada and have at certain times in the past influenced both the physical shape of the country and the balance of political power (see Nicol, Chapter 5 in this volume).

It was suggested that regionalism and the emergence of regional identities in Canada are a result of different patterns of exploration and settlement compounded by dissimilar patterns of economic development. Regionalism and provincialism grew during World War I and the economic boom of the 1920s. While Ontario and Québec prospered anger over the effects of high tariffs (and continuing debates about abolishing payments made to railways to offset the cost of shipping Prairie grain east, or 'Crow Benefits') among Prairie farmers and dissatisfaction over freight rates in the Maritimes helped solidify a growing sense of self-awareness and shared destiny in each of those regions. British Columbia also began to grumble. The bone of contention was again the economic costs of isolation or the costs of overcoming the distance between British Columbia and Canada's heartland of Ontario and Québec. British Columbia bemoaned the negative impacts of high tariffs on manufactured goods produced in the heartland, inadequate railway freight subsidies and the costs associated with the construction of the Canadian Pacific Railway.

By World War I (1914-1918) a number of overlapping and sometimes contradictory regions and regional identities were identifiable in Canada including 'the Maritimes' (Nova Scotia, Prince Edward Island and New Brunswick), 'the Prairies' (Alberta, Saskatchewan and Manitoba), British Columbia, 'the old provinces of Canada' (Ontario and Québec) and English-speaking vs. French-speaking Canada. By the

Great Depression (1929) provincialism and regionalism had produced successful political organisations including the Social Credit Party (SCP) in Alberta, the Co-operative Commonwealth Federation (CCF) in western Canada more generally and in Québec the *Union Nationale*, a party that promoted the idea of French-Canadian unity. The bases of each party's political power differed slightly, and for the CCF it was the impact of the Great Depression on wheat prices; but in essence each felt that other regions received special treatment at their expense.

In more recent times regional alliances and the regional bases of political power in Canada have solidified into the successful but short-lived Reform Party (western Canada) and the very successful Bloc Québécois and Parti Québécois in Québec. Squabbling and bickering between the provinces and territories and the remaining regionally based political parties continues. Championed in the recent past by the now defunct Reform Party, Alberta complains about the transfer of wealth from its oil enriched coffers to just about every other province and territory. Just about every region and when politically acceptable, every political party except the Bloc Québécois, complains about the transfer of resources to resource rich Québec. On the periphery, both literally and figuratively, 'the north' is ignored and therefore spared perhaps the worst of the bickering. Since John Diefenbaker made northern development a national priority during the 1950s and early 1960s federal resources have flowed northward. Development was and still is seen as an economic necessity and a matter of national security. So profound has been the impact of regionalism on Canada that Canada may be described as a string of regionalized societies that differ significantly in ethnicity, political culture and economic orientation.

Before ending our discussion of how distance and geography have influenced Canada we have to consider Canada's location relative to the United States. For Canada and Canadians there is no and there has never been any escaping the effects of sharing a border with the United States. The highly unbalanced distribution of power between the United States and Canada means that while Canadians worry about the effects on Canadian culture, political life and international relations, Americans do not even think about Canada. Sir John A. McDonald worried about the influence of the United States on Canada and Canadians as early as Confederation. The level of concern increased during the 1920s and 1930s as radio broadcasts from the US reached increasing numbers of Canadians. Fears about the power of US media and recognition of the need to promote the formation of a Canadian identity contributed to the passage of the 1958 Broadcasting Act and the establishment of the first Canadian cultural content quotas.

The influence of the United States on Canada is nowhere more keenly felt than in the international arena. The American Revolution determined first that the United States would not be British and secondly, that Canada would empathically be British with a parliamentary system modelled on that of Great Britain and a constitutional monarch. One of the most immediate consequences of the American Revolution was the northward movement of Loyalists. The arrival of Loyalists in Atlantic Canada, Québec and Ontario greatly affected the settlement and political structure of these regions. Later, the American Civil War served as an impetus for Confederation. Increasingly, Canada's proximity to the United States is seen as problematic. The United States feels and expects Canada's compliance with US foreign policy initiatives. Such cooperation began innocently enough with the establishment of Permanent Joint Board on Defence during World War II. At the start of the Cold War Canada somewhat unhappily found that participation in new defence initiatives, specifically the North American Air Defence Command (1958) and the Distant Early Warning Line (DEW Line) (see Petrov and Laidler, Chapter 11, and Nicol, Chapter 5, in this volume), was compulsory rather than voluntary. With each extension of the idea of joint or continental defence Canada has lost a little more control over its sovereignty. The instigation of new border control initiatives after September 11th has, in the opinion of some, further enhanced Canada's subservience to the United States and sped up the loss of Canadian sovereignty.

Purpose and Structure

For many geographers, particularly those educated in or from Canada a job in an academic institution in the United States teaching Geography inevitably leads to one specific assignment: teaching a course on North America. Such courses are listed in university course catalogues across the United States under various names including but not limited to: The Regional Geography of North America, The Regional Geography of the United States and Canada and The Geography of North America. For many students such courses are their first introduction to both geography as an academic discipline and the geography of the continent and country they call home. For a few students these courses will ignite a hitherto unknown interest in geography and some will go on to become geography majors. For most however their brief encounter with geography will never be anything more than a necessary evil; a way of fulfilling a graduation requirement.

Almost without exception the instruction of regional courses on North America generates considerable frustration. Much of the frustration generated comes from the fact that substantial numbers of American students have to be convinced that Canada actually exists as a separate sovereign nation, that North America includes Canada and having realized that Canada actually exists, that it is not simply a land of snow, ice, polar bears and penguins! The level of frustration is heightened by the coverage most standard textbooks on North America provide about Canada. Replete with colour photographs and flowery prose these texts are written to appeal to undergraduate students enrolled in American academic institutions. The vast majority of the maps, images, data and examples discussed describe the United States. Despite the sophistication of the data, diagrams, maps and images, discussions of Canada are superficial and all too often clichéd. In these other texts Canada remains of only secondary importance; in Michael Hart's words "a source of cold weather ... not much else" (1998).

For instructors, the superficial treatment of Canada provided poses a serious, often time-consuming problem: how to provide balanced coverage of Canada in their courses. The easy solution is to require students to purchase two texts, one on North America and second on Canada. While undergraduate students typically complain about the cost of textbooks a few minutes spent researching prices of textbooks on Canada might lead one to conclude that students' complaints are not without merit.

For instructors wishing to include additional information about Canada in their regional courses, the alternative solution to requiring the purchase of two textbooks is to look for information and data form other sources. The Internet has certainly made this task easier and less arduous, but the task can still be daunting. Moreover, the problem of finding accurate reliable information remains.

At a very fundamental level this text is an attempt to address both these problems. This book provides instructors teaching regional geography courses on North America a comprehensive source of accurate up-to-date information about Canada. We also hope that this volume can be used to supplement and complement the content of regional textbooks on North America.

The organization of this text is guided by two principles: the structure of published introductory regional texts on North America and the need to include information that complements that found in these introductory texts while also touching upon interesting contemporary geographical topics and issues (for example Chris Mayda's discussion of various environmental issues and renewable energy in Chapter 2).

Part I of the book contains chapters that focus on specific topics. These topical discussions reflect the contemporary structure of academic geography and the practice of what is often described as *systematic geography*. Systematic, or occasionally 'scientific' geography, is concerned with the formulation of general principles that can be used to describe the physical (physical geography, Chapter 1) and human (human geography) worlds and increasingly, the interaction between the two. Both physical and human geography are divided into specialized fields of study. Human geography for example includes population geography (Chapter 3), economic geography (Chapter 4) and urban geography (Chapter 6).

Part I of this text presents a systematic geography of Canada with chapters covering physical geography (Meserve), rural Canadian renewable energy – the environment (Mayda), political (Nicol), economic (Hecht), population (Bélanger) and urban (Bélanger and Jébrak). Chapter I by Peter Meserve (Fresno City College) on physical geography sets the stage for later discussions of political geography, Canada's economy, population and environmental issues presented by Heather Nicol (Trent University), Al Hecht (Wilfrid Laurier University), Hélène Bélanger and Yona Jébrak (Université du Québec à Montréal) and Chris Mayda (Eastern Michigan University). In many ways the order of the chapters in Part I is conventional. However, in the case of this volume the order has additional meaning. As we alluded to in a previous section of this introduction, distance, the terrain and the climate influenced Canada's evolution as a country and facilitated the existence of distinct often politically powerful regions. The character of Canada's economy, its reliance on the extraction and processing of primary or staple products is a product of geology and northerly location. The emergence of a manufacturing sector is at least in some part a product of Canada's proximity to the vast markets of the United States. The distribution of Canada's population and regional patterns of population growth and decline are historically products of the environment. Canadians and the urban areas they inhabit cluster along the US – Canada border because of the cold and the terrain. The chapter on rural Canada and renewable energy picks up on a number of ideas mentioned in other chapters including the ever-changing nature of Canada's economy. As well as being 'hewers of wood and carriers of water' Canadians are also producers of energy; most famously petroleum from the Alberta tar sands and hydroelectricity in Québec. And yet, more than 80% of the Canadian population lives in cities and towns (Chapter 6). Chris Mayda (Chapter 2) describes how Canada and many rural communities are implementing 'green' energy initiatives in light of global warming. While fear of the consequences of global warming is driving much of the interest in green energy, potential economic gain (increase in the number of green jobs) is also seductive. As a result, many Canadian communities and businesses are implementing environmentally friendly initiatives.

Part II of this book follows a different approach. The chapters in Part II focus on regions within Canada. Within geography the study of regions is usually termed regional geography. As a way of studying geography, regional geography dates from the mid-1800s. In its simplest form regional geography involved listing and learning about a particular part of the world. Over time, more sophisticated approaches developed including Richard Hartshorne's areal differentiation. Some geographers argue that regions are not just physical but social and cultural entities. As such regions are variable and great care must be taken to justify their identification and study. After a while, regional geography was criticised for its lack of theory and emphasis on description. From the 1960s on, as new cohorts of geographers were trained and 'social scientists' in general adopted more scientific approaches to the study of social phenomena, regional geography declined in importance and became consigned to geography's past.

Whether or not academic geographers no longer practise regional geography, regional geography courses remain popular in many American universities and regional geography remains for many students the only type of geography they ever study. For this reason alone a plethora of geography books describing the individual regions of the world have been and continue to be published. As stated previously, this volume is designed to complement a subset of these textbooks, regional textbooks on North America.

The contents of Part II are arranged so that readers can roam at will from east to west, or begin in *Atlantic Canada* (Chapter 7) and end with Chapter 11 on *The North: Balancing Tradition and Change*, by Gita J. Laidler (Carleton University) and Andrey Petrov (University of Northern Iowa). With the exception Chapter 10 by Douglas C. Munski (Memorial University) on *The Prairie*, all the regionally focused chapters begin by discussing some aspects of the physical geography of the subject region. Subsequent sections typically focus on history (often settlement history), urban concepts, population composition and change and finally the economy. Douglas C. Munski's Prairie chapter takes a slightly different approach by adding into the mix sections on various themes, for example the transformation of agriculture on the Prairies and the potential for energy production presented by the natural resource endowment of the region.

While the content of each chapter varies, it is possible to identify several recurring themes. The role Canada's harsh climate and unforgiving terrain is highlighted in Chapter 7 on Atlantic Canada by Vodden and Catto and again in Cameron's Chapter 9 on Ontario, in Chapter 8 on Québec and of course in Chapter 11 on the North. In all these chapters the environmental conditions found in Canada are universally conceptualised as impeding development. However, the environmental conditions in Canada also encouraged and fostered economic development through the exploitation of fossil fuels (see Chapter 2) and the harvesting of timber and fish (Chapter 12 on British Columbia and Chapter 7 on Atlantic Canada). The struggles to first explore Canada and then the unite the distant parts of British North America through Confederation and the creation of a national economy are discussed at some length by David Rossiter (Western Washington University) in Chapter 12 on British Columbia and the chapters on the Prairies – 10 (Douglas Munski, University of North Dakota) and Atlantic Canada – 7 by Kelly Vodden and Norm Catto (Memorial University). Uneven development is another important theme. All the chapters in Part II describe the patterns of uneven settlement, economic activity, population growth and urbanisation apparent in each region of Canada. In Chapter 8 on Québec Isabelle Maret, Paul Lewis, and Sandra Breux (Université de Montréal) go one step further when the authors argue uneven development and regionalism are fundamental to understanding the modern province of Québec. Finally, the changing status of Canada's First Nation peoples is discussed in several chapters. For many, debates about the standing of First Nation's peoples and Metis in Canada are recent in time and crystallise around the formation of Nunavut (see Chapter 11, Laidler and Petrov). Other chapters on Atlantic Canada, British Columbia and the Prairies remind us this is not the case and that these debates date back to the earliest days of English and French exploration, Confederation and of course, the Riel rebellions.

PART I

CONCEPTS IN CANADIAN GEOGRAPHY

CHAPTER 1

Canada

Physical Geography, Landscapes and Lithology

Peter MESERVE

Fresno City College

Canada's land is the product of approximately 4.5 billion years of geological activity. The largest surface features – such as the Canadian Rockies – are evidence of tectonic processes involving massive lithospheric plates moving slowly around on Earth's surface. (One exception to this rule is Hudson Bay.) Smaller scale landforms like Ontario's "Ten Thousand Lakes" (read the license plates) are the products of erosional and depositional activities that until recently were dominated by the glaciers of the Pleistocene (Late Cenozoic) Ice Age. Underlying all these features are some of the oldest – as well as newest – rocks on the Earth; not only were 4.28 billion year-old greenstones found near Hudson Bay in 2008, but in 2010 strong evidence was found of 4.5 billion year-old mantle underlying Baffin Island. Canada's bedrocks – and the fossils and mineral deposits they contain – are both an historical record and the key to the location of the resources that have formed modern Canada.

The simplest way to divide Canada into geologic regions is to distinguish between mountainous areas and those with more level landscapes, but subdividing these allows more useful regions to be determined. Canada is immense – the second largest country on earth – and while every attempt to establish its landform regions can be disputed, the following divisions can be defended on the basis of their locations, lithology (bedrock), and geologic history.

There are three mountainous regions formed by different tectonic (mountain-building) sequences, located on different boundaries, and eroded over different time periods. In the western provinces lie the Canadian Rockies and the rest of the *Canadian Cordillera*, the *Arctic Cordillera* covers the far northern islands, and in the east are the *Appa-*

lachians. Both the Cordillera and the Appalachians are extensions of continental mountainous regions and have much in common with their American extensions.

Of the many possible geologic divisions of the lower-lying parts of Canada, the largest and most important region – actually the geologic core of the entire North American Continent – is the *Canadian Shield* surrounding (and underlying) Hudson Bay. On the western edge of the Shield – and bordering the western mountains – are the *Interior Plains* (also called the *Interior Platform*) which is the Canadian portion of the Great Plains. To the southeast of the Shield, alongside the Appalachians, are *Lowlands* adjacent to the Great Lakes (especially Lakes Huron, Ontario, and Erie) and to the St. Lawrence River. Finally in eastern Québec, abutting the St. Lawrence Lowlands, is the *Grenville Region*, which has over millions of years been eroded from a mountainous area to form an irregular but far more level region.

Mountains

Canada's mountains, like all major mountains on earth, are the result of tectonic convergence. Over the past millions (and billions) of years Canada, as part of what is now the North American Plate, collided with other plates as they converged at rates exceeding two inches per year. These extremely slow and protracted collisions took millions of years to complete; the collision on Canada's Pacific coast still continues. Through the compression and folding of the crust (often composed of sedimentary rocks from offshore areas) as well as by igneous activity, large areas were elevated in mountain-building events called orogens. From their initial uplifts, however, mountains have been subject to the sculpting and lowering influences of rivers, glaciers, and other processes creating occasionally spectacular landforms.

1. Western Cordillera

The tallest, youngest, and most impressive mountainous region in Canada, the Western Cordillera, parallels the Pacific coast and forms most of British Columbia and Yukon Territory. After the supercontinent Pangaea began to split apart about 250 million years ago, what is now North America moved westward, colliding with the smaller tectonic plates in the way. Several major mountain-building events (e.g., the Laramide and Sevier orogenies) contributed to the build-up of the Cordillera, and at times islands were sutured onto the coast creating unique geologic areas called terranes; Wrangellia is sometimes used to refer to all these added terranes.

Like any feature this large, the Cordillera is geologically complex and is formed of many smaller ranges and regions. A series of ranges exists starting in the Yukon Territory with the St. Elias Range, which is still being uplifted (slowly) and which contains Mt. Logan (5,959 metres tall) in Kluane National Park.

Figure 1. Major Rock Categories: Sedimentary, Sedimentary and Volcanic, Volcanic, Metamorphic, Intrusive

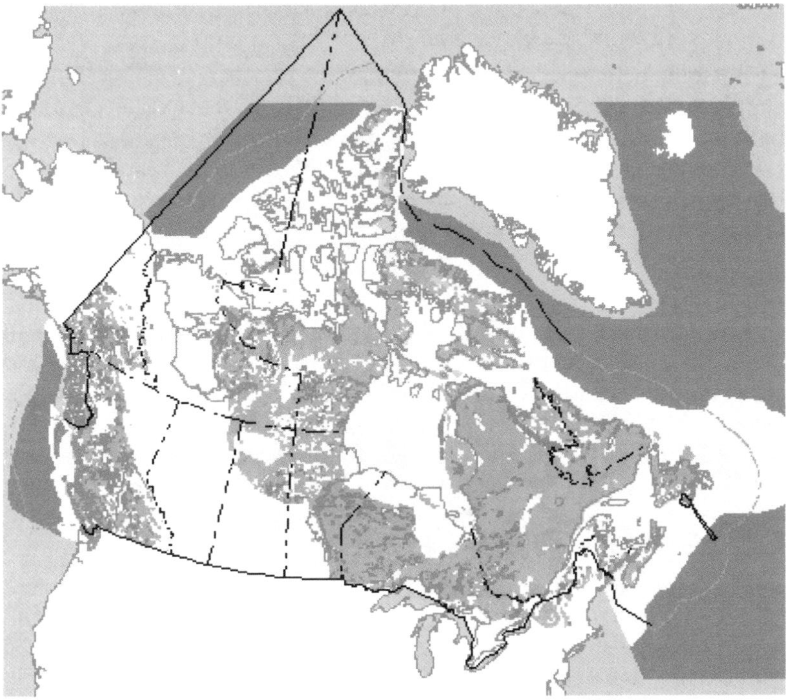

Source: Geological Map of Canada – Map D1860A [CD-ROM], 1997[1].

Extending southward along the coast almost 1,600 kilometres to the United States, lie the Coast Mountains and the Insular Mountains, noted for their spectacular fiords and glacial landforms. East of these ranges in central British Columbia is the rugged Interior Plateau – which includes the Okanagan and Kootenay Lakes – and even farther east, separated from the plateau by the Rocky Mountain trench, lie the Canadian Rockies, home of Banff National Park.

1 From http://atlas.nrcan.gc.ca/ and: Canada. Geological Survey of Canada. 1981. Geology and Canada. Adapted from Prospecting in Canada 4th edition by A.H. Lang, published in 1970.

The rocks of the Western Cordillera reflect the different processes that formed them and therefore appear quite fragmented and scattered, yet some are the sources of valuable resources. Sedimentary rocks from ancient seabeds and coastlines uplifted to form some of the Canadian Rockies. Not only do they contain tremendous coal deposits – especially in the East Kootenay region – but in Yoho National Park the Burgess Shale contains some of the world's most amazing fossils. Igneous rocks emplaced near Bonanza Creek, Yukon Territory; Cariboo Plateau, British Columbia; and Yale (Fort Yale), British Columbia contained enough gold to initiate major gold rushes.

Tectonic activity continues in this region and is responsible for several natural hazards. Active fault-lines, such as the Queen Charlotte fault, lie off the coast of British Columbia and remain capable of generating severe earthquakes. The Garibaldi volcanic belt in south central British Columbia contains numerous (technically) active volcanoes, though threat levels are very low.

2. Appalachians

The older and consequently far more eroded Appalachian mountain range along the east coast of Canada extends southward to Georgia and segments can even be found in Scotland and Norway. All these ranges were the result of several large and smaller tectonic plates converging to form Pangaea. Overall, the mountain forming-process is referred to as the Caledonian Orogeny, but one part of this convergence, the Acadian orogeny, occurred around 350 million years ago and formed most of Maritime Canada and the adjoining parts of Québec.

As in the west, different segments of the Appalachians have different names and geology; the many bays and gulfs in this region separate and divide many of the ranges. In the northwest of New Brunswick lie the Miramichi Highlands which cross the Gulf of St. Lawrence and in Québec become the Chic-Chocs. In Nova Scotia uplifted sedimentary rocks formed the Cobequid Hills which contained some of the most important coal deposits in the east (and where the Springhill mining disasters occurred). The northern end of the Appalachians is in Newfoundland where the Long Range Mountains and Avalon Peninsula, the namesake of the mini-continent which was squeezed into the Appalachians, are found.

The waters of the Maritimes are geologically significant in their own right, not least because of the tremendous tidal ranges. The Bay of Fundy competes for the world title (with Ungava Bay) for the greatest diurnal range, nearly 17 metres; in addition, the tides create Reversing Falls on New Brunswick's St. John River. Nearby on Dear Island is the "Old Sow", perhaps the world's largest whirlpool.

3. Arctic Cordillera

The northern-most islands of Canada's Arctic archipelago, in Nunavut and the Northwest Territories, were formed as part of the Innuitian orogeny around 200 million years ago. Several smaller ranges comprise this region, with the surprisingly named United States Range farthest north on Ellesmere Island. Although the climate has limited exploration of this region, coal and metal resources have been discovered.

The Shield and Surroundings

Lower-lying areas of Canada are the result of either extreme weathering and erosion of (former) mountain ranges or of the deposition of sedimentary rocks (often on areas of extreme erosion). There are still hills, valleys, and even cliffs in the non-mountainous regions of Canada but most of the landforms are relatively low-lying. While the (geologically) recent Ice Age is responsible for much erosion and for the thousands of lakes, the Canadian Shield has been eroding for literally billions of years in some areas.

4. Canadian Shield

The Canadian Shield is the largest exposure of ancient (Precambrian) rock on earth; it covers roughly half of Canada (while underlying another quarter) and contains rocks more than 4 billion years old. During those billions of years the shield would have grown by convergence, the associated mountain building, and by the suturing of exotic terranes (much like what happened to the Western Cordillera). Incredible amounts of overlying rock were later removed which in turn exposed rocks and mineral deposits dating back nearly to Earth's origin. As with other large geologic regions, subdivisions abound in the Shield; Natural Resources Canada divides the Shield into seven provinces, each differentiated by structure.

Within the Shield igneous and metamorphic rocks dominate since they formed at great depths before being uncovered. One result of this is that there are very few fossil fuel resources from this area, but an incredible amount and diversity of other mineral resources. Nickel and copper are found near Sudbury, Ontario; diamonds and gold near Yellowknife, NWT; uranium near Tulite, NWY; and copper and zinc near Flin Flon on the Manitoba/Saskatchewan border.

Near the centre of the Shield lies Hudson Bay, the centre of Canadian drainage and the heart of historic Rupert's Land. During much of the 2.5 million years of the Pleistocene Ice Age, Hudson Bay was buried by perhaps three kilometres of ice. As the glacial ice flowed outwards it shaped the surface not only of central Canada but also of the central

United States down to Kansas and Missouri; the hydrology of much of Canada is still the legacy of the Ice Age. The excessive weight of the ice is the primary reason that Hudson Bay exists; the crust sank isostatically in this area (just as a canoe sinks down when weighted down) and is very slowly rising in response to the melting of the glaciers.

The abundance of lakes in Ontario, Manitoba, and Québec is the direct result of Ice Age erosion. Hundreds of thousands of lakes, many relatively small and connected by streams, were eroded out of the bedrock while the Great Lakes were deepened considerably by the glaciers.

Review question

- Describe the general spatial patterns in Canada of the major rock categories. Relate these patterns to those of geologic history and resources.

CHAPTER 2

Rural Canadian Renewable Energy

Chris MAYDA

Eastern Michigan University

Introduction

The twenty-first century is shaping up to be the renewable energy century, much as the twentieth century was all about fossil fuels. Every developed or developing country in the world has embraced some form of leaving fossil fuel. Even oil soaked Abu Dhabi is building Masdar, a green city based on solar power.

Behind the twenty-first century push for renewable energy sources are elevated greenhouse gas (GHG) emissions that have caused climate change. The increase in GHG is linked to human use of fossil fuels as the prime energy source. In an effort to ease elevated GHG emissions attention has turned to other sources of energy and the more efficient use of non-renewable energy sources.

Canada has been especially affected by climate change, and in fact considered a bellwether developed country because of its northern clime, which along with its high standard of living leads Canadians to use more per capita energy than any other country: 18% more than the United States and more than twice the European or Japanese standard. Climate change has made many Canadians aware of and working to reduce their carbon footprint[1]. Though behind European efforts to find alternative sources of energy, many Canadian provinces have been focusing on alternative energy sources to replace fossil fuels and to reduce GHG emissions.

However developing alternative energy sources does not address a major foundation of Canada's fossil fuel dependent economy, resource extraction. Canada is heavily invested in mining enterprises in the far

---

1 Allcut 1945.

north as well as the oil sands of Alberta, the source of about one-quarter of all US oil. While Canada is often perceived as aware and actively addressing its GHG emissions, the oil sands are a major glitch in reducing its carbon footprint. And there are more challenges.

Canada is transmitting and storing the new energy, however a major impediment is the limitations of the current grid. The current infrastructure and transmission grids are aging and in need of refurbishing and should include transmitting alternative sources of energy. Solar and wind power locations are usually not in the urban areas where most power is needed, but instead in rural areas. Transmission over large areas is problematic. Most renewable energy sites produce relatively small amounts of energy when compared to coal burning or nuclear plants. Gathering the energy and then transmitting it over large distances make the cost of electrical transmission prohibitive.

Canada has implemented programs to begin addressing the infrastructure problems. For example, the $2.05 billion Infrastructure Canada Program (ICP), established in 2000 has as its first priority to improve the green municipal infrastructure, including renewable energy services. Though alternative energy is only a small part of the overall program, it is a start and hopefully as the industry grows more attention and money will increase the ability to spread alternative energy across Canada.

Since the 1990s the provinces favoring alternative energy sources--Ontario, Québec, Prince Edward Island, Alberta and British Columbia--have adopted progressive programs in wind, solar, and biomass, most of which are active in the rural areas.

This chapter will discuss the sustainable and alternative energy efforts of three rural Canadian areas: The Okanagan Valley in British Columbia, Cowley Ridge – Pincher Creek in Alberta, and Prince Edward Island (PEI). Each location has unique attributes. The Okanagan Valley is the sunniest part of Canada and has utilized this asset by investing in solar power. The area outside of the Crowsnest Pass of Southern Alberta is the windiest in Canada, and by the early 1990s the region was already investing in wind farm installations. PEI now called the "Green province" has addressed its lack of traditional energy sources and become a leader in biomass and wind energy.

The first section of this chapter will briefly introduce Canadian energy, climate change in Canada, and the country's reaction to the Kyoto Protocol and other programs. This will be followed by an overview of Canadian fossil fuel use and alternative energy programs, including a brief introduction to biomass, solar, and wind power. Then each of the case areas and the governmental policies that support them will be discussed.

Canadian Energy

Geography, climate, and a dependence on natural resources for its economy have contributed to the high demand for and energy use in Canada. These factors include Canada being the second largest country in the world, its location in northern latitudes, its widely dispersed 33 million people, and large distances between cities, all of which add up to a high per capita use of energy. Canada has a much different carbon footprint than the United States. While Canada does depend on fossil fuel for its transportation sector, almost two thirds of its electricity is provided by hydropower. Canada is the third highest per capita producer of hydropower in the world[2]. Canada is blessed with abundant fresh water having the third most in the world[3].

Annually, Canadian rivers discharge about 9% of the world's renewable freshwater. Canadian freshwater produces 13% of all hydropower in the world and hydropower produces two-thirds of Canadian electricity. The most important project is the James Bay Project, where Hydro-Québec built power stations on the La Grand River off of Hudson Bay, which generates about half of electrical power consumption in Québec as well as electricity to other parts of Canada and the northeastern United States. As the largest hydropower system in North America, it generates 34,600 MW (2005).

While continuing to generate new hydropower options Hydro-Québec has begun investing in wind generation and plans to raise 3,000 MW of wind power in the province. Hydro Québec is committed to reducing GHG emissions. While hydropower is a long standing alternative energy source in Canada there has been a concerted effort in several provinces to invest in other alternative energy sources, such as wind, solar, biomass, and tidal. While they continue to increase, these sources supply only about 0.5% of total Canadian electricity and remain far below the production in other developed and developing countries (Table 1). Alternative sources of energy are not meant to replace existing energy systems, but to compliment them.

Alternative sources, especially wind, compliment hydropower because each has its optimum time of operation at different times of the year. Hydropower is at its peak during spring through fall, whereas wind reaches its peak during winter. Dams can be shut and store reserves while wind is at its maximum, and then opened when winds die down.

---

[2] The highest electricity production from hydropower is Iceland (24,423.46 kWh), followed by Norway (23,687.105 kWh), and Canada (10,658.414 kWh) (2004).

[3] Brazil is the country with the most renewable fresh water. Canada is third after Russia.

An additional advantage to building a robust renewable sector is to build a new renewable based manufacturing sector in Canada.

Table 1. Cumulative installed wind power 2000-2007

	2000	2001	2002	2003	2004	2005	2006	2007
Germany	764	8,754	11,994	14,609	16,629	18,415	20,622	22,247
Spain	567	3,337	4,825	6,203	8,263	10,027	11,623	15,145
US	469	4,275	4,685	6,372	6,725	9,149	11,575	16,818
India	402	1,456	1,702	2,125	3,000	4,430	6,270	8,000
China	346	402	469	567	764	1,260	2,604	6,050
Denmark	2,300	2,417	2,880	3,110	3,117	3,128	3,136	3,125
Canada	137	198	236	322	444	684	1,460	1,770

Source: World Watch, Canada Wind Energy Association.

The new manufacturing sector is rising in the European Union where natural resources are minimal and renewable energy is essential to remain solvent. While Canada rests on its hydropower abundance, jobs are still disappearing and alternative options and transmissions standards remain constant instead of progressive.

The Canadian administration in 2010 has adopted similar tactics to the 2000-2008 Bush administration in the United States, which acted on fear of change and a lack of understanding of the employment opportunities possible when working with the environment rather than conquering it. The government has only slowly adopted an attractive stimulus package and government subsidies similar to the fossil fuel industry, as is now provided in several EU countries.

On the other hand, the recent flurry of alternative energy investment and subsidies has blurred an honest assessment of what renewable energy can and cannot do. Some renewable energy sources have become popular and trendy. Prices are high and investments may be made based more on subjective reasons than objective facts. Non-renewable resources must be curbed, but trying to maintain an equally consumptive and wasteful standard of living dependent on renewable resources alone may not be possible.

Each of these issues is a large subject unto itself, but need to be understood so that the entire alternative and traditional grid can be seen in perspective. The problematic issues will be discussed in this chapter in relation to the areas discussed.

Climate Change in Canada

The first thing to know about climate and climate change is that climate is not static but ever changing. In the past couple of hundred years climate has been relatively stable. Interestingly, the time period that we have good climate records are the same as the latest stable period. So when scientists began to see average climatic patterns change, they had no hard data to compare it to. All scientists could do was speculate about the complex interactions of variables in a changing climate – plant proliferation, continental collisions, biogeochemical processes – and theorize how they all played a role in the climatic evolution of Earth. These were difficult enough to theorize, but as human impact on Earth increased the equation became more complex still.

Climate has covered a range of conditions over time. Hot or cold periods alternate over the four billion years of Earth's existence. The current interglacial period, beginning about 12,000 years ago, has been the period of complex human development. Human development is tied to the carbon cycle and photosynthesis, which has generally kept Earth at temperatures advantageous for human occupation. In the carbon cycle plants absorb greenhouse gases, such as carbon dioxide, sequester excess carbons in biomass, and release oxygen into the air. The biomass eventually decays, some of it turning into coal, some returning to the atmosphere or soils. The carbon contained in animals also decays and transforms to carbon fossil fuels.

Somewhere between 30,000 and 12,000 years ago the Americas were settled by the indigenous groups who were decimated when Europeans began to settle in the sixteenth century. Following initial French occupation the British dominated the Canadian landmass after 1763.

The British initiated the Industrial Revolution, and became the dominant global power as Canada evolved from a colony to a country. By the mid nineteenth century industry in the developed countries, including Canada, depended increasingly on fossil fuels. Developed countries, including Canada evolved from resource extraction to manufacturing to post industrial economies, but Canada's path was somewhat different by remaining largely dependent on natural resource extraction as a primary source of income. As the easily available resources in temperate areas were mined, the resources farther north, such as in Canada, became the new resource discovery area. Today Canada north is a major mining site for gold, zinc, oil, gas, diamonds and various other precious resources.

While climate change has been a source of worry for a few scientists throughout the twentieth century, by the 1970s many scientists acknowledged that climatic changes were happening[4]. Called global warming, though far more complex than warming alone, Earth's climate was changing and Canada was among the first to register large changes, and

---

4 Bryson, R.A. and Murray, T.J. 1977, *Climates of Hunger*. Madison, Wisconsin: university of Wisconsin Press.

many of the changes were a warming trend. Few people listened or cared about their findings, largely because the developed world was in a period of rapid industrial and consumer expansion. Focused on bigger and better technologies to continually bypass the unintended consequences of development, humanity eased into the twenty-first century largely ignoring the increasingly erratic weather patterns.

Although several greenhouse gases were involved in climate change, there was one, carbon dioxide (CO_2) that was rising very fast and was directly related to fossil fuel consumption. Carbon dioxide along with several other greenhouse gases (which include water vapour, methane, nitrous oxide, ozone, CFCs) trap solar radiation and cause the greenhouse effect that raises global temperatures. Carbon dioxide levels were rising and accelerating, from 280 ppm (parts per million) prior to the Industrial Revolution, to 389.7 ppm in December 2010. Not only were CO_2 levels rising, but they were accelerating from 1 ppm per year in 1950 to 1.8 ppm annually in the 1990s, to 3 ppm from 2007 to 2008. Increased levels of CO_2 were causing the Earth's atmosphere to warm quicker than anticipated by scientists, largely due to a positive feedback loop that amplified the original impact. The more CO_2 in the atmosphere the more it increased warming. In Canada average temperatures have been rising since the 1950s, with southern Canada increased 1.0-2.7 °F (0.5-1.5 °C), and western Canada registering the largest increase, up to 10° F (6 °C). Warming is expected to continue at about 0.4 °F (0.2 °C).

By the time of the Rio Earth Summit in 1992 many leading scientists were convinced of the effects of CO_2 on climate and the United Nations established the Intergovernmental Panel on Climate Change (IPCC) to bring attention to the issue. Some of the effects – deforestation and burning of fossil fuels – are especially applicable to Canada. The development of Canada's natural resources could accelerate worldwide climate change.

Climate change has already affected Canadian ecosystems and agricultural zones, in an unevenly distributed and seemingly haphazard hodgepodge of changes that create havoc for plant communities, insects, animals, and human life. The changes are, of course, not haphazard, but complex relationships are changing within ecosystems that are constantly seeking balance to survive, and results in radical shifts that ill-fit the linear studies of modern science. Linearity worked when the climate was somewhat stabilized, but has little place in a world of rapid change. Climate change will continue to alter a long list of patterns that humans have come to depend on to define their world. The changes include: temperatures, precipitation patterns, extreme weather events, rising sea levels, shoreline erosion, habitat and species losses, melting of permafrost and ice sheets, increased forest fires, disease and infestation outbreaks, and shifting ecosystems. Studies of Canadian landforms, ecosystems and ecological dynamics show that abrupt shifts in ecosystems states may and in some cases have already triggered rapid changes, especially in the north[5].

Among the changes already recorded are natural responses such as insect infestation, shifting forest areas, and human impacts such as the extraction of oil sands. There has been a large increase in pine beetles throughout western Canada over the past 50 years due to changing climatic patterns. Milder winters allow the beetles to survive and destroy more trees the following year. The continental expanse of boreal forest is an important part of the filtering process of plants converting CO_2 to sequestered carbon and oxygen. The hugely successful and profitable Alberta oil sands are releasing CO_2, but also increasing the use of fossil fuels to mine, extract and ship the resulting oil. These processes are increasing Canadian GHG emissions by 40%.

GHG emissions increased 70% since 1970, and will continue to grow unless actions are taken to halt the increases. Since about 1988 most scientists have taken climate change seriously and many countries have begun to address the issue and its effects. Understanding it included a growing awareness of the fragility of Earth systems and how only a few degrees of temperature change could have dire consequences. The rapidity of the changing climatic conditions was recognized and climate change became a *cause célèbre* by the millennium.

The Kyoto Protocol, meant to stabilize GHG emissions, was ratified by 183 countries. Canada ratified in 2002, but by 2006 a newly elected Conservative government backed out of the treaty. The goal of the Kyoto treaty was to reduce emissions to 1990 levels by 2012, but Canada was well above those levels by 2002 and was also concerned because the United States did not ratify the treaty. Canada feared attaining the goals might harm its economy.

Nonetheless, Canada did set a pace of adopting alternative energy sources. Canada adopted its own climate change plan to reduce GHG

---

5 Manon D. Fleury, Dominique Charron, David Waltener-Toews, Abdel Maarouf, *Ecosystem Approaches to Climate Change and Infectious Diseases Research*, Presentation IDRC-CRDI, 2003; Zicheng Yu, "Holocene carbon accumulation of fen peat lands in boreal western Canada: A complex ecosystem response to climate variation and disturbance", *Ecosystems*, 9: 8, 2006; Daniel Scott, Jay R. Malcolm and Christopher Lemieux, "Climate change and modelled biome representation in Canada's national park system: implications for system planning and park mandates", *Global Ecology and Biogeography* 11: 6, 2002; Frederick J. Wrona, Terry D. Prowse, James D. Reist, John E. Hobbie, Lucie M.J. Lévesque, and Warwick F. Vincent, "Climate Impacts on Arctic Freshwater Ecosystems and Fisheries: Background, Rationale and Approach of the Arctic Climate Impact Assessment (ACIA)", *Ambio: A journal of the Human Environment*, 35: 7, November 2006.

emissions by 20% from 2006 levels by 2020, less than that required by Kyoto but more than the plan adopted by the Obama administration in the United States. Canada has also implemented a tax incentive plan for renewable fuels.

In April 2007 the Harper government instituted its program for Canadian climate change, Turning the Corner[6]. The target was a 20% reduction from 2006 levels by 2020, and 60 to 70% by 2050, which is much less than what scientists believe is necessary to avoid dangerous repercussions. Other countries, most notably in the EU are setting goals that far outreach even Kyoto. For example, in October 2008 EU legislators voted to reduce emissions by 20% by 2020, and UK scientists announced their belief that GHG emissions in Britain needed to be cut 80% by 2050, including all aviation and shipping emissions that were previously excluded. Renewable power generation is expected to be boosted[7].

In Harper's "*Turning the Corner*" plan all sectors of the economy are targeted to reduce emissions by 18%. Electric utilities are encouraged to reduce emissions by replacing fossil fuels with non-emitting devices. Included within the program is a renewable energy plan that invests $1.48 billion to renewable sources constructed between April 2007 and March 31, 2011. Projects registered under this program include many of the renewable projects discussed in this paper including wind farms in Alberta and PEI. However, critics of the plan believe that subsidies are minimal and are holding back Canada's ability to capitalize on the manufacturing and job creation possibilities that generous tax incentives have attracted in the EU and some states in the United States[8].

---

6 Canada online: http://www.ec.gc.ca/doc/virage-corner/2008-03/541_eng.htm.

7 UK emissions need to be reduced by at least 80% as a reasonable contribution to a global strategy of cutting global emissions by around 50% by mid-century. Reductions of this scale are required to limit the expected global temperature increases to around 2 °C de above pre-industrial levels, and to reduce the chances of exceeding 4 °C to a very low level (e.g. to less than 1%). Temperature rises above 2 °C are likely to have a major and increasing impact on human welfare and the natural environment. Temperature rises above 4 °C could be catastrophic. www.theccc.org.uk/downloads/.

8 The German *Renewable Energy Source Act* (2000) allows utilities to pass on the price of renewable and therefore turn a profit. Companies, Canadians among them, are flocking to Germany to take advantage of feed-in tariffs. These tariffs are not capped, ease connection to the renewable grid, offer a premium price to producers, and offer long term contracts that reduce over time as efficiency and technologic innovations drop prices. In 2007 German investment in solar alone was 5.7 Euros, and employed 250,000 people in alternative energy fields.

Canadian GHG Emissions and Alternative Energy Programs

Canada has been relying on fossil fuels for its transport and heating needs. In order for Canada to reduce emissions to retard climate change it is necessary to adopt renewable energy programs. But investment support from the government – similar to the incentives given fossil fuels such as the oil sands – is necessary to build this new industry. Climate change is real and greenhouse gas emissions are rising much more rapidly than previously anticipated. Support, such as the feed-in tariffs used in the EU to stimulate companies and economies may be a useful stimulus in Canada[9]. The structure of the Canadian electrical grid is also problematic because of fragmentation. Each province has its own policy and is not encouraged to think beyond its own borders, which not only affects sharing power but transmission goals. However, that is beginning to change. A few provinces are looking beyond or working with each other, such as New Brunswick and PEI that have begun to sell wind power to the northeastern US. But still Canadian commitments are much smaller than in the EU, and they further lack the commitment to intellectual capital that can help recreate lost manufacturing jobs.

Despite the lack of tax and incentive support Canadian renewable energy has grown rapidly since the millennium and promises to continue. Wind power alone quintupled from 2005 to 2010 (table 2).

Lacking a federal policy several provinces – Ontario, Québec, Manitoba and British Columbia – have adopted their own emission standards that do not endorse the continued reliance on oil sands. In Atlantic Canada, Nova Scotia, New Brunswick and Prince Edward Island, are also taking action to address growing GHG emissions.

Among the renewable resources adopted by the provinces, including Alberta home to most of the oil sands, are several innovative and supported renewable energy programs meant to halt the continued reliance on fossil fuels. What follows is an introduction to the various renewable energy sources and a sampling of their application in British Columbia, Alberta and Prince Edward Island.

Solar Power

Ultimately, our energy and power comes from the sun. Solar energy runs the planet and is the fulcrum of other energy sources. The amount of solar energy reaching Earth in a year is much more than available

---

[9] http://pubs.pembina.org/reports/FITariffs-factsheet.pdf.

non-renewable energy. Yet solar power is one of the least efficiently used energy sources in our quiver.

Table 2. Canadian Wind power capacity, 2000 through February 2010

MW	2001	2002	2003	2004	2005	2006	2007	2008	2009	2010
■	198	236	322	444	684	1460	1770	1876	3319	3359

Source: Canadian Wind Energy Association.

Transforming sunlight into electricity and heat can be done in a number of ways and for different purposes. There is active technology that increases available energy, such as photovoltaic panels, and passive technology which reduces the amount of energy used from other sources. In passive solar, space is designed to circulate air and optimize available solar energy. Generating electricity has been done with photovoltaic solar panels, which are often mounted on rooftops. Solar thermal power uses large mirrors to capture and concentrate the sun's energy and convert it to usable forms for power. Their basic method of operation is to boil water under the sun's concentrated heat, which generates high-pressure steam that drives turbine generators. Solar thermal plants produce power for industrial applications. Because of the conversion of sunlight to steam, thermal plants can provide energy 24 hours a day.

Other solar energy uses include solar heating, detoxification of water and air, and passive energy. Solar energy to heat water and space with solar collectors is becoming more popular and available. For example in Oshawa, Ontario a solar wall was constructed in a General Motors battery plant in 1991. The metal wall draws in heat and ventilation fans which distribute the warmed air through the building.

Water heating alone accounts for one quarter of household energy use. Prices for solar heating are falling as more people research solar power and more cities and towns offer incentives. Solar detoxification processes cause chemical reactions using sunlight to breakdown contaminants in water and air. Passive solar energy uses building design and placement to maximize sunlight for space heating and cooling and receiving light.

Wind Turbines

The windmill is an iconic symbol, such as the windmills of the Netherlands that have reclaimed land in the low country for hundreds of years and the rural North American plains and prairie windmills that grinded grains and pumped water.

Wind is a renewable source that is also dependent on solar energy. Less than 2% of all solar power is converted into wind energy. Wind is created by the uneven distribution of heat. At its simplest wind power generation from wind turbines occurs when wind turns the tower rotor blades, which turn a generator that produces electricity. Cables transfer the electricity to transmission lines that carry it to homes or industry.

Wind turbines have been growing in size, because the higher the turbine the more stable and constant the wind speeds and the freer they are from obstacles, such as buildings, trees, or rock formations, which can decrease the efficiency of the turbines of create irregular wind flows, turbulence. The size varies between 600 kW per turbine in 2000 to 3 MW towers. The increase in size allows each tower to generate more power. The doubling of wind speed generates eight times the power.

The current models of wind turbines are horizontal towers that rise up more than 300 feet high (100 meters), which is the length of a football field. The propeller blades are attached to the nacelle, a box that is deceptively small when viewed from below, but dwarfs a car (Figure 1). As the propellers turn they power a generator, which passes the electrical current to cables that carry the current to transmission lines on the power grid.

Figure 1. A wind turbine nacelle dwarfs the size of a car. This nacelle in PEI is made by Vestas the Danish company that is the world's largest wind turbine maker

Photograph by Chris Mayda

Wind is not as reliable as conventional power. Wind farm figures are often given as capacity but capacity can vary. Wind is not a constant and so the energy is less than the ideal "capacity," the sum of the generator ratings and the number of hours in a year. A capacity factor needs to be recognized in order to truly understand the possibilities in any system. Average capacity factors are between 20 and 40%. A typical wind farm in windy Alberta for example can harvest wind only 35% of the time. The intermittent wind is still hampered by a lack of battery storage facilities. The lack of batteries to store the electricity makes wind power only good instantaneously. The inconsistency and lack of storage of wind power has made it necessary in some places to build additional fossil fuel based power stations to take up the slack when power is not wind generated.

Canada has the right geography for good wind power. Canadian wind resources can meet 20% of its electricity demands. The best wind energy is generated over fetches, an unobstructed distance, and off shore turbines, which generate more power than on shore due to the strong ocean winds. Canada has a very long shoreline and uses its large Prairie region for wind power generation.

In 2010 the installed capacity in Canada was 3,359 MW which was enough to power almost a million homes. This is equal to about 1.1% of Canada's total electricity demand. Wind farms are distributed throughout the country (figure 4)[10].

Figure 2. Canadian wind farm capacity, 2010

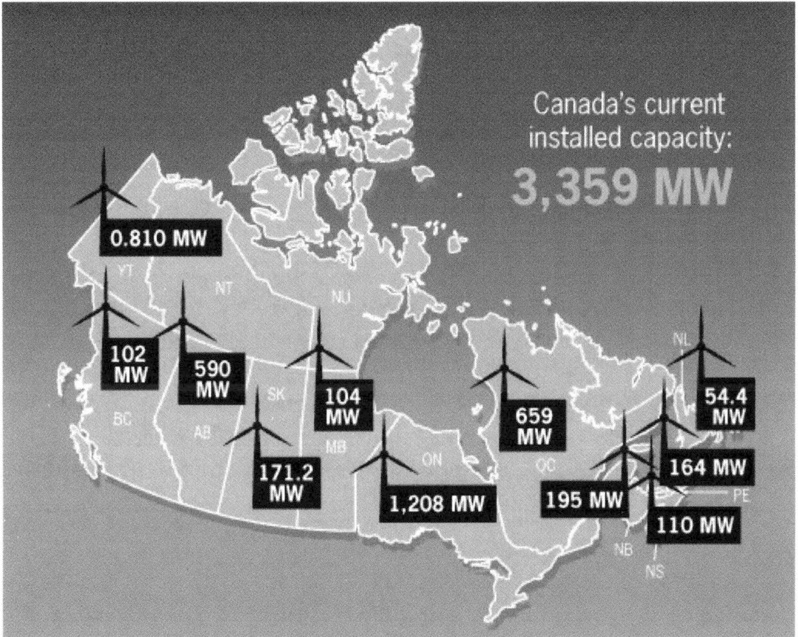

Source: Canadian Wind Energy Association.

Across the globe countries are building wind turbines, with Danish, Spanish and German companies in the lead. Vestas, the largest wind turbine company in the world is a Danish company. The largest wind power company in the United States is GE Energy. Canada has no large wind power companies, although there is one manufacturer, AAER out of Québec that began to supply the North American market in 2008. However strong demand has created a several year backorder for turbines. Wind farm installed generation has increased almost tenfold since 2000. The Canadian wind power market has grown quickly with several investment companies including:

- Acciona: Largest wind developer in the world. 172 wind farms in 10 countries. Also manufactures generators.

---

[10] http://www.canwea.ca/images/ and http://www.canwea.ca/farms/.

- Canadian Hydro: Has 7 wind farms in Alberta and Ontario. New projects in Ontario and Québec
- ENMAX: Major Calgary, Alberta energy supplier for past 100 years.
- Greengate Power: Calgary, Alberta based wind developer of 1,590 MW in Alberta and Ontario
- Greenwind Power: Vancouver company developing wind farms in southwest Alberta
- Shear Wind: Halifax, Nova Scotia company founded in 2005 invested in Alberta and Nova Scotia. Five wind farms projected.
- Suez Renewable Energy, North America: Wind energy and biomass generator of electricity. Suez international headquartered in Belgium.
- Suncor: Integrated energy company headquartered in Calgary, Alberta. Major oil sands extractor.
- TransCanada: major electric and gas supplier. A Calgary based division of TransAlta Energy which is also involved in traditional energy sources. The division focuses on wind generation. It operates three wind farms in Alberta and is involved in the construction of several more in Alberta and in other provinces.
- Ventus: Ontario wind farm developer purchased by Suez Renewable Energy in 2007. Had developed 25 wind energy projects in six provinces in Eastern Canada.

Biomass

Biomass is a renewable energy derived from plants. The sources can be primary where energy production is the primary purpose, such as 'energy crops' as sustainable forests or sugar beets. Secondary sources are a by-product of another process; recycling waste wood, forest residue, manure, and agricultural crop waste. Low water content fuels (wood, chicken litter, straw and husks) can be burned directly, whereas the fuels with higher water content need to be processed by gasification or anaerobically digested to produce a usable fuel.

Biomass energy is considered carbon dioxide neutral because it merely burns the CO_2 that plants have photosynthesized over their growth cycle, thereby reducing GHG emissions, whereas burning fossil fuels releases CO_2 that has been stored for millennia and therefore increases the CO_2 content in the atmosphere. Biomass also has low sulfur content which results in better air quality and reduced acid rain.

Prior to the nineteenth century introduction of abundant and inexpensive fossil fuels Canadian energy was dependent on biomass, especially wood. Throughout the twentieth century most heating energy was fossil fuel based. However, with the rise in energy costs attention has turned to alternative sources of energy. The Canadian boreal forest, accounting for 10% of the world's forest, is a natural for biomass energy. In Canada biomass conversion is returning as an important source of alternative energy supplying 6% of energy, second after hydroelectricity.

Many Atlantic Canadians, continue to burn wood for home heating, and pulp and paper mills in British Columbia burn wood waste products for energy. After World War II fossil fuel dependent individual building heating systems were popular. In the 1980s as gas prices began to rise, Canada and especially Atlantic Canada began to burn woodchip and sawmill waste in biomass-fired large plants, as are practiced in the Nordic European countries[11].

Provincial Alternative Energy Projects in Rural Areas

British Columbia

British Columbia is best described as a mountainous landscape where the rugged topography has made eastern access difficult. Dividing the mountain systems are valleys, one of which, the Okanagan Valley is within the Interior Plateau of the province and divides the Cascade Ranges from the eastern ranges and the Rocky Mountains system. The valley is glacially carved. When the glaciers retreated they gouged out the valley and left sand and gravels behind; good soil for growing the many irrigated fruits the area is known for.

At the southern end of the Canadian portion of the Okanagan Valley is the town of Osoyoos (pop. about 5,000) and the only desert in Canada, a fact that the locals are proud of. Its unique geography has made it a favorite of tourists and retirees who arrive in droves to see the Canadian desert and feel the 100 °F (38 °C) summer days. The summer averages about 86 °F (30 °C) high and 60 °F (16 °C) lows, while winter, though cold, are not nearly as cold as other parts of Canada 24 °F (-5 °C).

The case is made in Osoyoos that it is the upper Sonoran Desert. Some do not agree, but regardless it is a warm and arid region that receives about 10 inches (250 mm) of precipitation, and has desert type vegetation. Osoyoos is known for Antelope brush, a deciduous shrub that can be found from British Columbia south into Arizona. Growing among perennial bunchgrasses Antelope brush is the indicator species of the Osoyoos Desert but Antelope brush and its ecological community,

---

[11] Malmo, Sweden, http://www.iclei-europe.org/sitemap/; Finland forest sector eyes biomass business, http://goo.gl/wuDKN; Kenth Hasselgren, "Use of municipal waste products in energy forestry: highlights from 15 years of experience", *Biomass and Bioenergy*, 15: 1, July 1998, 71-74.

are one of Canada's most endangered ecosystems[12]. It is similar to other desert plants in that it protects habitats from something many Canadians flock to – sunshine – the Okanagan receives about 5.5 hours per day or 2,000 hours annually, about 600 more hours than either Halifax or Ottawa. Available solar energy is determined by the number of hours, the height of the sun, and the cloud cover.

In August 2008 the Canadian government created the ecoENERGY for Renewable Heat Initiative that encourages Canadians to invest in residential solar water heaters. EcoENERGY also supports institutional investment in solar power. Three provinces have been highlighted for the project – Alberta, PEI, and British Columbia. The program subsidizes about 20% of installing solar water heat. There are several solar projects in the Okanagan.

The Desert Society is located on 67 acres of land about 2 miles (3 km) from Osoyoos, and their mission is to protect the tender desert environment and ecosystem, and to inform the public about Canada's desert. Because of the remote location away from power lines the Desert Society's Desert Centre has not had electric power. In March 2008 the Desert Centre in Osoyoos received power for the first time, thanks to rooftop solar panels. Now the centre, far from the town and the grid, can have lights, run a vacuum and make evening presentations, all taken for granted amenities. A local rancher provides well water for the centre's use until a waterline from Osoyoos is installed. The photovoltaic system received $20,000 government funding from Western Economic Diversification Canada, as well as an additional $23,000 from the Centre.

Burrowing Owl, an exclusive winery located on near perfect soils and climatic conditions north of Osoyoos, received awards for its solar heating system, but has also been a champion of the namesake of the winery, a small owl that lives in old badger holes. Burrowing Owl received a $16,000 rebate from Fortis Power a Kelowna based utility, to bring energy efficient options like geothermal heating, energy efficient irrigation system, and solar water heating for guests at the hotel and for winery processing. The total cost of energy efficient items was $181,000 in 2006 but the winery is expecting to save $20,000 annually.

---

12 Antelope-Brush ecosystems. Province of British Columbia Ministry of Environment, Lands and Parks. March 1995.

Figure 3. Desert Centre in Osoyoos, B-C. The centre did not have electricity prior to the installation of solar panels in 2008

Photograph by Chris Mayda.

Other installations include solar heating at a hotel in Kelowna, and the University of British Columbia Okanagan. Solar powered buildings include several residences and apartment buildings in the valley.

British Columbia has reached beyond confederation standards to achieve carbon neutrality by 2010. All public buildings are now LEED gold or equivalent and GHG emission reductions are geared to be 33% below 2007 levels by 2020.

Alberta

The Rocky Mountain crest separates Alberta and British Columbia. Dropping down from the Rockies Alberta enters into the wide-open Prairie region. The leeward side of the Rockies brings year round winds, most notably the dry Chinook winds of winter that blow through Southern Alberta onto the Prairie. To the east of Crowsnest Pass in Southern Alberta is the windiest area of Canada.

Wind has been a source of energy generation since the first pioneers settled here in the late nineteenth century. Harnessed wind power provided power to grind grains, pump water for livestock, and a few ingenious farmers were providing small amounts of electric power for lights and radio into the 1950s when electricity was not yet available in the rural areas of the province.

The wind has been used for large-scale wind turbine generation since 1993 when 25 lattice turbines were built on Cowley Ridge.

Figure 4. "Egg beater" Darrieus turbines near Pincher Creek were early experiments in Alberta wind turbines in the 1990s. No longer operating because unable to withstand extreme wind conditions

Photograph by Chris Mayda.

Alberta deregulated the electricity industry in 1999. Once customers could choose their power providers wind power has been on the rise. Customers agreed to pay about $5 monthly to receive wind power. By 2010 Alberta wind farms generated more than 4% or 1,000 MW of electricity across the prairie from Cowley Ridge to Taber. The total energy created by the wind farms is about equal to a good-sized coal burning power plant.

Building a wind farm begins with an agreement made between wind farm producers and local land owners who are in possession of windy areas or ridges (Figure 5). Before a wind farm is established the site is assessed and designed to maximize the wind power. Most farmers lease their land though some decide to become producers themselves. Farmers can continue to farm because the relatively small footprint of the wind towers, about 5% of the land, and the minimal effects to crop production or livestock grazing. Farmers are paid royalties that average thousands of dollars annually. Wind power production in the southern prairie of Alberta takes a few years to install.

Figure 5. Pincher Creek area of southwestern Alberta, where farmers are able to capitalize on windy ridges for wind turbines

Photograph by Chris Mayda.

First the wind energy of several sites must be analyzed at measuring stations for at least one year. Then solitary towers are built at several spots, which are compared for the best wind generation. While wind farms are present as far north as Edmonton, the largest number of wind farms is located between Cowley Ridge and Pincher Creek in south-western Alberta (Figure 6).

Figure 6. Wind farm map, Alberta. The largest number of wind farms is between Cowley Ridge and Pincher Creek near Crowsnest Pass in southwestern Alberta

Map by Doug Rivet.

Alberta is investing into wind turbines heavily and is a leading megawatt producer. However, Alberta produces the highest GHG emissions in Canada because of its oil sand production and will continue to increase emissions until at least 2020.

Table 3. Alberta Wind Farms, 2010

Southern Alberta Wind farms (Company)	Date Installed	Location	# Turbines	Total inst. capacity (MW)
Blue Trail Wind Farm (TransAlta)	2009	Fort Macleod	22	66.0
Castle River (TransAlta)	1997 2000	Cowley Ridge	1 59	0.6
Cowley Ridge (Canadian Hydro)	1993-2000	Cowley Ridge	57	21.4
Cowley Ridge (Canadian Hydro)	2001	Cowley Ridge	15	19.5
Chin Chute (Suncor/Acciona)	2006	Chin Chute	20	30.0
Kettles Hill (Phase 1) (Phase 2) (Enmax)	2006 2007	Pincher Creek	5 30	9.0 54.0
(Lundbreck JV)	2001	Lundbreck	1	0.6
Macgrath (Suncor/Acciona)	2003	Macgrath	20	30.0
McBride Lake (Enmax, Transalta)*	2002		114	75.0
McBride Lake East	2001	Fort McLeod	1	0.66
Old Man River (Alberta Wind Energy)	2007	Pincher Creek	2	3.6
Sinnot (Canadian Hydro)	2001	Pincher Station	5	6.5
Soderglen (Nexen/ Canadian Hydro)	2006	Soderglen	47	70.5
Summerview (Transalta)	2002 2004	Pincher Creek	1 38	1.8 68.4
Taber Wind Farm (ENMAX)	2007	Taber	37	81.4
Tallon Energy (Talon Energy)	2004	Pincher Creek	1	0.75
Taylor (Canadian Hydro)	2004	Macgrath	9	3.38
Vestas Prototype (TransAlta)*	2004	Edmonton	1	3.0
Waterton (Transalta)	1998	Hillspring	6	3.78
Weather Dancer	2001	Pincher Creek	1	0.9
Projected Wind Farms				
Finavera Ghost Pine	2010+	Three Hills		75.0
Glenridge Phase I (Shear Wind)	2010+	North of Medicine Hat		200.0
Black Spring Ridge Chigwell Ponoka Radar Hill Wintering Hills Halkirk (Greengate)	2010+ 2010+ 2010+ 2010+ 2010+ 2010+	Lethbridge Lacombe Ponoka Red Deer Drumheller Stettler		300.0 150.0 150.0 100.0 150.0 150.0
Legacy ridge Energy	2010+	Pincher Creek		20.0
Old Man Stage 2	2009+	Pincher Creek		46.0
Waterton Phase I	2010+	Waterton		150.0

(TransAlta/Vision Quest)				
Wind Power River View	2010+	Pincher Creek		115.0
Windrise Power	2010+	Fort McLeod		99.00

*Northern Alberta. Source: Canadian Wind Energy Association.

Prince Edward Island

Prince Edward Island (PEI) is the smallest province in Canada, located in a cold-temperate climate zone off the coast of Nova Scotia and New Brunswick. It is a crescent shaped island located in the southern Gulf of St Lawrence in the c-shaped area called the Magdalen Shallows, which the Gaspé Peninsula and Cape Breton Island semi-surround. PEI's low elevation is especially susceptible to climate change, which has led it to adopt energy tariffs that embraced renewable energy. The island is now nicknamed the "Green Province" having reached its target goal of 15% renewable energy in 2007, three years ahead of schedule. The province's commitment to biomass and wind will be discussed.

Though GHG emissions on the island were about 10% above 1990 levels (May 2008), the PEI Renewable Energy Act requires a 5% reduction of 2004 levels by 2010, and thereafter that every public utility obtain at least 15% of all electricity from renewable energy sources[13]. In October 2008 the island government outlined the paper "Island wind energy, securing our future: the 10 point plan" that expressed a goal of 500 megawatts by 2013. Many islanders wish to reduce its carbon footprint by building alternative energy systems. The island has been experimenting with several sources of renewable energy including biomass energy and wind power.

PEI lacks fossil fuel sources, and has the most expensive energy costs in the provinces. PEI has been largely dependent on coal and oil from New Brunswick. After extensive environmental assessment, a submarine cable was built and began transmission across the Northumberland Strait in 1993, but within a decade cable transmission neared capacity. As petroleum costs rise, local energy production has become increasingly important for PEI. The island has therefore turned to natural gas and renewable forms of electricity – solar, biomass, wind turbines. By 2006 biomass accounted for 6.5% of power, and the active construction of wind turbines promises an additional 10% of electricity by 2010. Though the government is endorsing and subsidizing much of the new building in alternative energy, there are those who find fault with what they consider blind endorsement for the "wind craze."

13 PEI online: http://www.gov.pe.ca/law/statutes/pdf/R-12-1.pdf.

Figure 7. Prince Edward Island sits off the coast of Nova Scotia and New Brunswick. At opposite ends, winds are abundant

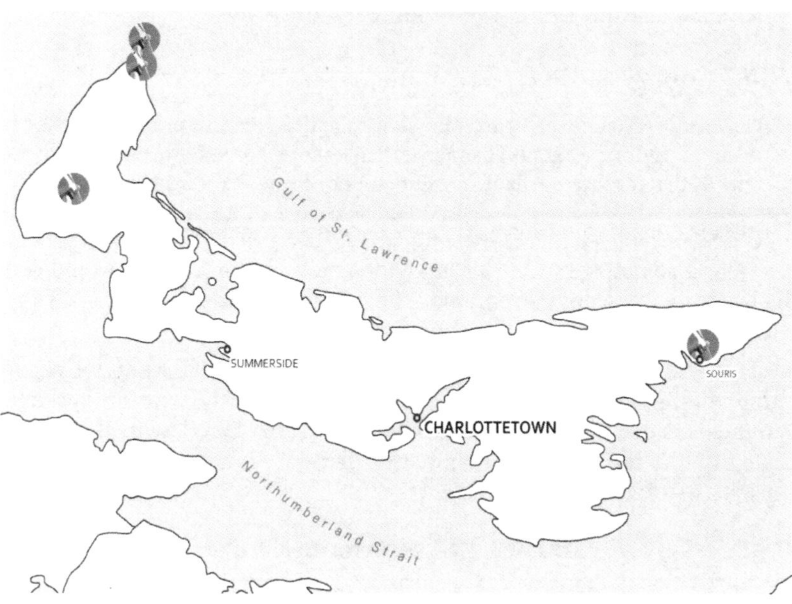

Map by Doug Rivet.

Bioenergy

Since the early 1980s Charlottetown has heated water and air in various institutional buildings by periodically adding bioenergy produced steam heat. This is done in a district wide distribution rather than each building or institution having its own power generation. In several places across Canada and a few in the United States, the district distribution uses biomass fuels. Use of bioenergy expanded in the late 1990s to more than 60 customers plus government, hospital and university buildings. The cost is about 10% less than fuel oil (the traditional heating source on the island). 90% (45% solid waste, 45% sawmill residue) of the Charlottetown district is fuelled by bioenergy, which saves more than four million gallons of heating oil annually. Added benefits include

- self-sufficiency – Charlottetown burns tonnes of waste material and imports millions of gallons less heating oil (which came from New Brunswick)
- more money remaining in the local economy (about 70 cents of each dollar remains)

- waste materials are no longer a liability but an asset
- reduced GHG emissions
- improved air quality
- reduced landfills (the waste is burned in the plants).

Wind Energy in PEI

> Our wind resource in Prince Edward Island is one of the best in North America and we want all Islanders to have the opportunity to share the benefits of that resource. – Pat Binns, Premier of Prince Edward Island, 2005[14]

PEI has major wind farms located at opposite ends of the island.

Taking advantage of the strong winds blowing through the Magdalen Shallows the crescent extremities – East Point and North Point – are the locations of two large wind farms (Figure 7).

The first wind farms were operated by the public PEI Energy Corporation, but more recent ventures, such as at West Cape are private ventures located on private land. Revenues are shared with those impacted by the turbines. The major wind farm areas have a total installed capacity of 166.56 (2010) (Table 4).

Table 4. PEI wind farms, 2010

PEI Wind Farms (Company)	Date Installed	Location	Number of turbines	Total installed capacity (MW)
North Cape (PEI Province)	2001-2004	Tignish	16	10.56
Aeolus Wind Farm	2003	Norway	1	3.0
Vestas Prototype (Transalta, Vestas)	2004	Tignish	1	3.0
West Cape Phase 1 (Ventus) Phase 2 (Suez Renewable)	2007 2009	West Cape	11 44	19.8 79.2
Norway (Suez)	2007	Tignish	3	9.0
Eastern Kings (PEI Province)	2007	Elmira	10	30.00
Summerside Wind Project	2009	Summerside	4	12.00

Source: Canadian Wind Energy Association.

The PEI government has adopted wind farms for energy price stability, independence, and to meet the Renewable Energy Act mandates. Designated areas for wind farm projects under the Renewable Energy Act must have an average wind speed of 7.5 meters per second and meet other necessary approvals to precede development. Both extremities are

[14] Renewable Energy Access Newsletter, June 15, 2005. In Canada, record RFP programs are yielding similar record demand. Developers, wind-based IPPs, and turbine manufacturers are scrambling to meet utility demand, while financiers and investors are becoming more receptive every day.

located on a promontory of the island's red sandstone cliffs. However, the largest and most recently constructed wind farms are along the roads leading to West Point and in the town of Summerside.

The North Cape Wind Farm has been online since 2001. North Cape is the location of the federally and provincially funded Atlantic Wind Test Site and the Wind Energy Institute of Canada. The Wind Energy Institute has researched, developed and tested wind turbines since 1980. The site's location takes advantage of one of the most windy and harsh climatic spots in the country. The corrosive marine environment (the Institute is exposed for 300 degrees to the Gulf of St. Lawrence) and the occasional freezing spray and frost make it a trying but ideal turbine testing site to establish standards and reliability. The 16 turbines are a mismatched set of wind turbines developed by different manufacturers and tested onsite. They have an installed capacity of 10.56 MW which supplies about 3% of the total energy on the island, enough power for more than 5,000 homes when running at full capacity.

The provincial government built Eastern Kings Wind Farm on the eastern extremity in Elmira and qualifies for up to $9 million in federal funds for 10 years. Installed in 2007 it has 10 turbines with 30.0 MW installed capacity.

Privately held Suez Renewable Energy NA has invested $250 million partially subsidized by the provincial government for PEI wind farms. Norway Wind Park was installed in 2007 and has 3 turbines with 9.0 MW installed capacity. Norway Wind Park electricity is for the domestic market. West Cape Wind Farm has been built in two phases. The first built in 2007 has 11 turbines with 19.8 MW installed capacity. Phase two added an additional 44 turbines during the summer of 2008. The 55 wind turbines have a total capacity of 99 MW (Figure 8). Nearby Summerside contracted to purchase 9 MW of power from the West Cape Wind Farm, but the remainder of the power generated by the West Cape will be sold to the New England Power Pool export market. It is the first wind power generated in Canada sold to the United States. Summerside completed its own wind farm in 2009.

The West Wind project has met with some local opposition. Though PEI residents generally agree with renewable energy plans, some find fault with environmental issues with turbines killing bats and birds, sound pollution, and others have fought additional power lines necessary to transmit power.

Beyond the reduction of GHG gases the wind farms economically benefit the island. When more than 80% of power was from off island sources millions of dollars were leaving the island, but on-site wind farms add jobs and more money stays within the local economy.

Figure 8. West Cape wind turbines are installed on privately held farms

Photograph by Chris Mayda.

Conclusion

Northern climes have been the first to be heavily affected by climate change. Canada needs to address climate change or the country could face environmental, ecological and economic turmoil in the near future. With this in mind alternative energy, the production of jobs, and energy efficiency are keys to a successful future. Canada has an abundance of hydropower energy that already provides two-thirds of its electricity, but the opportunity to further retract from fossil fuel dependence and to create a new economy points the country toward wind, solar, biomass and other alternative energies. Provincial support in some provinces has been continuous and in 2007 some federal support was also given to alternative energy projects. While it is essential to shift power production the process cannot be done blindly, but requires transparent, objective, and progressive judgment.

Review Questions

- Canada has been turning its focus to more renewable energy but there are some major snags in their way. Name two of the major problems.
- How are wind and hydropower complimentary sources in Canada?

- How did Canadian energy policy change from Kyoto Protocol to "Turning the Corner"?
- What are some of the shortfalls of wind turbines?
- What are the different conditions that allow wind turbines to be productive in Alberta and PEI?

References

Allcut, E.A. (1945), "A Fuel Policy for Canada", The Canadian Journal of Economics and Political Science/Revue Canadienne d'Économique et de Science politique, 11: 1, 26-34 Feb.

Bailie, Alison, Stephen, Bernow, William, Dougherty, Benjamin, Runkle, Marshall, Goldberg, *The Bottom Line on Kyoto: economic benefits of Canadian Action*, The David Suzuki Foundation, April 2002, http://www.davidsuzuki.org/files/kyotoreport.pdf.

Bhatti, Jagtar S. Rattan Lal, Michael J. Apps, Mick A. Price (eds.) (2005), *Climate change and managed ecosystems*, CRC.

Blackford. B.L. (1978), "Wind-driven currents in the Magdalen Shallows, Gulf of St Lawrence", *Journal of Physical Oceanography*, 8: 4, 653-664.

Castaldo, Joe (2008), "Alternative energy: Out of juice?", *Canadian Business Online*, October 27.

David Suzuki Foundation and the Pembina Institute (2005), *The Case for deep reductions: Canada's role in preventing dangerous climate change*, The David Suzuki Foundation, http://www.davidsuzuki.org/files/climate/Ontario/Case_Deep_Reductions.pdf.

Province of British Columbia Ministry of Environment, Lands and Parks (1995), *Antelope-Brush ecosystems*, March.

Stephenson, T.A. and A. (1954), "Life between tide-marks in North America: IIIA. Nova Scotia and Prince Edward Island: Description of the Region", *The Journal of Ecology*, 42: 1, Jan., 14-45.

Torrie, Ralph, Richard, Parfett, Paul, Steenhof, *Kyoto and beyond*, The David Suzuki Foundation and The Canadian Climate Action Network, September 2002, http://www.davidsuzuki.org/files/Kyoto_Beyond_LR.pdf.

US Department of Energy: Energy Efficiency and Renewable Energy (2008), Annual Report on U.S. Wind Power Installation, cost, and performance trends: 2007, May.

Internet sources

Canadian Wind Energy Association http://www.canwea.ca.

Environment Canada: climate change http://www.ec.gc.ca/climate/overview_science-e.html.

U.S. Department of Energy, Energy Efficiency and Renewable Energy http://www.eere.energy.gov/.

CHAPTER 3

Population

The Canadian Mosaic

Hélène BÉLANGER

Université du Québec à Montréal

Introduction

The world's population is over 6.7 billion, and one half (3.3 billion people) live in urban areas (United Nations, 2011). The world's population continues to grow and it is estimated that it will exceed 10 billion people in 2100. Population growth will take place mostly in developing countries. Industrialised countries, such as Canada, face an aging population and low natural growth, which could result in a zero population growth or a contraction unless the population is maintained or increased by means of immigration.

Population studies examine the evolution, structure and spatial movements of specific groups of people that are defined in some way as being a distinct "population" (Inuit, French-Canadians, Protestants, etc.). The main disciplinary approaches in the study of population are demography and population geography. Demography describes the cultural and social structures of the population and its renewal by means of different movements (births and deaths; immigration and emigration). Population geography has the same objective but is more concerned with spatial aspects. Undoubtedly, the distinction between these two disciplinary approaches is blurred. Other disciplines take specific interest in population studies as well. Sociology, for example, has a great interest in population movement and its implication on the evolution of societies. The field of urban studies is interested in populations that live in urban centres – how they share space, issues and relationships concerning the built environment, symbolism and so on.

This chapter will address population questions in the Canadian context. The author does not have the intention of presenting an exhaustive study of the Canadian population. Rather, this text is an introduction to the main characteristics of the Canadian population: the evolution of the Canadian population through time, its movements, its socio-professional and cultural characteristics, and how this is related to Canada's physical environment.

Studying population using demographic and population geography approaches means also using numbers. In order to reduce redundant information about data sources, it must be specified that, unless otherwise noted, all data come from Statistics Canada census and demographic reports. The student will find all the catalogue numbers in the bibliography at the end of this chapter.

For students who are exploring the field of population studies for the first time, many of the concepts that appear in this text will be new. For this reason, terminology specific to this and similar fields appears in italics. Explanations and definitions of these concepts are given directly in the text.

Population Distribution

Canada is a large country of almost 10 million square kilometres with an estimated population, in 2008, of more than 33 million inhabitants. The national population density is very low with only 3.69 people per square kilometre. However, this figure hides important density variation throughout the country. The majority of Canadian population is concentrated in four provinces: Québec, Ontario, Alberta and British Columbia. And, due to its physical environment and climate, 83% of the population lives on only 13% of the territory (figure 1). This area, in the southern part of the country in close proximity to the United States of America, includes Canada's three largest cities: Toronto, Montréal and Vancouver.

Physical Environment and Climate

Physically, Canada consists of several large physiographic sets such as the Canadian Shield, the prairies and many mountains. The Canadian Shield is an old rock formation and it is the largest physiographic area of the country. The soil in this area is rich in minerals but is poor for agricultural activities. Surrounding the Canadian Shield are the prairies, where many types of rocks and soil are to be found. The very deep strata of the hinterland ground provides gas, oil and coal, and the low lands of the Great Lakes and the St. Lawrence Valley concentrate the most productive agricultural lands of the country.

Figure 1. Population density, Canada, 2006

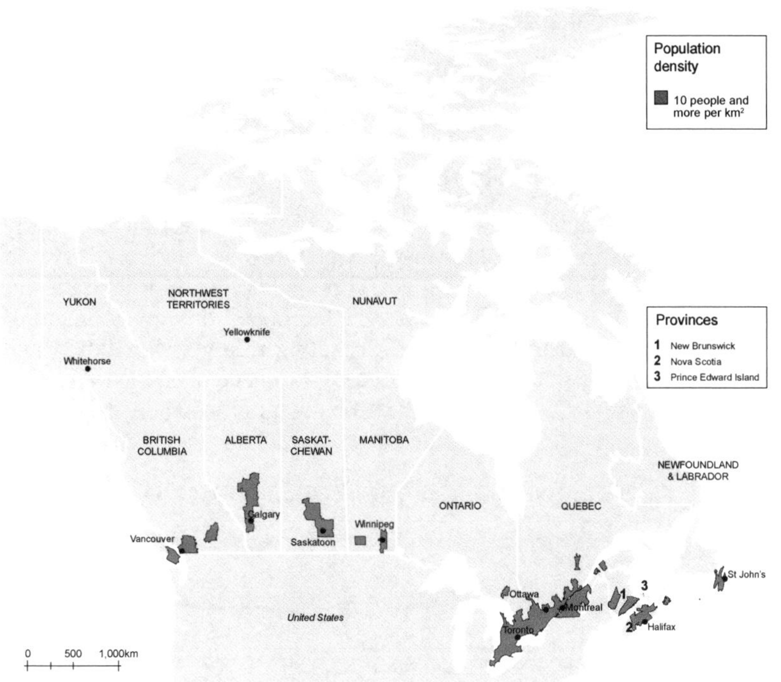

Source of data: Statistics Canada. 2006 Census.

The physiographic characteristics of Canada influence settlement patterns. In the far north (for example, the city of Yellowknife), the climate is Arctic with the temperature averaging -32 °C during January, the coldest month (see Environment Canada 2005, 2008). Close to the Pacific Ocean (for example, the city of Vancouver), the climate is temperate – here, the temperature is mild during the coldest month (around zero degrees), but with precipitation averaging 1,167 millimetres per year. In the hinterland, the weather is Continental[1]. Nevertheless, Canada remains a cold country. However, with an average high temperature around 26 °C and precipitation between 360 to 940 millimetres (between the city of Montréal and the city of Regina), the southern part of the country enjoys long, warm summers which favour human settlement as well as the development of a productive agricultural economy.

1 Being at distance of oceans means they will not influence the climate. A continental climate is thus cold during the winter and hot during the summer.

Natural resources were very important in the settlement and development of communities, and many communities continue to depend on these resources. For example, on the East Coast, many communities developed fisheries; in the St. Lawrence Valley, agriculture; in the Prairies, energy source extraction (gas, oil and coal); and wood extraction on the West Coast.

Urban and Rural Development

Colonization began in the St. Lawrence Valley and expanded rapidly towards the Ontario Peninsula, where agriculture was relatively easy and the hydrographical network facilitated communication and the transport of goods. Later on, these prime locations facilitated the development of commercial and industrial activities, attracting a larger population and contributing to the urbanization process. Canada's industrialization began during the 19th century around the Lachine canal in the city of Montréal.

Figure 2. Urbanization rate, Canada, 1901-2006

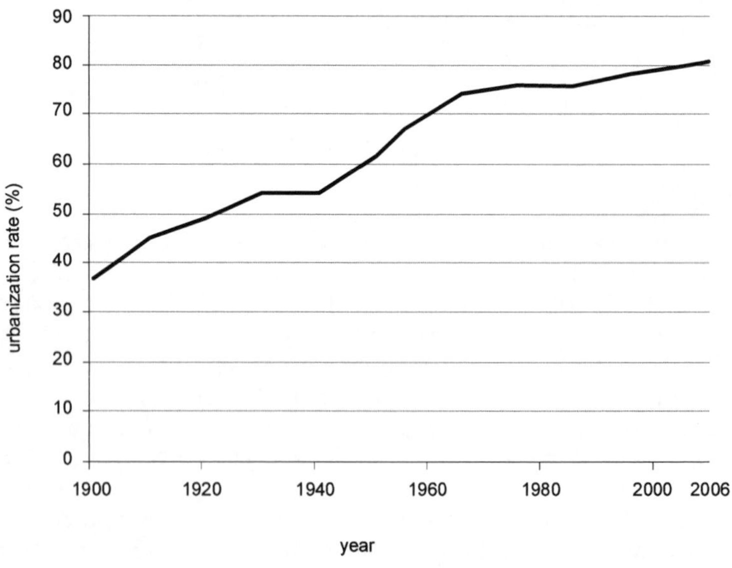

Source of data: Statistics Canada.

Mostly rural at the beginning of the 20th century (see figure 2), the urbanization rate increased steadily during the last century: the majority of the Canadian population became urban during the 1920s. Even though agricultural activities form the economic base of many communities, urbanization occurred on some of the best agricultural land in the

St. Lawrence Valley and the Ontario Peninsula. Today, it is estimated that 80% of the Canadian population is urban and lives in an area of 23,000 square kilometres.

With its physiography and climate, only 7% of Canada's total area is suitable for agricultural activities. According to the latest agricultural census (in 2006), 675,867 square kilometres were being used for agriculture, 2,400 less than 20 years ago. Today, more than 14,300 square kilometres of agricultural land are urbanized. The urban footprint is increasing while Canada faces an increasing demand for agricultural land due to population and economic growth.

Population Growth

Jean Talon, Canada's first official statistician and first "intendant" of Justice, Police and Finance of New France, completed the first Canadian census in 1666, registering 3,215 inhabitants in the colony (see Statistics Canada 2008j). Among them 3,190 were concentrated in the three major settlements: Québec, Montréal, and Trois-Rivières. The first national census took place four years after Confederation, in 1871.

Canada's population reached one million during the first half of the 19th century[2]. The population reached five million in 1895, ten million in 1929, 15 million in 1954 and 20 million in 1966. However, as shown in figure 3, there were two important events that influenced population growth between 1851 and 2006. During the Depression of the 1930s, the birthrate dropped considerably, and after World War II, birthrates increased substantially.

Between 1946 and 1961 population growth reached an average of 2.6% per year; this period is called the Baby Boom. The Baby Boom was the result of the conjunction of specific economic and social circumstances: the Depression resulted in a slump in the birthrate (i.e., a delay in starting new families or having additional children due to economic constraints), however economic growth in the 1940s, combined with the end of World War II, resulted in a rapid increase in the birthrate. In addition, young adults tended to marry at a younger age and have a larger number of children (which some demographers attribute to an attempt by young men to avoid conscription). The Baby Boom continued after World War II as couples tended to have children sooner after marriage. Since the middle of the 1960s, the growth rates have decreased slowly to reach an average of 1% per year between 2001 and 2006, with a total rate of 5.4% over a five-year period. According to the

[2] Between 1867 and 1901 population was "estimated". See Statistics Canada. CANSIM table 051-0001.

most recent estimates, Canada had the second highest growth rate, following the United States, among the G8 countries (Britain, Canada, France, Germany, Italy, Japan, Russia and the United States of America) (CIA, 2008).

Figure 3. Total population growth, Canada, 1851-2006

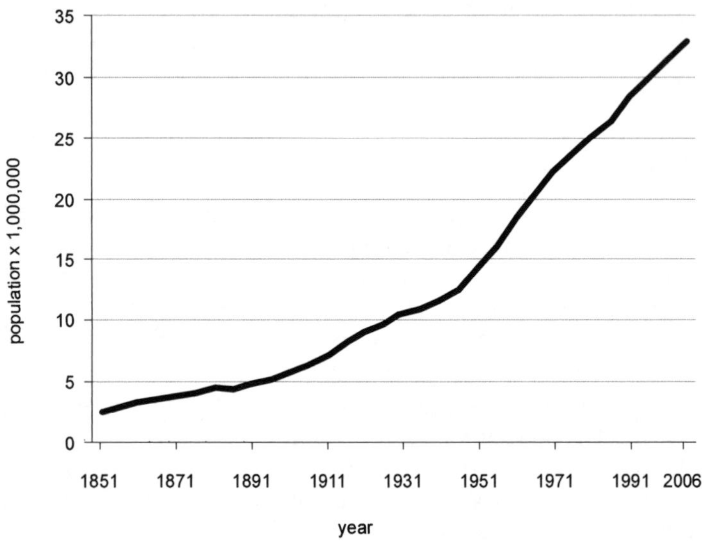

Source of data: Statistics Canada. 1851 to 2006 censuses.

Population growth is the result of what demographers and geographers call four different population *movements*: birth, death, immigration, emigration. The two first movements give the *natural increase* and the last two movements give the *migratory increase.*

Growth = (births – deaths) + (immigrants – emigrants)

For a long period of time (since Confederation in 1867), population growth in Canada was predominantly due to natural increase. More recently, migratory increase has played a larger role in Canada's population growth. Since the middle of the 1990s, migratory increase has become more important than natural increase, as shown by figure 4. The last year that the Canadian total fertility rate replaced the previous generation in terms of overall population numbers was in 1971.

Figure 4. Total population growth, natural increase, migratory increase, Canada, 1981-2007

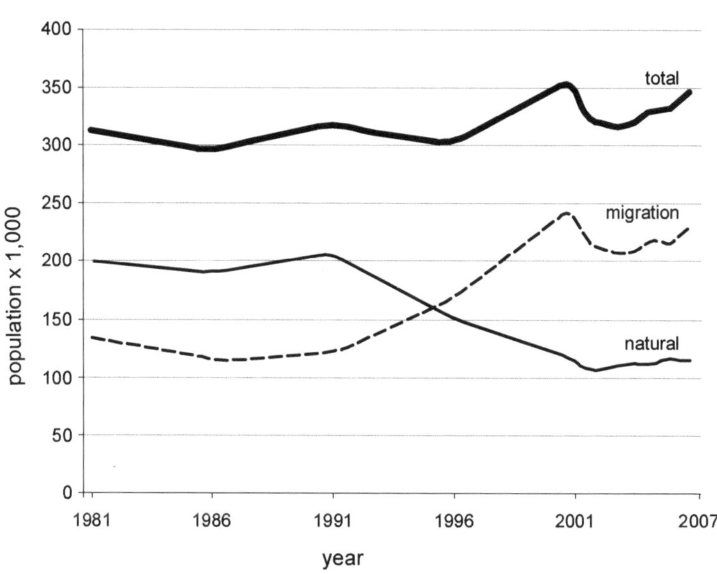

Source of data: Statistics Canada.

The *total fertility rate measures* the number of children a woman would give birth to, during her childbearing years (evaluated to be approximately between the age of 15 and 49). In order to replace the previous generation, the total fertility rate has to be of a minimum of 2.1 children per women. Why 2.1 children? Because of the gender differences in the number of births between girls and boys: for every 100 girls who will be born, 105 boys will be born. This means, that to replace 100 women, we need 205 children. An easy calculation gives a figure of 2.05. The difference between 2.05 and 2.1 is explained by the mortality among girls – Not every girl will reach her fertility age. In Canada, the total fertility rate was 4.6 in 1901, far surpassing the rate required to replace the previous generation and assure population growth. But, since the beginning of the 21st century, the total fertility rates were at their lowest levels since the 1920s, with variations between 1.51 and 1.56. These rates are comparable to the European rate (1.45 in 2005, See United Nations, 2007) as a whole but lower than in the United States (estimated at 2.1 in 2008. See CIA 2008).

There are variations at the regional level, with a total fertility rate of 1.34 in Newfoundland and Labrador and of over 2.7 in Nunavut. In the case of Nunavut, the high total fertility rate is explained by the fact that

85% of the population is Aboriginal, a population group that has a higher fertility rate than non-aboriginal groups. There are also differences between rural and urban areas, where rates are generally lower in urban areas. However, precautions have to be taken in the interpretation of these differences – As the figure 11 shows, internal movements of working age populations between provinces could have an impact on the total fertility rate as well. The movement of young adults in search of job opportunities transfers potential births from one region to another. Some demographers (Caron *et al.*, 2007) explain the distinction between rural and urban area fertility rates as the result of *suburbanization* and *rurbanization*, or the movement of young couples to suburban or rural areas when the time comes to start a family.

Structure and Composition of Population

The age structure of a population is illustrated using an age pyramid, a diagram illustrating of the distribution of the population according to age and sex. On the 'x' axis, the population is shown; on the 'y' axis, the ages are shown.

Figure 5. Age distribution, Canada, 1956-2006

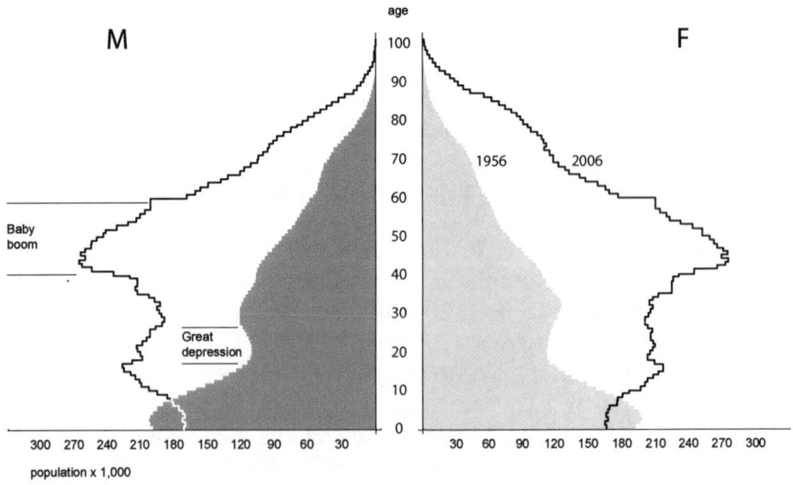

Source of data: Statistics Canada. 1956 and 2006 Censuses.

In an age pyramid for 1956, in the middle of the Baby Boom, the pyramid has a large base, showing the size of this new generation (see figure 5). The noticeable hollow before that period shows the population born during the Depression. Forty years later, the base of the pyramid is shrinking, showing the decreasing fertility rate. In parallel, Canada witnessed an increasing life expectancy at birth, which is today one of the highest of the industrialized world. In 1971, the life expectancy at birth was 69.6 years for men and 76.6 for women. In 2004, it increases to reach 82.6 years for men and 86.6 years for women, and the difference between men and women is shrinking. With a decreasing fertility rate and an increasing life expectancy at birth, Canada's population is aging. This is well demonstrated by the increase of the median age which was 27.2 years old in 1956 and 39.5 years old in 2006 – an increase in median age of 12 years in just a 50-year period.

Figure 6. Total fertility rate and age structure by provinces and territories, Canada, 2006

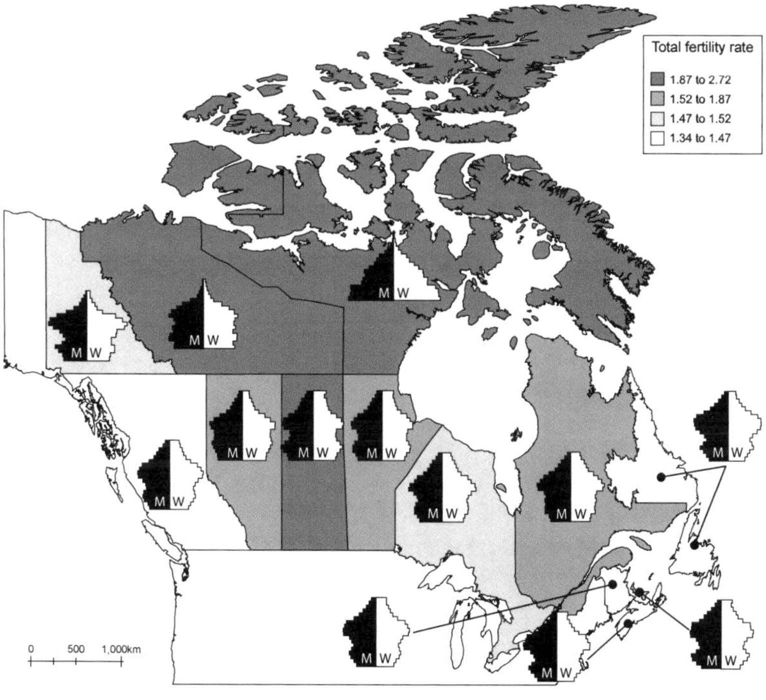

Source of data: Statistics Canada.

The study of the dependency ratios also indicates that the population of Canada is aging. The dependency ratio gives the proportion of young people (or elderly people or both) versus the population of potential workers, using age as an indicator. The population is considered being potentially active in the workforce between the ages of 15 and 64 years old. Not surprisingly, the dependency ratio of young people (less than

15 years) decreased from 32.5 in 1956 to 17.7 in 2006. These figures mean that for every 100 potentially active adults, there were 32.5 children less than 15 years in 1956 and 17.7 in 2006. As for the dependency ratio of elderly people, it has increased substantially from 7.7 in 1956 to 13.7 in 2006. This indicates an aging population. However, differences are noticeable throughout the territory as illustrated by the pyramids in figure 6. Nunavut, with its pyramid showing a large base, still has a young population structure. At the opposite, Manitoba is slowly taking a more rectangular shape and Newfoundland/Labrador a more obelisk shape.

All things being equal, an aging population means that we can expect the mortality rate to increase in the near future, simply because an aging population has a higher risk of death.

Household and Family Composition

Before industrialization and urbanization, when most households were involved in farming, large families were desired. Children were a source of labour and thus, could increase the family production and income (see Milan, 2000). In an urban environment, once child labour became illegal, children came to represent an expenditure since they did not earn money and parents had to invest in the child for him or her to become autonomous. Having fewer children allows parents to invest more in education and well-being.

Industrialisation and Urbanization also represent changes in the lifestyle of the population. From the middle of the 19th century, an increasing part of the female population began to join the labour force temporarily at first but in a more permanent fashion with time. Even so, the birthrate remained relatively high until the 1960s. Regional differences in birthrate could be attributed to social and religious practices, with family planning more accepted among the Anglo-Protestant women than the French-Catholic ones (see for example Appleby, 1999).

Since the 1960s, the number of marriages has decreased. For example, between 1981 and 2002, marriages decreased by 23%. Divorces have also been on the rise since 1968, the year divorce law was adopted. Divorce law facilitates a process which had been previously less common and more difficult. With an easier process and more acceptance of divorce by Canadians, it is now estimated that one in three marriages will end in divorce before the spouses reach the age of 30.

Another transformation in Canadian society is the increasing number of couples with common-law status. Favoured by younger couples, it is often a prelude to a legal marriage. However, more and more couples are choosing to keep common-law status instead of getting married.

There is great variation between regions regarding the marital status of its population according to the last census. In 2006, the proportion of the population of 15 years and over that is legally married varies from 31% in Nunavut to 54.3% in Newfoundland and Labrador. As for the couples choosing common-law status, Nunavut has the highest percentage (22.7%), followed by the province of Québec (19.4%). In the province of Québec common-law couples have represented an increasing percentage of the population since the 1960s. In the case of the Northwest Territories and Nunavut, common-law status may be related to cultural practices. For the government of Canada, a marriage ceremony can be celebrated by a religious or a civil celebration but a formal contract must be signed if the marriage is to be legally recognized. Even if the government recognizes traditional marriage celebrations, the high proportion of common-law status couples may rely on the absence of contract between spouses in some Aboriginal communities.

Same-sex Unions

Canada was among the first countries in the world to recognize same-sex marriages. It was a long fight since the decriminalization of homosexuality by the federal government in 1969. The Government of Québec included the sexual orientation into the Québec Human Rights Code in 1977, a decision that was followed by other provinces and territories (with the exception of Alberta, Prince Edward Island and the Northwest Territories). Many attempts were made to include the recommendation of including sexual orientation in the Canadian Human Rights Act between 1979 and 1996, the year Bill C-33 passed. In 2000 same-sex couples gained the same benefits as other common-law couples. The first provincial government to rule in favour of the recognition of same-sex marriages was the government of Ontario in 2002, followed by the government of British Columbia in 2003 and the government of Québec in 2004. Same-sex marriage was finally recognized by the federal government in 2005. The subject still divided the Canadian population and the following Conservative Government tried to reopen the debate but the vote on this issue was defeated in the House of Commons by the opposition parties (see among others Hurley, 2007). In 2006, there were 45,345 same-sex couples representing less than 0.6% of all couples and 16.5% of them legally married. By itself, the Toronto region concentrates 21% of the declared same-sex couples in Canada, whereas Montréal and Vancouver are home to 18% and 10%, respectively.

With this changing dynamic in marital status, the single-parent family profile has also changed. In the beginning of the 20th century, most single-parent families were the result of the death of one spouse.

Table 1. Marital status, provinces, territories and Canada, 2006

	Never legally married (single)	*Com-mon-law*	Legally married, not separated	Separated but legally married	Di-vorced	Widow
Newfoundland and Labrador	30.9%	*(7.7%)*	54.3%	2.3%	5.6%	6.9%
Prince Edward Island	31.3%	*(7.2%)*	52.0%	3.5%	6.3%	6.9%
Nova Scotia	32.2%	*(8.9%)*	49.5%	3.6%	7.6%	7.1%
New Brunswick	32.3%	*(10.1%)*	50.0%	4.0%	6.7%	7.0%
Québec	43.2%	*(19.4%)*	37.5%	2.1%	10.6%	6.5%
Ontario	31.6%	*(7.0%)*	51.9%	3.5%	6.8%	6.1%
Manitoba	33.3%	*(7.2%)*	50.2%	2.8%	6.8%	6.9%
Saskatchewan	32.9%	*(7.4%)*	50.8%	2.5%	6.6%	7.3%
Alberta	34.0%	*(8.6%)*	50.7%	2.8%	7.7%	4.8%
British Columbia	32.1%	*(8.2%)*	50.4%	3.2%	8.3%	6.0%
Yukon Territory	43.7%	*(16.1%)*	38.9%	4.0%	9.6%	3.9%
Northwest Territories	52.0%	*(18.8%)*	36.3%	3.0%	5.8%	3.0%
Nunavut	61.4%	*(22.7%)*	31.0%	2.2%	2.4%	3.0%

Source of data: Statistics Canada, 2006 census.

Today, divorce parties the cause of single-parent families. Some divorced parents remarry or form common-law unions that create other types of reconstituted families. Moreover, the increasing number of births outside unions and the growing number of childless couples indicates that the traditional family, composed of a couple with children, is becoming less common.

In sum, families tend to have fewer children or no children at all. With fewer children, family (and household) size is shrinking. Family composition is also changing, taking more complex forms than the traditional couple with children. The Canadian total fertility rate is not sufficient to maintain population size. Today, the migratory increase is already more important than the natural increase in terms of population growth. The Canadian population is aging and the need for immigration will become more important when, eventually, the mortality rate will increase due to the population age structure.

Table 2. Family types, provinces, territories and Canada, 2006

	Couples	Single-parent families
Newfoundland and Labrador	84.5%	15.5%
Prince Edward Island	83.7%	16.3%
Nova Scotia	83.1%	16.9%
New Brunswick	83.7%	16.3%
Québec	83.4%	16.6%
Ontario	84.2%	15.8%
Manitoba	83.0%	17.0%
Saskatchewan	83.4%	16.6%
Alberta	85.6%	14.4%
British Columbia	84.9%	15.1%
Yukon Territory	79.3%	20.7%
Northwest Territories	78.6%	21.4%
Nunavut	72.4%	27.6%
Canada	84.1%	15.9%

Source: Statistics Canada, 2006 census.

Ethnocultural Characteristics and Immigration

It was the French who established the first colony on land that would later become Canada. During the French occupation, population growth was slow; there were few colonists with very few women among them. Various programs and incentives, such as the attribution of free lands, distribution of crops, materials, were necessary to attract colonists. Population growth began to increase significantly once Canada was handed over to Britain. The handover resulted in the arrival of successive waves of immigrants, mainly from Scotland and Ireland. The new arrivals rapidly colonized Newfoundland, Nova Scotia and Prince Edward Island and significant numbers also settled in the province of Québec (Careless, 1997). However, descendants from the French colonists were not assimilated by the British. Today, these two cultures are still distinguishable in many aspects of Canadian daily life.

Over the last 100 years, Canada has welcomed 13.4 million immigrants. Today, 6.2 million Canadians are first generation immigrants. This represents nearly 20% of the Canadian population. Canada is second only to Australia (22%) with regard to the proportion of immigrants in its population (Boyd and Vickers, 2000). With a low fertility rate and an aging population, immigration represents an important part of population growth. The role immigration will play in maintaining or increasing population growth may become more significant in the near future.

Why Migrate?

Why do people migrate to and from a country or specific regions within a country? Three categories of factors have an impact in the decision to migrate (*push/pull factors*): political factors, economic factors, and environmental factors (Rubenstein, 2004). Political factors include the political systems (both of the country of origin and the receiving country). For example, political refugees are pushed by the political system of their country and are attracted by the Canadian democratic system. This category of factors mainly significant for international migrants, although Canada represents an example where political factors may also play a role at the regional level. In the province of Québec, the election of a political party dedicated to obtaining independence from Canada may act as a push factor towards other regions of the country for some Québecois. Economic factors are related to employment and wealth. A lack of jobs in the country or region of origin may push migrants while the employment possibilities in the receiving country or region may attract them. Finally, environmental factors include the environmental risks such as floods, droughts, earthquakes, or less dramatic environmental factors such as topography or weather.

Migration has negative and positive impacts on both the region of origin and on the destination region. For the region of origin, emigrants may decrease the pressure on the job market and on infrastructures (health, education, welfare, etc.). However, the region of origin looses its investment in people (such as health and education), and the future economic contribution of the emigrant. For the destination region, immigrants contribute to the economy without any investment prior to this contribution. At the same time, immigrants add pressure both on the destination region's economy and infrastructures.

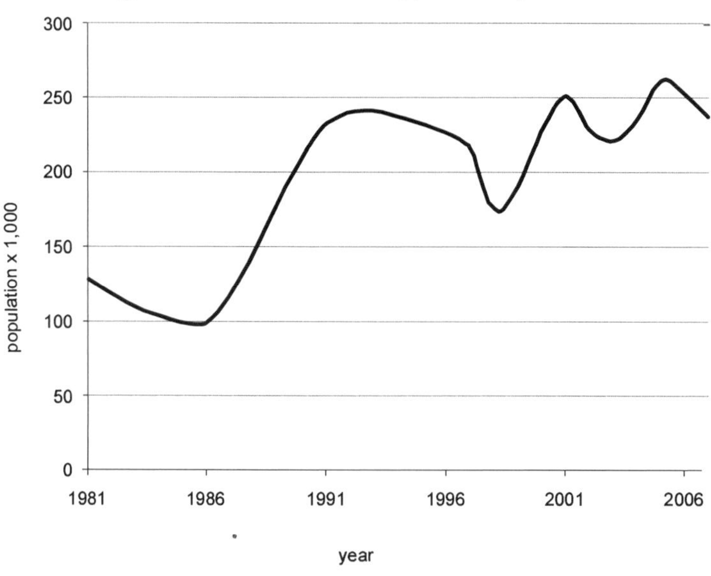

Source of data: Statistics Canada. Catalogue No. 91-209-X.

The most common type of immigrant in Canada is the economic immigrant. Economic immigrants have represented between 24% and 29% of newcomers during the last ten years. The second type of immigrant is the family immigrant (family reunion). In 2007, one in four immigrants admitted to Canada was a family immigrant. The third type of immigrant is the refugee immigrant. They represented 12% of the newcomers in 2007 (Statistics Canada, 2008i).

The Canadian Mosaic

With two predominant cultures resulting from colonization, one in five people foreign-born, and more than 1 million aboriginal people, Canada is a mosaic of different cultures. More than 200 ethnic identities were declared in the 2006 census. Immigrants and aboriginal people tend to concentrate in specific regions of the country with 85% of immigrants residing in four provinces: 38% in Ontario; 24% in Québec; 13% in British Columbia and 10% in Alberta.

Composition and Distribution

The majority of Canadians consider themselves to be of Canadian ethnic origin, or they declared a regional identity – a growing tendency, even more so among Québecois. The other main ethnic groups declared were British, French, Scottish, Irish, German, Italian, Chinese, First Nations and Ukrainian (Statistics Canada, 2008i). In 1981, the majority of newcomers emigrated from Great Britain. Today, Chinese, Indians, Filipinos and Pakistanis form the most significant group of newcomers; most certainly, this will have an impact on the Canadian mosaic.

Figure 8. The Canadian Mosaic: Distribution of immigrants by provinces and territories, 2006

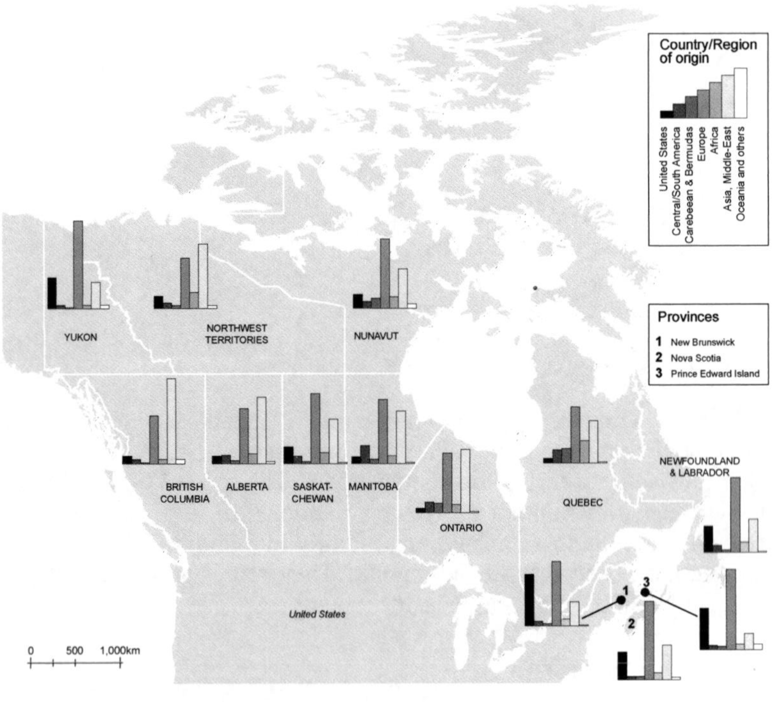

Source of data: Statistics Canada. 2006 census.

Aboriginal People

Canada's aboriginal population now exceeds 1 million inhabitants and represents 4% of the country's population, placing Canada in the second position among countries regarding the size of its aboriginal population. New Zealand occupies the first position with 15% of its population claiming aboriginal descent. With a young population and a higher fertility rate, the population growth rate, over the last ten years, of aboriginal population was almost 6 times higher than the rate of the non-aboriginal population (45% compared to 8%, between 1996 and 2006).

Canada's aboriginal people come from three distinct groups: Inuit, First Nations (also called North American Indians) and the Métis. Inuit concentrate mainly in Northern Québec (along the north coast of Labrador) and in Nunavut and the Northwest Territories (in an area referred to as the Inuit Nunaat, the Inuit homeland). The Métis are the descendants of the unions between aboriginals and European ancestors, with succeeding generations the result of unions between Métis people. The Métis eventually formed their own communities with common cultural practices, thus becoming a distinct aboriginal group (Métis National Council, 2008). The Métis witnessed a population growth rate of 91% during the last ten years, the most significant increase among aboriginal groups. Nearly 86% of Métis are living in an area that stretches from Ontario to the Pacific Ocean, with the greatest concentration in Alberta (22%). As for the First Nations, 23% of them are living in Ontario and another 19% in British Columbia.

The aboriginal population forms 4% of the Canadian population, however, as shown in figure 9, there are important regional variations. In Nunavut for example, 85% of the population is Aboriginal, 50% in the Northwest Territories and less than 1.5% in the province of Québec.

Aboriginal populations follow the general tendency to urbanize, even if the aboriginal population has a significantly lower urbanization rate than the overall Canadian population. Ten years ago, 50% of the aboriginal population was already urban – this proportion grew to 54% in 2006 (and the national average is 80%). Aboriginal concentrate in cities such as Winnipeg, Edmonton, Vancouver or Toronto with populations that vary from 26,575 in Toronto to 68,380 in Winnipeg.

Other Visible Minorities

In 2007, visible minorities[3] accounted for 16% of the Canadian population (five million people) and their growth rate is more important than the total Canadian population growth rate. The significant increase of population identified as visible minorities is more related to ethnic origin of newcomers than on their fertility rate. Three new immigrants out of five are part of visible minorities.

Different groups compose visible minorities. The importance of each group varies, the most important one being south Asian (including ascendancy with the Indian subcontinent), followed by Chinese and Black.

3 For Statistics Canada, visible minorities refer to "persons, other than Aboriginal peoples, who are non-Caucasian in race or non-white in colour".

For this last group, half of it originates from the Caribbean and 40% from Africa.

Figure 9. Aboriginal population, provinces and Canada, 2006

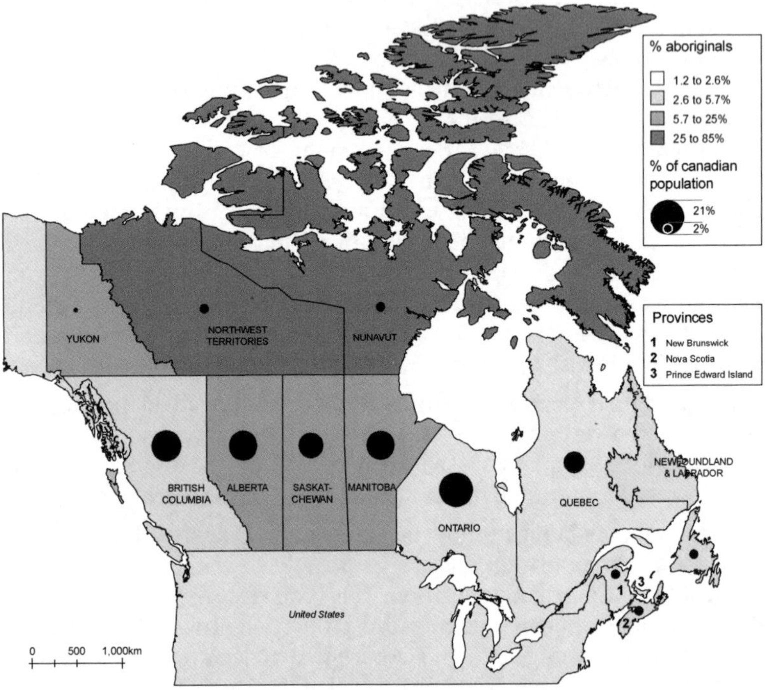

Source of data: Statistics Canada, 2006 Census.

Visible minorities concentrate in certain areas of Canada – 25% of Canada's visible minorities concentrate in British Columbia, 23% concentrate in Ontario and 14% in Manitoba. Only 8% of Canada's visible minorities live in the Atlantic Provinces (Nova Scotia, New Brunswick, Newfoundland and Labrador and Prince Edward Island) combined. In addition to concentrating in certain provinces, 96% of Canada's visible minorities live in metropolitan areas.

Language

With a changing ethnic profile, it is not surprising that Canada has witnessed an increased number of Canadians who have a non-official language as a mother tongue. In ten years, the population whose mother-tongue is English decrease of two percentage points and French mother-tongue 1 point to other languages. Today, Canadians claim over 200 different languages as a mother tongue; among these languages are the two official languages (English and French) and Aboriginal languages as well as German, Italian, Dutch, Chinese, Spanish and Arabic. More than six million Canadians have a different mother tongue than the official languages and this number will probably continue to increase alongside increased immigration from non-English or non-French speaking countries and/or communities.

There are only three regions of the country where the number of official language speakers is increasing: Nunavut, the Northwest Territories and Saskatchewan. In these regions, an increase in the use of English has resulted in the loss of the use of aboriginal languages. In fact, Norris (1998) noted that ten languages have been completely lost during the last century for reasons such as marriage between language groups, urbanization, the strong presence of official languages and historical attempts to assimilate young aboriginals into the main culture[4]. Facing this dramatic loss of culture, aboriginal groups now try to instil and celebrate their cultural practices among their youth – for example, Inuktitut, the language of the Inuit, is now part of Inuit school curriculum.

Religion

With increasing ethnic diversification, it could be assumed that the cultural practices in Canada are also diversifying. According to some investigation made by Statistics Canada[5] the first generations of immigrants are more susceptible of keeping their traditions and customs and to find them important in their daily lives. It is even more important among recent immigrants. These traditions and customs include clothing, food, music and religious practices.

Among the non-aboriginal population, almost 44% of Canadians are Catholic, 29% are Protestant and close to 17% claim to have no religious affiliation. The regional differences highlight the differences between the double colonization (French and British). In the province of Québec, 83% of the population are Catholic while in the neighbouring province of Ontario, the percentage drops to 35% (equal to the percentage of Protestants). Only 6% of Québecois claim to have no religious affiliation, but this percentage rises to more than 36% in British Columbia (2001 census).

---

[4] In June 2008, Prime Minister Stephan Harper made, on behalf of Canadians, a Statement Apology for the Indian school system put in place to assimilate young aboriginals. The statement can be read on the ministry of Indians and North Affairs (http://www.ainc-inac.gc.ca).

[5] One investigation was made in 2002 among more than 42,000 people aged 15 years and over (excluding aboriginals – subject to another study (Catalogue number 89-593).

Despite the fact that more than 80% of Canadians claim some religious affiliation, only 50% of the adult population participate actively in their religion (Statistics Canada, 2003a).

Figure 10. Population's mother tongues by provinces and territories, 2006

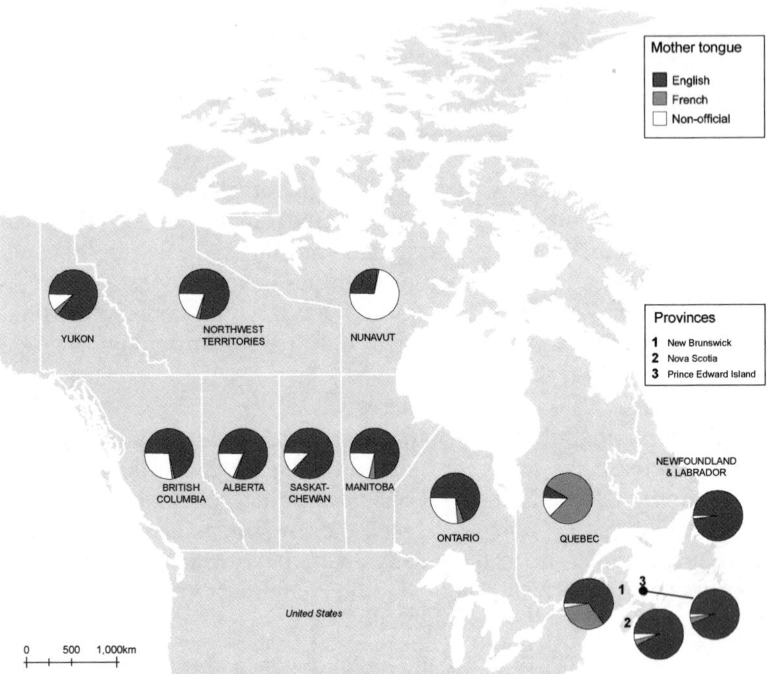

Source of data: Statistics Canada, 2006 Census.

Canadians also have a tendency to keep religious affiliation and practice private (i.e., not participate in public rites) (Warren and Schellenberg, 2006). As a group, Canadians are becoming more private with respect to religious practice. Traditional, more public displays of religious affiliation to Catholic or Protestant churches are on the decrease.

The two predominant cultures often identified as the "two solitudes", are composed of the descendants of French Catholics and English speaking Protestants. The first group is found in the largest numbers in the province of Québec, while smaller French speaking communities and individuals are dispersed throughout the country. The other "solitude" constitutes the majority of the population of Canada, but is a minority in Québec.

After centuries of Catholic Church control over the everyday life of Québecois[6], 1960 marked a change in the society. Considered by some as a catch-up period in the process of modernization of Québec society, the election of a more progressive government in 1960 gave Québecois the tools to develop modern state institutions covering health, education, legal rights for women, etc. These initiatives modernized Québec society and also recognized a uniquely Québec identity. With the creation of the *Office de la langue française* in 1961 and the adoption of bills such as Bill-22, the declaration of French as the official language in Québec, and Bill-101, the Charter of the French language. These bills had a profound impact on the English community. In the presence of economic decline and a perceived cultural gap, many made the decision to move to other regions of the country.

Human Development: Wealth and Poverty

Education

Education (or schooling) is an interesting indicator of a population's productive capacity and its integration into a national or regional economy. In Canada, the level of education, measured by the average years of schooling for the population 15 years and over, is increasing. In 1990, the average level of education in Canada was 12 years – this number almost increased to 13 years in only a ten-year period. But what about the highest level of education reached among the adult population, the types of training they have received, and their activity in the labour market?

Not only did the education level increase, but the gender gap decreased and is almost inexistent today. In 1971, only 3% of the Canadian women aged 15 years and over had a university degree, less than half the percentage of men who had a university degree. Thirty years later, only one percentage point separates women from men (15% vs. 16%).

As for the adult population, aged 25 to 64, only 15% of them have not earned a certificate or diploma, and more than 60% of them completed postsecondary education. Canada occupies the 6th position among the 30 member countries of the Organisation for Economic Co-operation and Development (OECD) with respect to the percentage of adults in the population who have earned a university degree. However, there are regional differences that reflect the overall economic well-being of the populations. As shown in the following figure, the highest proportion of adults without any certificate or diploma live in Nunavut (46%), fol-

[6] The Catholic Church tended also to impose its view on women fertility: *à la grâce de Dieu* (translation: as many as God wants us to have) was the leitmotiv.

lowed by Newfoundland and Labrador (26%). As for the percentage of a population that has earned a university degree, Ontario (31%) and British Columbia (30%) have a greater percentage than the national average (28%).

The gender gap in the level of education may have almost disappeared; women and men still tend not to choose the same fields of study. The most popular fields of postsecondary studies – business, administration, and marketing – are the same for both sexes although its popularity is higher among women. One-in-four women choose this field of study (postsecondary studies) compared to only one-in-six men. In second and third position, women choose health and welfare programs (about 20%) and education (more than 10%). As for men, they choose techniques and mechanics (around 10%) and engineering (less than 10%).

Table 3. Education level, population 25-64 years old, Provinces, territories and Canada, 2006

	No certificate or diploma	University degree
Newfoundland and Labrador	26%	18%
Prince Edward Island	19%	22%
Nova Scotia	19%	25%
New Brunswick	21%	20%
Québec	17%	26%
Ontario	14%	31%
Manitoba	20%	24%
Saskatchewan	19%	22%
Alberta	15%	27%
British Columbia	12%	30%
Yukon Territory	15%	26%
Northwest Territories	23%	23%
Nunavut	46%	15%
Canada	15%	28%

Source of data: Statistics Canada, 2006 Census.

A higher level of education is generally associated with a better *employment rate*, lower *unemployment rate*, higher income, and lower gender disparity. The employment rate is the proportion of the active population (which exclude students, people living in institutions and people who have retired from the work force for diverse reasons) who has a job the week prior to the census. The unemployment rate is the proportion of the active population who does not have, but are in search of a job. As shown in table 4 below, for Canadians who have not completed their secondary education ("high school") and who are between 25 and 64 years of age, the percentage of those unemployed is 9.3% for men and 11.15% for women. Those who have earned a university degree have a lower unemployment rate –, only 4.2% of men and 4.0% of women in this group are unemployed. And the gender disparity is also lower in terms of income; a woman with a university degree earns 85 cents for each dollar a man earns (in 2005) while a woman with less than 9 years of schooling earns only 70 cents (Statistics Canada, 2006).

Table 4. Employment and unemployment rate among adults 25-64 years old, without a secondary certificate and with university degree, Canada and OECD, 2003

			Without high-school certificate	With university degree
Employment rate	Canada	Men	63.5%	86.7%
		Women	43.1%	79.2%
	OECD	Men	68.7%	89.0%
		Women	44.5%	79.4%
Unemployment rate	Canada	Men	9.3%	4.2%
		Women	11.2%	4.0%
	OECD	Men	12.2%	3.1%
		Women	13.2%	3.9%

Source of data: OECD 2008.

Labour Force Composition and Regional Mobility

Activity

Over the last 30 years, Canadian women have increased their education level and their participation in the labour market. The *activity rate* was of 46% (women aged 15 years and over) thirty years ago and has now reached 63%. And the gap between men and women is shrinking with the double occurrence of increased participation in the labour market by women and a slight decrease in participation by men. Despite this shrinking gap, there is still a difference of 10 percentage points (63% for women, 73% for men).

The reasons explaining the men's activity rate decrease are related to two economic slowdowns: a recession in the 1980s and another in the 1990s. Manufacturing jobs are more sensitive to economic cycles and massive layoffs occurred during these recessions. In 1982, 208,000 jobs were lost, and 315,000 were lost between 1990 and 1992 (Statistics Canada, 2008f). Since women tended to be more active in the service sector, their employment was less vulnerable to economic downturns.

The gender gap still exists, and is highest in Alberta where the activity rate of women is 68% (the highest among the provinces), while it is close to 81% for men. Alberta also has the highest activity rate (74% according to Statistics Canada estimation for 2007, CANSIM table 282-0002). Alberta also boasts the lowest unemployment rates (less than 4% for both men and women). The provinces of Newfoundland and Labrador and Prince Edward Island have the highest unemployment rates, with rates of 14% and 10% respectively.

With an activity rate of nearly 68% and an unemployment rate of 6%, the Canadian economy is going well overall, with some regions doing better than others.

Alberta has a booming economy related (in part) to energy extraction which employs nearly 223,000 workers. But, as in other regions of the country, the general tendency is towards the *tertiarisation* of its economy. The tertiarisation of the economy is the evolution of an economy based on primary activities (extraction, agriculture, fisheries and so on) or secondary activities (often called blue collar activities which include manufacturing jobs, construction and so on) towards an economy based on service activities (often called the white collar activities). For many reasons – such as the cost of manpower, the high value of the Canadian dollar in comparison with other currencies, and poor productivity rates – many businesses delocalise their manufacturing activities to countries where labour is much less expensive. In Canada today, three out of four employees are working in the service sector (which also includes low-end jobs such as in restaurants), a gain of 10 percentage points in 30 years.

But, the tertiarisation of the economy does not mean the end of the primary or the secondary sector of the economy. They are still dynamic sectors, especially in some regions. For example, the primary sector constitutes no less than 13% of the Saskatchewan manpower while in Alberta it occupies 10% – the national average is only 4%. The manufacturing sector is more dynamic in Québec and Ontario (14% each). However, these figures do not highlight the manpower movement between regions.

Table 5. Activity, employment and unemployment rates, population 15 years and over, provinces and Canada, 2007

	Activity rate			Unemployment rate			Employment rate		
	Men	Women	Both	Men	Women	Both	Men	Women	Both
Newfoundland and Labrador	63.7	55	59.2	14.9	12.1	13.6	54.2	48.3	51.2
Prince-Edward Island	71.8	65	68.2	11.2	9.2	10.3	63.7	58.9	61.2
Nova-Scotia	67.8	59.8	63.7	9.3	6.7	8.0	61.5	55.8	58.6
New Brunswick	68.3	59.9	64	8.8	6.2	7.5	62.3	56.2	59.2
Québec	70.6	61	65.7	7.9	6.4	7.2	65	57.2	61
Ontario	72.6	63.5	68	6.8	6	6.4	67.7	59.7	63.6
Manitoba	75.4	63.6	69.4	4.5	4.3	4.4	72.1	60.9	66.4
Saskatchewan	76	63.5	69.7	4.3	4.1	4.2	72.8	60.9	66.8
Alberta	80.5	67.6	74.1	3.3	3.7	3.5	77.8	65.1	71.5
British Columbia	71.7	61	66.3	4	4.5	4.2	68.9	58.2	63.5
Canada	72.7	62.7	67.6	6.4	5.6	6	68	59.1	63.5

Source: Statistics Canada, Catalogue 71.001.X. CANSIM table 282-0087.

With less dynamic economy based on primary or secondary activities in some regions, it is not surprising that there are important regional migrations from some provinces to others. According to the last census, 3.4% of the active population moved from one province or territory to another during the last five years, which represents more than half a million workers.

As figure 11 shows, Alberta, with its booming economy not only in the energy extraction sector but also in construction and public administration, attracted the largest proportion of the population.

Figure 11. Regional mobility between provinces and territories, Canada, 2005

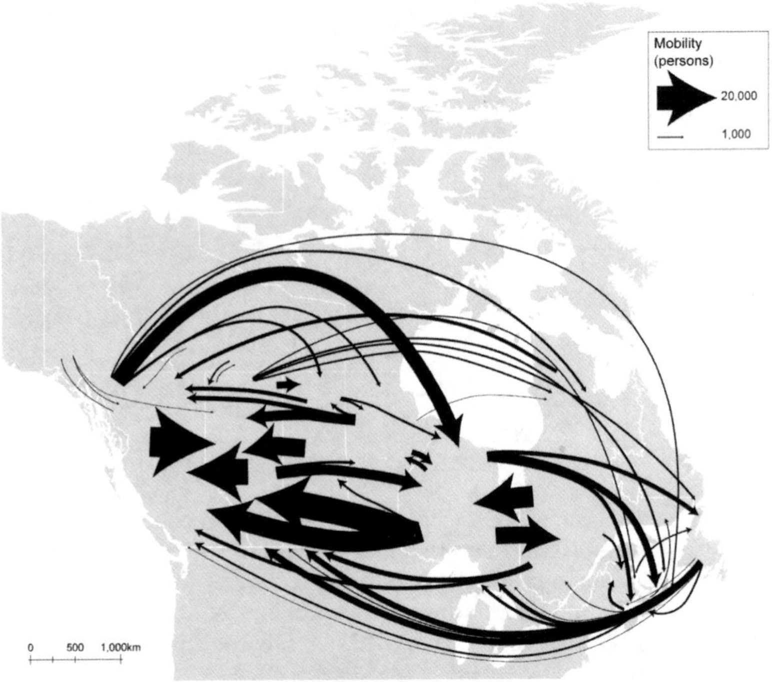

Source of data: Statistics Canada, Catalogue No. 91-209-X.

Personal and cultural characteristics (such as the language spoken) as well as the economic base of the regions will influence the migration process. Migration movements have an impact both on the receiving communities and on the communities of origin. Because the migrants tend to be more educated, this could represent an important loss of fiscal taxes for the community of origin. For the receiving communities, it could represent a gain in fiscal taxes, and it may be the only way to facilitate population growth when they are facing a very low fertility rate.

Socio-professional Profiles

With 2.6 million workers, sales clerks and sales representatives continue to be the most popular profession for both men and women in 2005. With a flourishing economy, demands for goods and services are on the rise which explains the dynamism of this sector. Despite an increasing level of education for both men and women, and despite a growing economy, the other most common jobs are of low end and traditionally gendered: truck drivers, concierges, mechanics and so on for men; cashiers, nurses, secretaries for women. Only 48% of adults with a university degree from a program of less than 3 years have a skilled occupation, almost 20 percentage points below OECD average (OECD, 2008).

Poverty

The United Nations 2006 Human Development Report classifies Canada in 6th position among all countries regarding its level of human development (UNDP, 2007). Unsatisfied with the use of basic indicators such as the Gross Domestic Product, The United Nations composed an indicator: the *Human Development Index (HDI)*. The HDI looks at three dimensions of human development: "living a long and healthy life (measured by life expectancy), being educated (measured by adult literacy and enrolment at the primary, secondary and tertiary level) and having a decent standard of living (measured by purchasing power parity, PPP, income)" (UNDP, 2007). This indicator, even if incomplete, allows studying the human development evolution in a country and facilitates basic comparison between nations. However, this indicator is rather poor for evaluating the integration of the population in the society.

What Is Poverty?

There are two basic ways of evaluating poverty within a population. The first one is *absolute poverty*. In situations of absolute poverty, the population is not able to meet its basic subsistence needs. This is why the use of an indicator such as the HDI for industrialised countries puts them at the highest rank among all countries in terms of human development. However, this could hide important inequalities inside the country. The second way of evaluating poverty is by looking at the *relative poverty*, the people's integration into the society in which they live. In this case, the problem of how to evaluate this level of integration remains. There is no real consensus on the measure of poverty in Canada.

Indices of Poverty and Wealth

Various types of indicators exist for evaluating poverty. Some indicators are exclusively based on income while others take into account the living environment. Depending on the poverty index used, the level of poverty could be very low or relatively high in Canada. For example, a well-known economist from the Fraser Institute, a right-wing think-tank, developed an index, which, despite its pretension of giving a complete basket of goods and services considered as essential, gives a poverty line closer to absolute poverty than the relative poverty. Using this index could cut the poverty rate by three quarters if compared with the *low-income cut-off* used by Statistics Canada. The low income cut-off is used by Statistics Canada, not as a poverty index but rather as an indicator of "a difficult situation" faced by families or by individuals with regard to an established standard of living. Canada does not officially have a poverty index.

The low income cut-offs were determined for the first time 40 years ago, based on an investigation on income and family spending. It was then evaluated that the average family used approximately half of its income for basic needs (food, housing and clothing). A cut-off was arbitrarily determined 20 points above the median proportion of income spent. In other words, a family was considered in a difficult situation if it spent 70% or more of its income for basic needs. Other investigations were done more recently for evaluating the family spending but the cut-offs is still established at 20 points above and now take also into account family size and where they live.

The use of the low-income cut-off as an indicator shows a general tendency of improvement of family situations. In 1995, one family unit in five was below the cut-offs point while it was one in six ten years later. In the context of a dynamic economy, Canada witnessed a decrease in poverty after 25 years of increase. It took more than 10 years for the income to adjust after the recession in the 1990s.

Table 6. Low-income units, family types, Canada, 1995 and 2005, in proportion

	1995	2005
Families of two persons or more	12.1%	7.4%
Families of elderly people	3.3%	1.6%
Couples (other than elderly) without children	8.4%	6.4%
Families (couple) with children	10.9%	6.7%
Single parent women-run families	52.7%	29.1%
One-person households	37.3%	30.4%

Note: Income after taxes, based 1992

Source of data: Statistics Canada

There are some types of families, individuals, and groups who are more at risk of facing a period of poverty – single-parent families, one-person households, Aboriginals, disabled persons, and so on. Their availability for work, the economic burden of having a family living on a small income and problems of integration into the workforce explain these higher risks. But even among more at-risk groups, the general situation has improved. For example in 2005, among the single-parent, woman-run families, 29% are below the cut-off line, 24 percentage points lower than 10 years earlier. The decrease can be explained by an increase in the participation of women in the labour market. As for other at risk groups, 30% of one-person households were below the cut-off line (in 1996) compared to 36% of visible minorities (in 1996) and 31% of disabled people (in 1996) and Aboriginals.

The case of aboriginal groups is particularly dramatic. 43% of aboriginals live below the low income cut-off line, the highest rate among the different at-risk categories. Even if poverty seemed generalized among these communities, there are spatial distinctions that can be made. Communities living in British Columbia or in the South of Ontario have faced a less dramatic situation than the communities living far north. That being said, the wealthiest aboriginal communities tend to be comparable of the poorest non-aboriginal communities (Armstrong, 2001).

Conclusion

Canada is a vast country where people concentrate in a small part of its territory. Canada's population is aging, but different waves of immigration continue to contribute to its population growth. Canada is a cultural mosaic: aboriginal groups which form a minority, the descendants of the European conqueror and successive waves of immigrants – first from European countries but now from every part of the world.

Canadian society evolved according to demographic movements. There are cultural transformations such as less dominant Catholic and Protestant churches; this has allowed for the evolution of a more secular society where the rights of women, homosexuals, aboriginals and immigrants are recognized. Canada is a mosaic in many ways: a cultural, economical, physical and economic montage.

Review Questions

- What did and still does influence human settlements in Canada?
- Why the Baby Boom was so important in Canada?
- Can you explain why a total fertility rate of less than 2.1 is not sufficient to maintain a population size?
- Calculate the dependency ratio of elderly for the province of Newfoundland and Labrador and for Nunavut in 2006. To answer this question, visit Statistics Canada website and search the summary tables for age distribution by provinces and territories.
- How can you explain the aging Canadian population?
- What could be the pull factors for immigrating to Canada?

- Which Canadian metropolitan area concentrates the highest proportion of visible minorities? To answer this question, visit Statistics Canada website and search the summary table by metropolitan areas.
- In your opinion, why an index of poverty measuring the absolute poverty is less useful in a country like Canada?

Further Readings

Battle, Ken (2003), *Minimum Wages in Canada: A Statistical Portrait with Policy Implications*, Ottawa: Caledon Institute of Social Policy.

Behiels, Michael (2004), *Canada's Francophone Minority Communities*, Montréal: McGill-Queen's University Press.

Bissoondath, Neil (1995), *Selling Illusions: The Cult of Multiculturalism in Canada*, Toronto: Penguin.

Conrad, Margaret and Alvin, Finkel (2002), *History of the Canadian Peoples*, Vol. II, 3rd ed., Toronto: Addison Wesley.

Edwards, John (ed.) (1998), *Language in Canada*, Cambridge, U.K.: Cambridge University Press.

Jackson, Robert J. and Doreen, Jackson (2006), *Politics in Canada: Culture, Institutions, Behaviour and Public Policy*, Toronto: Pearson – Prentice Hall.

Kaplan, William (ed.) (1993), *Belonging: The Meaning and Future of Canadian Citizenship*, Montréal: McGill-Queen's University Press.

Knox, Paul and Steven, Pinch (2006), *The Cultural Landscape: Urban Social Geography. An Introduction*, Edition, Prentice Hall.

McBride, Stephen (1992), *Not Working: State, Unemployment and Neo-Conservatism in Canada*, Toronto: University of Toronto Press.

Roach, Ruth, *et al.* (1993), *Canadian Women's Issues*, Toronto: Lorimer.

Rubenstein, James M. (1996), *An Introduction to Human Geography*, 8th ed., Upper Saddle River: Prentice Hall.

Smith, Miriam (1999), *Lesbian and Gay Rights in Canada: Social Movements and Equality-Seeking*, Toronto: University of Toronto Press.

Trigger, Bruce G. and Wilcomb, E. Washburn (eds.) (1996), *The Cambridge History of the Native Peoples of the Americas*, Cambridge. U.K.: Cambridge University Press.

Webber, Jeremy H.A. (1994), *Reimagining Canada: Language, Culture, Community and the Canadian Constitution*, Montreal: McGill-Queen's University Press.

Websites

Canada e-Book. Statistics Canada: http://www43.statcan.ca/r000_e.htm

Same-sex marriage in Canada. The right of gay and lesbian couples across the country. CBC news in depth story. Interactive: http://www.cbc.ca/news/interactives/map-samesex-marriage/

Statistics Canada: http://www.statcan.ca

The Aboriginal Canada Portal: http://www.aboriginalcanada.gc.ca/acp/site.nsf/en/index.html

The Atlas of Canada. Government of Canada. Natural resources Canada: http://atlas.nrcan.gc.ca

The Canadian Encyclopedia: http://www.thecanadianencyclopedia.com

The Companion Website for The Cultural Landscape: An Introduction to Human Geography 8th Edition: http://wps.prenhall.com/esm_rubenstein_humangeo_8/

The virtual library of the Canadian Parliament: http://www.parl.gc.ca/common/Library.asp?Language=E

Urban Poverty Project 2007. Canadian council on social development: http://www.ccsd.ca/pubs/2007/upp/

References

Almey, Marcia (2006), *Women in Canada: Work Chapter Updates*, Statistics Canada, catalogue No. 89F0133XWE, [Online – http://www.statcan.ca/english/freepub/89F0133XIE/89F0133XIE2006000.pdf].

Appleby, Brenda Margaret (1999), *Responsible parenthood: decriminalizing contraception in Canada*, Toronto: University of Toronto Press.

Armstrong, Robin (2001), "The Geographical Patterns of Socio-economic Well-being of First Nations Communities", *Agriculture and Rural Working Paper* Series No. 46, Statistics Canada, catalogue No. 21-601-MIE, [available online – http://www.statcan.ca/english/research/21-601-MIE/21-601-MIE2001046.pdf].

Baldwin John R. and W. Mark Brown (2004), *Four Decades of Creative Destruction: Renewing Canada's Manufacturing Base from 1961-1999*, Analytical Paper, Statistics Canada, catalogue No. 11-624-MIE – No. 008.

Bélanger, Alain and Éric, Caron Malenfant (2005), "Ethnocultural diversity in Canada: Prospects for 2017", *Canadian Social Trends*, No. 79, Statistics Canada, catalogue No. 11-008, [available online – http://www.statcan.ca/english/freepub/11-008-XIE/2005003/articles/8968.pdf].

Boyd, Monica and Michael, Vickers (2000), "100 years of immigration in Canada", *Canadian Social Trends*, No. 58, Statistics Canada, catalogue No. 11-008, [available online – http://www.statcan.ca/english/freepub/11-008-XIE/2000002/articles/5164.pdf].

Canadian Council on Social Development (2007), Poverty by Geography. Urban Poverty in Canada, 2000.

Careless, James Maurie, Stockford (1997), *Canada: A Celebration of Our Heritage*, Mississauga: Heritage Publishing House.

Caron Malenfant, Éric *et al.* (2007), *Demographic Changes in Canada from 1971 to 2001 Across an Urban-to-Rural Gradient*, Research Paper, Statistics

Canada, catalogue No. 91F0015MIE – No. 008, [available online – http://www.statcan.ca/english/research/91F0015MIE/91F0015MIE2007008.pdf].

CIA (2008), *The World Factbook*, [available online – https://www.cia.gov/library/publications/the-world-factbook/fields/2002.html].

Environment Canada (2005), *Climate Normals 1961-1990*, Climate Information Branch, Canadian Meteorological Centre.

Government of Canada, Natural Resources (2008), *The Atlas of Canada*, [online – http://atlas.nrcan.gc.ca/site/english/maps/peopleandsociety/population].

Hofmann, Nancy (2001), "Urban Consumption of Agricultural Land", *Rural and Small Town Canada Analysis Bulletin*, Vol. 3, No. 02, Statistics Canada, [available online – http://www.statcan.ca/english/freepub/21-006-XIE/21-006-XIE2001002.pdf].

Hurley, Mary C. (2007), "Sexual Orientation and Legal Rights", Parliamentary Information and Research Services, *Current Issue Revue*, No. 92-1E, [available online – http://www.parl.gc.ca/information/library/PRBpubs/921-e.pdf].

Métis National Council (2008), Website – http://www.metisnation.ca/.

Michalowski, Margaret *et al.* (2005), *Projections of the Aboriginal Populations, Canada, Provinces and Territories, 2001 to 2017*, Statistics Canada, catalogue No. 91-547-XWE, [available online – http://www.statcan.ca/english/freepub/91-547-XIE/91-547-XIE2005001.pdf].

Milan, Anne (2000), "One hundred years of families", *Canadian Social Trends*, No. 56, Statistics Canada, Catalogue No. 11-008, [available online – http://www.statcan.ca/english/freepub/11-008-IE/1999004/articles/4909.pdf].

Norris, Mary Jane (1998), "Canada's Aboriginal Languages", *Canadian Social Trends*, winter, Catalogue 11-008.

OECD (2008), *Education at a Glance 2008*, OECD Indicators, [available online – http://213.253.134.43/oecd/pdfs/browseit/9608041E.PDF].

Osberg, Lars (2000), *Schooling, Literacy and Individual Earnings*, Statistics Canada, catalogue No. 89F0120XIE.

Rubenstein, James M. (2004), *The Cultural Landscape: An introduction to Human Geography*, 8th edition. Prentice Hall.

Statistics Canada (2001), *History of the Census of Canada*, [online – http://www.statcan.ca/english/census96/history.htm].

Statistics Canada (2002), *General Social Survey – Cycle 15: Family History*, Catalogue No. 89-575-XIE, [available online – http://www.statcan.ca/english/freepub/89-575-XIE/89-575-XIE2001001.pdf].

Statistics Canada (2002), *General Social Survey – Cycle 15: Changing Conjugal Life in Canada*, Catalogue No. 89-576-XIE, [available online – http://www.statcan.ca/english/freepub/89-576-XIE/89-576-XIE2001001.pdf].

Statistics Canada (2003a), *2001 Census: analysis series. Religions in Canada*, Catalogue No. 96F0030XIE2001015, [available online – http://www12.statcan.ca/english/census01/products/analytic/companion/rel/pdf/96F0030XIE2001015.pdf].

Statistics Canada (2003b), *Ethnic Diversity Survey: portrait of a multicultural society*, Catalogue No. 89-593-XIE, [available online – http://www.statcan.ca/english/freepub/89-593-XIE/89-593-XIE2003001.pdf].

Statistics Canada (2003c), *The Canada e-book*, Catalogue No. 11-404-XIE, [available online – http://www43.statcan.ca/r00-e.htm].

Statistics Canada (2005), *The Canadian Labour Market at a Glance*, Catalogue No. 71-222-XIE, [available online – http://www.statcan.ca/english/freepub/71-222-XIE/71-222-XIE2006001.pdf].

Statistics Canada (2006), *Women in Canada. A Gender-based Statistical Report*, Fifth Edition, Catalogue No. 89-503-XIE, [available online – http://www.statcan.ca/english/freepub/89-503-XIE/0010589-503-XIE.pdf].

Statistics Canada (2007), *Population and dwelling count highlight tables, 2006 census*, Catalogue No. 97-550-XWE2006002, [available online – http://www12.statcan.ca/census-recensement/2006/dp-pd/hlt/97-550/Index.cfm?Page=INDX&LANG=Eng].

Statistics Canada (2008a), *Aboriginal Peoples, 2006 Census*, Catalogue No. 97-558-XWE.

Statistics Canada (2008b), *Births 2006*, Catalogue No. 84F0210X.

Statistics Canada (2008c), *Canada at a Glance*, Catalogue No. 12-581-XWE, [available online – http://www.statcan.ca/bsolc/english/bsolc?catno=12-581-X].

Statistics Canada (2008d), *Canada's Changing Labour Force, 2006 Census*, Catalogue No. 97-559-XWE.

Statistics Canada (2008e), *Canadian Demographics at a Glance*, Catalogue No. 91-003-X, [available online – http://www.statcan.ca/english/freepub/91-003-XIE/91-003-XIE2007001.pdf].

Statistics Canada (2008f), Canadian Economic Observer/L'observateur économique canadian, Catalogue No. 11-010-XIB.

Statistics Canada (2008g), *Deaths 2005*, Catalogue No. 84F0211X.

Statistics Canada (2008h), *Education, 2006 Census*, Catalogue No. 97-560-XWE.

Statistics Canada (2008i), *Ethic Origin and Visible Minorities, 2006 Census*, Catalogue No. 97-562-XWE.

Statistics Canada (2008j), *Jean-Talon: 1625-1694*, [online –http://www.statcan.ca/english/about/jt.htm].

Statistics Canada (2008k), *Labour Force Information*, September 14 to 20, Catalogue No. 71-001-X, [available online – http://www.statcan.ca/english/freepub/71-001-XIE/71-001-XIE2008009.pdf].

Statistics Canada (2008l), *Language, 2006 Census*, Catalogue No. 97-555-XWE.

Statistics Canada (2008m), *Population 1867 – 1970*, CANSIM table No. 075-001.

Statistics Canada (2008n), *Report on the Demographic Situation in Canada. 2005 and 2006*, Catalogue No. 91-209-X, [available online – http://www.statcan.ca/english/freepub/91-209-XIE/91-209-XIE2004000.pdf].

Statistics Canada (2008o), *The Canada Year Book*, Catalogue No. 11-404-XWE, [available online – http://www65.statcan.gc.ca/acyb.r000-eng.htm].

Statistics Canada (2008p), *Community Profiles, 2006 census*, Catalogue No. 92-591-XWE, [available online – http://www12.statcan.ca/census-recensement/2006/dp-pd/prof/92-591/index.cfm?Lang=E].

United Nation Development Program (2007), *Human Development Report 2007/2008*,
http://hdr.undp.org/en/media/HDR_20072008_EN_Complete.pdf.

UNFPA, the United Nations Population Fund (2011), State of World Population 2011. People and possibilities in a world of 7 billion, [available online – http://foweb.unfpa.org/SWP2011/reports/EN-SWOP2011-FINAL.pdf].

Warren, Clark and Grant, Schellenbert (2006), "Who's religious?", *Canadian Social Trends*, No. 81, Statistics Canada, catalogue No. 11-008, [available online – http://www.statcan.ca/english/freepub/11-008-XIE/2006001/main_religious.htm].

Warren, Clark (2006), "Interreligious unions in Canada", *Canadian Social Trends*, No. 82, Statistics Canada, Catalogue No. 11-008, [available online – http://www.statcan.ca/english/freepub/11-008-XIE/2006003/main_interreligious.htm].

CHAPTER 4

The Changing Canadian Economy

From Resources to Knowledge

Alfred HECHT

Wilfrid Laurier University

Introduction

Some people have described Canada as the only 'modern underdeveloped' country. The abundance of raw materials corresponds to the situation of many developing countries, while its high standard of living and the advanced structure of its economy make it part of the developed world. It is a member of the group of eight most developed countries, the G8, but it is the smallest member in both population and gross domestic product (GDP). From abroad, its economy is seen as still being highly dependent on the extraction of primary recourses. Relatively speaking, this view is true. However, it is not true in absolute terms, especially if one looks at GDP of output by the various sectors and employment in them. Resources are still strong in the export field where they still contribute around 30% to the total value of exports. However, over the years, increased productivity in the extraction of recourses and the profit from them has allowed Canada to develop strong niches in the auto, rail, aircraft and other transport related industries. These are now Canada's main export sectors.

Within the last two decades of the 20th and 21st centuries, Canada also developed a strong internationally oriented high tech sector with strong service and production components. However, by far the greatest sector, as measured by employment, is the service sector. Of the group of eight, Canada, after the United States, has the largest fraction of employment in this sector. Its present strength goes back to the trading mindset of the fur trade and the Hudson Bay Company era. Strong governmental and private service institutions are found in the fields of education, banking, health and retail. Large variation over space exists in nearly all sub areas of the Canadian economy. Largely they are the outgrowth of the historic settlement pattern and the different natural resource bases of the regions and provinces. Overarching inter-business linkages between Canadian enterprises and those in the USA make the Canadian economy highly dependent on the USA. This is strength and a weakness at the same time.

Table 1. Canadian GDP Absolute and in Percent, by Industries, 2008 and 2011[1]

	July 2008	% of	Oct. 2011	% of
Millions of chained dollars (2002)	$ (000,000)	Economy	$ (000,000)	Economy
All industries	1,238,091	100	1,275,001	100.00
Goods-producing industries	369,565	30.01	361,668	28.37
Agriculture, forestry, fishing and hunting	25,809	2.10	28,705	2.25
Mining and oil and gas extraction	56,843	4.62	59,004	4.63
Manufacturing	181,482	14.74	162,473	12.74
Construction industries	74,342	6.04	77,190	6.05
Utilities	31,089	2.52	34,296	2.69
Services-producing industries	869,617	70.61	909,927	71.37
Transportation and warehousing	57,450	4.67	60,126	4.72
Information and cultural industries	45,020	3.66	45,578	3.57
Wholesale trade	73,334	5.95	71,164	5.58
Retail trade	75,165	6.1	77,468	6.08
FIRE & RLMCE - see definition below	247,860	20.13	266,101	20.87
Professional, scientific and technical services	58,236	4.73	61,824	4.85
ASWMRS – see definition below	31,216	2.53	30,594	2.40
Public administration	69,519	5.65	76,406	5.99
Educational services	60,842	4.94	63,737	5.00
Health care and social assistance	79,358	6.44	84,837	6.65
Arts, entertainment and recreation	11,690	0.95	11,408	0.89
Accommodation and food services	27,365	2.22	27,341	2.14
Other services (except public adm.)	32,562	2.64	33,343	2.62

Note: adjusted to annual rates, FIRE & RLMCE = Finance/insurance/real estate/ renting/leasing enterprises: ASWMRS = Administrative and support, waste management and remediation services.

However, the great overall diversity in the Canadian economy (see Tables 1 and 2) bodes well for the future of Canada in the 21st century.

---

1 Statistics Canada; "Gross Domestic Products at Base Prices, by Industries (Monthly)" http://www40.statcan.ca/l01/cst01/gdps04a.htm, accessed September 29, 2008 and January 8, 2012.

Table 2. Employment by Industry and in Percentages, 2007 and 2011[2]

	2007	%	2011	%
All industries (in 000s)	16,806	100.0	17,306	100.0
Goods-producing sector	3,976	23.7	3,805	22.0
Agriculture	335	2.0	306	1.8
Forestry, fishing, mining, quarrying, oil and gas	342	2.0	337	1.9
Utilities	138	0.8	140	0.8
Construction	1,131	6.7	1,262	7.3
Manufacturing	2,031	12.1	1,760	10.2
Services-producing sector	12,830	76.3	13,501	78.0
Trade	2,673	15.9	2,670	15.4
Transportation and warehousing	820	4.9	843	4.9
Finance, insurance, real estate and leasing	1,056	6.3	1,083	6.3
Professional, scientific and technical services	1,130	6.7	1,309	7.6
Business, building and other support services	699	4.2	677	3.9
Educational services	1,180	7.0	1,219	7.0
Health care and social assistance	1,835	10.9	2,092	12.1
Information, culture and recreation	776	4.6	784	4.5
Accommodation and food services	1,074	6.4	1,093	6.3
Other services	722	4.3	759	4.4
Public administration	865	5.1	971	5.6

In July of 2008, at the end of the second quarter, the Canadian economy had an overall, expenditure based GDP of $1,616 billion, but in 2002 dollars, it was only $1,238. By October 2011, it had barely increased, less than 3%, in large part because of the world financial crisis, which started in 2008 and still had not returned to normal by the end of 2011. The major change over this short time period was in the manufacturing sector, which decreased about two percentage points, 14.74 to 12.74 of the total Canadian economy. The relative increases, to compensate for this decrease in manufacturing, are distributed broadly over the whole economy, but were picked up mainly by the various service sector industries.

The picture for overall Canadian employment in 2007 and 2011 (Table 2) reflects the same relative distribution of employment by industrial sectors for the Canadian economy as did the GDP. However, a major difference is in the proportion of people in the goods producing sector, being 23.7% for 2007 and 22.0% for 2011, about 6% less than the GDP percentage, indicating that this sector is more efficient than the service sector.

---

2 Source: http://www40.statcan.gc.ca/l01/cst01/econ40-eng.htm, Statistics Canada, "Employment by Industry", accessed January 8, 2012.

Evolution of the Canadian Economy

Canada has followed a unique path to its present day high living standard. This path is best illustrated by the Staple Growth Theory which was first proposed by Harold Innis[3] in 1930. The theory suggests that Canada became wealthy through the export of its abundant natural resources called staples. In fact, many people around the world still associate an image of Canada's economy with the extraction and export of natural resources. This image has been so strong that the last 50 years, when compared to other sectors of the economy, the Canadian economy has often been described as being made up of 'hewers of wood and carriers of water'. In comparison to other members of the G8, the 'resource image' is probably true. However, in Canada, the importance of natural resources has been strongly declining.

Figure 1 shows the Staple Growth Theory in diagrammatic format. One can see that Canada has harvested and exported different resources over time. It began with fish in the early 1500s and was followed by fur, lumber, wheat, forest products, minerals, and energy.

Figure 1. Importance of Various Resources to the Canadian Economy over Time

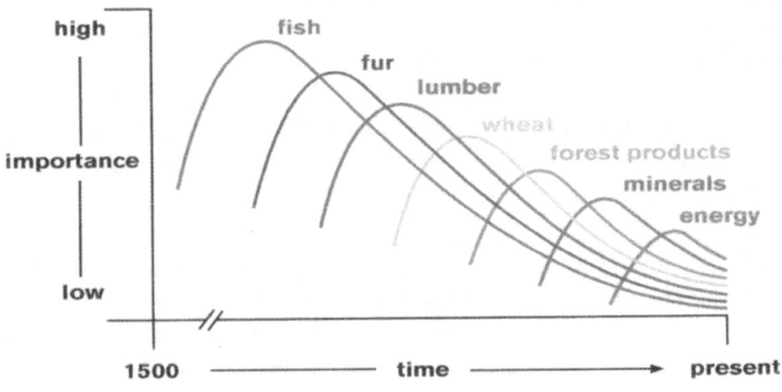

Source: Constructed by Alfred Hecht.

---

3 Innis, Harold A. and Contributor Arthur J. Ray, The Fur Trade in Canada, 1999, Toronto: University of Toronto Press. (First published in 1930).

What resource will be next is debatable. However, over time the importance of natural resources in the Canadian economy has declined, especially when viewed in terms of their value in the total economic output. In the western world, the importance of labour and capital in the creation of goods increased substantially during the industrial revolution, with devastating impact on the value of resources in final or finished products (Figure 2).

Figure 2. Changing Value of Manufacturing Inputs over Time

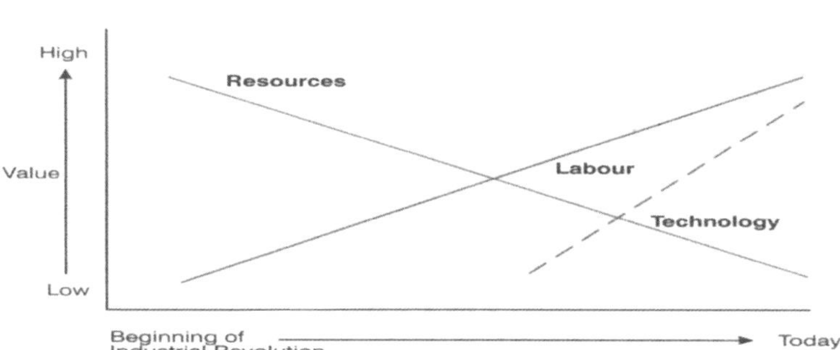

Source: Constructed by Alfred Hecht.

The Canadian Fish Resource Economy Past and Present

Fish was the first major export commodity from the present Canada territory. It started at the beginning of the 16th century, after the rich fishing grounds of the east coast of Canada were re-discovered. Fishermen initially came in the spring from present day France, Spain, Portugal and England and returned to Europe[4] in the fall loaded with salted barrels of codfish, caught mainly on the rich Grand Banks fishing grounds. Since filling the barrels took place on the ships, going on land was not required. How significant was this operation? At the beginning of the 17th century up to 300 ships per year joined the cod harvest from England alone. In part, they replaced earlier fishing fleets from the Iberian Peninsula. Between 1530 and 1600, Basque whalers from France and Spain had contributed significantly to the export economy of Canada's east coast.

4 Newfoundland and Labrador Heritage Web Site Project, "European Migratory Fishery", see www.heritage.nf.ca/exploration/efishery.html, accessed September 29, 2008.

When it was discovered that drying the cod first and then transporting it to Europe made more economic sense, drying operations on land became the norm and with it the first settlements.

Nevertheless, by 1800 the cod fishing industry was in decline. Declining prices, catches, wars and other economic opportunities, as well as what is frequently referred to as the "Tragedy of the Commons"[5], were the main causes. The final 'death blow' for traditional cod fishing came in 1992 when Canada placed a moratorium on cod fishing off the east coast[6]. Fisherman had to switch to other catches to continue their livelihood. However, by 2005 the Canadian catch was down to 61% of what it had been in 1990.

Besides the east coast fisheries, Canada also has commercial fishing on the west coast. Here the most important fish is the pacific salmon. Of the present Canadian total catch, only a quarter[7] comes from the west coast. It has also declined going from 11,663,963 salmon fish caught in 2001 to only 7,950,200 at the end of the fishing season in 2007.

Table 3. Gross Domestic Product at Basic Prices, Primary Industries, 2006 and 2010[8]

	2007	2010
All Canadian industries, 2002 chained dollars	1,218,981	1,233,930
Percentage	%	%
Agriculture, forestry, fishing and hunting	2.31	2.31
Crop production	1.24	1.38
Animal production	0.41	0.41
Forestry and logging	0.44	0.38
Fishing, hunting and trapping	0.08	0.09
Support activities for agriculture and forestry	0.14	0.13
Mining and oil and gas extraction	4.79	4.45
Oil and gas extraction	3.48	3.25
Mining (except oil and gas)	0.72	0.66
Coal mining	0.07	0.07
Metal ore mining	0.30	0.27
Non-metallic mineral mining and quarrying	0.38	0.33
Support activities for mining/oil/gas extraction	0.59	0.61
Total Primary Industries	7.10	6.90

5 http://en.wikipedia.org/wiki/Tragedy_of_the_commons, "Tragedy of the Commons", accessed September 29, 2008.

6 "Crosbie Announces First steps in Northern Cod Recovery Plan", see www.stemnet.nf.ca/cod/announce.htm, accessed September 29, 2008.

7 www.dfo-mpo.gc.ca/communic/statistics/commercial/landings/sum0407_e.htm, Fisheries and Oceans Canada, "Commercial Landings Summary Tables", accessed September 29, 2008.

8 Statistics Canada, "Gross domestic product at basic prices, primary industries", http://www40.statcan.ca/l01/cst01/prim03-eng.htm, accessed January 10, 2012.

When one looks at the Canadian GDP for 2010, as seen in Table 1, it is apparent that the total output from the agriculture, forestry, fishing and hunting economies, was only 28,705 billion, 2.25% of GDP. Fishing, hunting and trapping, however, contributed only 0.09% of the total Canadian economy in 2010 as shown in Table 3. Clearly, this staple resource sector contributes little to the present total Canadian GDP as the above Staple Growth Model suggests. However, the Canadian commercial fish catch was quite important in terms total world catch. As late as 2005, it contributed 1.17%, to the total, but by 2009, it had declined to 1.07% (Table 4). In fact, Canada ranks 17th among the largest fishing nations in the world and is a major exporter of fish. On the other hand, its fish aquaculture is not as well developed as that of the world as a whole. Canada's aquaculture production is only 16% of its total fish catch while that of the world is 62%[9].

However, because the ocean fish catch is now highly restricted by the federal government, due to low stocks of the major fish species, almost all fishing villages have high unemployment rates both on the east and west coasts and are in depressed economic situations.

Table 4. Nominal Fish Catches, Canada and the World[10], in Tons*

Year	1990	1994	1998	2002	2005	2009
World Total	114,539	130,461	88,811	94,560	93,253	88,918
Canada, CDN	1,786	1,183	1,041	1,080	1,090	939
CDN % of World	1.56	0.91	1.17	1.14	1.17	1.07

* Note: Quantity in thousands of metric tonnes, live weight.

The Famous Fur Staple

As shown in Figure 1, the second major Canadian staple export commodity, from about 1700 to 1850, was furs. It had its roots in the fashion industry in Europe where furs, especially beaver hats, became a major fashion item at the beginning of the 17th century. To supply this demand for furs, the Canadian fur industry was born which, in turn, relied upon the Canadian Native Indian hunters and trappers to supply the furs. The latter had the expertise of hunting fur animals, processing the furs, and shipping them by canoes. Two large companies, the Hud-

9 ftp://ftp.fao.org/FI/STAT/summary/a-0a.pdf, UN-FAO, Fisheries and Aquaculture Department, "World fisheries production, by capture and aquaculture, by country (2009)", accessed January 12, 2012.

10 ftp://ftp.fao.org/FI/STAT/summary/a-0a.pdfUN-FAO, Fisheries and Aquaculture Department, "World fisheries production, by capture and aquaculture, by country (2009)", accessed January 11, 2012.

son Bay Company[11] and the North West Company[12], were the main buyers until they joined their operations in 1821. In contrast to the exploitation of fish, in this case the Aboriginals were very involved in the staple economy. One could go as far as to say that they were the very backbone of the industry. Exploitation of this staple also contributed tremendously to the exploration of Canada[13]. As the demand for fur pelts increased, especially beaver, the interior of the continent had to be exploited and with its exploration came the geographic knowledge of the country. Bundles of furs were carried by canoes, from the continental interior to the network of trading posts, over creeks, rivers, ponds, and lakes. Final destinations were the ocean ports on Hudson Bay or ports on the St. Lawrence River or on the east coast. However, as the demand for furs decreased in Europe, in the middle of the 19th century, the fur economy started to decline. Today, the worldwide animal rights movement also helps to keep the demand for furs down. In addition, fur animals raised in captivity compete strongly with the remaining traditional fur trade in Canada. Hence, by 2007 the total value of wild fur catch in Canada is less than $100 million and thus only a minuscule fraction of the total Canadian economy. Canada's Aboriginals, who live in the remote areas of the provinces and territories, produce much of this fur output. Here the trapping industry is still a significant factor in earning a living.

Canada's Timber/Lumber Resources

Canada has 10% of the world's forested area even though only 45% the country is wooded and only 56% of this acreage has some commercial value[14]. Most of the higher quality forests are located in the southern fringes of the country, the area first settled by Europeans. Less than 8% of the forests are in private hands. Only 23% of wood volume consists of broad leaf forests or hard woods, which were in high demand in England at the start of the 19th century. England was in great need of timber/lumber for their commercial and navy shipbuilding industry since its own forests were depleted by this time. Ships would come to

---

11 Hudson's Bay History Foundation, "Exploring the Fur Trade and the Hudson's Bay Company", see http://www.canadiana.org/hbc/intro_e.html, accessed January 29, 2008.

12 http://www.collectionscanada.gc.ca/2/24/h24-1601-e.html, Library and Archives Canada, "The North West Company", accessed January 29, 2008.

13 Bélanger, Claude, Quebec History, "Discovery and Exploration of Canada", see http://faculty.marianopolis.edu/c.belanger/quebechistory/encyclopedia/DiscoveryandexplorationofCanada.htm, accessed January 29, 2008.

14 http://www2.nrcan.gc.ca/cfs-scf/industrytrade/english/View.asp?x=11, "Forest Industry in Canada", Natural Resources Canada, accessed September 29, 2008.

Canada filled with immigrants and return home with lumber. The main sources of lumber were the Atlantic Provinces and the tributary region of the St. Lawrence River. Recently hard wood output contributed some 58% of all wood products in Canada but only some 20% of its exports. Most of the forestry export consists of pulp and paper products. Nevertheless, in terms of direct contribution to the Canadian economy, hard wood lumber output contributes only about 1.7% of GDP.

The Canadian Wheat Staple

To many people in the world, Canada's image is of a large, golden-coloured, waving wheat field. Wheat was the commodity that helped open the Prairies in the last decades of the 19th century. When the Canadian Pacific Railroad was completed in 1885, commercial wheat could be exported from the Prairies. Western Europe needed vast quantities of wheat and Canada could provide it. Between 1895 and 1915 the wheat acreage in the prairies increased fivefold[15] and the output and export between 1870 and 1915 increased over 20 times (Table 5) while exports amounted to nearly 70% of production in 1915.

Table 5. Canadian Wheat Production and Export 1870-1915

(thousands of bushels)		1 ton = 37.744 bushels
Year	Production	Exports
1915	393,543	269,158
1910	132,078	62,398
1905	107,033	47,293
1900	55,572	14,774
1895	55,703	10,760
1890	42,223	3,444
1885	42,736	5,157
1880	32,350	4,502
1875	26,093	7,940
1870	16,724	3,128

Source: Historical Statistics of Canada, "Section M: Agriculture".

By 2006, wheat had reached an output of 25 million tons (928,337,000 bushels)[16] of which 22.4 was produced in Western Canada

15 McInnis, Marvin, "Canadian Economic Development in the Wheat Boom Era: A reassessment", http://qed.econ.queensu.ca/faculty/mcinnis/Cdadevelopment1.pdf, see page 12, accessed January 30, 2008.

16 Statistics Canada, The Daily, August 23, 2007, "Estimates of Production of Principal Field Crops", http://www.statcan.ca/Daily/English/070823/d070823b.htm, accessed September 29, 2008.

and 12.9 million tons or 58% was exported[17]. Over this time wheat production had increased by a factor of 2.35. However, if one recalls that in 1915 Canada's population was only 8 million and in 2006 it was 33[18] then the output per person declined substantially. In addition, one must remember that the wealth per person in Canada had increased tremendously over this time span, resulting in a tremendous decline of the importance of wheat in the Canadian economy as the Staple Growth Model (Figure 1) had predicted.

The Forest Industry

In the beginning of the 20th century, a tremendous increase in newspaper circulation, education and business operations caused a phenomenal demand for paper. This demand could be met from the coniferous zones in the Canadian provinces. Here, trees were smaller and less useful for lumber, but were quite adequate for pulp and paper products. They were also easier to harvest by mechanical means, thus allowing the much maligned 'clear cutting' method of harvesting forests. Even today, the pulp and paper industry is still the main land user of the so-called mid-Canada land corridor (Boreal Forest). Nevertheless, when combined with the hard wood harvest in the southern part of the provinces, the output of the forest industry was, at basic GDP prices, only 0.49% of all Canadian industrial output (Table 6) in 2006 and only 0.38% in 2010, a substantial decline during the recession period covered by this table. Even if one adds the support activities, as shown in Table 6, the values do not increase much. Many people employed in this sector live in small communities, which unfortunately experience substantial fluctuations in their economies as world demand and prices fluctuate up and down.

Table 6. Gross Domestic Product at Basic Prices, Primary Industries, 2006-2010

	2006	2007	2008	2009	2010
	billions of chained dollars (2002)				
All industries	1,191	1,219	1,230	1,193	1,234
Forestry and logging	0.49	0.44	0.40	0.34	0.38
Support activities for agric. and forestry	0.15	0.14	0.14	0.14	0.13

17 http://www.cwb.ca/public/en/newsroom/releases/2007/080207.jsp, "Wheat Marketing Year Ends with Record Durum Exports", Canadian Wheat Board, accessed September 29, 2008.

18 Statistics Canada, "Estimated Population of Canada, 1605 to Present", http://www.statcan.ca/english/freepub/98-187-XIE/pop.htm, accessed September 29, 2008.

If we look at forest product exports as a proportion of total Canadian exports (Table 7), we see it was 6.32% in 2007 on a balance-of-payments basis. However, by 2010 during the financial worldwide downturn, it had declined to only 5.5%. In 2011, "Canada is the third largest exporter of forest products in the world, after Germany and the United States. The forest sector is also the third largest contributor to Canada's balance of trade, after energy and minerals"[19]. Of the $33.6 billion exported in 2007[20], by far the greatest proportion is shipped to the USA followed some distance by the EU, Japan, and China.

Table 7. Canadian Forestry Exports on a Balance-of-Payments Basis, 2007[21]

years	2007	2010
exports in millions	463,120	404,834
% of Canadian export	%	%
Forestry products (total)	6.32	5.40
Lumber and sawmill products	2.72	2.00
Wood pulp and other wood products	1.44	1.56
Newsprint/other paper/paperboard	2.15	1.83

However, the forests industry is not without its critics and problems. Much of the harvesting, especially for the pulp and paper industry, is done by 'clear cutting' the forest. On the positive side, in 2005, nearly 50% of the 1.1 million hectares harvested in Canada were replanted or reseeded. In the past, this had been left exclusively to natural regeneration. Unfortunately, in 2010, natural fires destroyed 3.2 million hectares and insects another 7.5. In contrast only 0.6 million were harvested[22]. Clearly defoliation of the forests by insects and the killing of trees by beetles in 2010 was a major problem in Canada, destroying the equivalent of 12.5 times the harvested area. In 2007, the mountain pine beetle destroyed major areas of forest on the Alberta British Columbia border[23]. Other major pests are the spruce budworm, spruce beetle, forest tent caterpillar and the large aspen tortrix.

19 Natural Resources Canada, "Industry" http://cfs.nrcan.gc.ca/pages/52, accessed January 10 2012.

20 http://canadaforests.nrcan.gc.ca/statsprofile/trade/ca, Natural Resources Canada, "Canada's Forests", accessed September 16 2008.

21 Statistics Canada: "Exports of Goods on a Balance-of http://cfs.nrcan.gc.ca/pages/52 – Payment Basis by Products", http://www40.statcan.gc.ca/l01/cst01/gblec04-eng.htm, accessed January 10, 2012.

22 Natural Resources Canada, "The State of Canada's Forests; Annual Report 2011", http://cfs.nrcan.gc.ca/pubwarehouse/pdfs/32683.pdf, page 25, accessed January 10, 2012.

23 http://www.srd.gov.ab.ca/forests/health/insects/default.aspx, Alberta, Sustainable Resource Development, "Forest Insects", accessed September 29, 2008.

Canada's Mineral Staples

With the coming of the Industrial revolution in the 19th century, the demand for minerals increased substantially and with Canada's large land mass and favourable geological structures, it became an important world supplier. Historically, the mining industry was the backbone of many rural and remote communities. Since the 1950s, Canada has become a major supplier of minerals to world markets, especially the USA. The height of the importance of the mining industry was probably during and after World War II (Figure 1). Nevertheless, even in 2006, Canada was one of the world's largest producers and exporter of minerals, producing more than 60 different minerals and metals. Yet the value, \$33.6 billion[24] of output in 2006, which includes metals, non-metals and coal, represents still less than 4.0% of the national GDP. Direct employment in the mining, smelting, refining and mineral processing manufacturing industries employment just over 388,000 people, that made up only 2.4% of Canada's total employment[25]. Understandably, Canada exports about 80% of its mineral output, which contributes about 16% to total exports (Table 10).

As can be expected, the bigger provinces have the larger share of mineral output: Ontario (27%), British Columbia (17%), Quebec (14%) and Saskatchewan (11%)[26].

The most important minerals mined by value were nickel, with shipments valued at \$6.2 billion, followed by copper at \$4.6 billion. The value of diamonds mined was \$1.6 billion. Leading non-metallic minerals were potash at \$2.2 billion and coal had shipments valued at \$2.2 billion[27].

In a comparison of world mineral production in 2009, Canada ranks second in uranium (20.5%), third in nickel (12.7%), fourth in zinc (6.6%), sixth in gold (4.3%), ninth in copper (3.3%), first in potash (26.0%), seventh in gypsum (3.6%), sixth in diamonds (8.8), fifth in salt (5.4) and fourth in gypsum (3.6%). In addition, even though Canada does not mine aluminum ore, it is the third largest producer of primary

---

24 Natural Resources Canada, "Canadian Minerals Yearbook, 2006, Statistical Report", http://www.nrcan.gc.ca/mms/cmy/content/2006/69.pdf, accessed September 29, 2008.

25 Natural Resources Canada, "Statistics and Facts on Minerals and Metals", http://www.nrcan.gc.ca/statistics/minerals/default.html, accessed September 29, 2008.

26 Natural Resources Canada, "Canadian Minerals Yearbook, 2006, Statistical Report", http://www.nrcan.gc.ca/mms/cmy/content/2006/69.pdf, Table 3, accessed September 29, 2008.

27 Natural Resources Canada, "Canadian Minerals Yearbook, 2006, Statistical Report", http://www.nrcan.gc.ca/mms/cmy/content/2006/69.pdf, Table 1, accessed September 29, 2008.

aluminum in the world[28]. Therefore, it is not surprising that Canada has a worldwide image of a resource rich country.

Nevertheless, world demand for minerals varies substantially from year to year. The result is that many of the mining communities go from plenty to famine in short periods, frequently counter to the national economic cycle. The Sudbury District in northern Ontario, for instance, had 8.75% of total employment in "agriculture and other resource-based industries" in 2001 while the province of Ontario had only 3.18%[29]. Furthermore, its unemployment rate of 12.5% was more than twice as high as the 6.1% value for the whole of Ontario. By 2006, conditions had improved slightly with the comparative values being 11.6% and 6.4% respectively[30]. If one compares conditions in the different provinces, the differences are also great. In 2007 for instance, in Newfoundland, a resource exploitation province, the unemployment rate was 12.4% while the value for Canada as a whole was only 6.3%[31].

Even though many of the mining centres are found in the peripheral settlements of the country, they do attract workers, in large part because the industry as a whole pays well. In 2007 for instance, workers in the mining, oil and gas extraction industries had an hourly pay of $30 while those in Canada as a whole had only $19[32].

Energy, Canada's Last Major Export Staple

Canada is blessed with an abundance of energy sources especially oil, natural gas, coal, waterpower, and nuclear power. In 2007, it ranked seventh in crude oil and third in natural gas production[33]. The country's proven oil reserves in 2007 stood at 179.2 billion barrels (Table 8) which placed it second to the reserves of South Arabia[34], With present

---

28 Natural Resources Canada, "Statistics and Facts on Minerals and Metals", http://www.nrcan.gc.ca/statistics-facts/minerals/902, accessed January 11, 2010.

29 Statistics Canada, 2001 Community Profiles, Sudbury District, http://goo.gl/Eh0CH, accessed 2012-01-20.

30 Statistics Canada, 2006 Community Profiles, Sudbury District, http://goo.gl/oPQgg, accessed 2012-01-20.

31 Statistics Canada. "Latest Release from the Labour Force Survey", Friday, January 11, 2008, http://www.statcan.ca/english/Subjects/Labour/LFS/lfs-en.htm, accessed February 5, 2008.

32 Statistics Canada, "Earnings, Average Hourly for Hourly Paid Employees, by Industries", http://www40.statcan.ca/l01/cst01/labr74a.htm, accessed September 17, 2008.

33 Canadian Association of Petroleum Producers, "Canada", http://www.capp.ca/default.asp?V_DOC_ID=603, accessed September 16, 2008.

34 Wikipedia, "Oil Reserves", http://en.wikipedia.org/wiki/Oil_reserves#Canada, accessed Feb. 6, 2008.

production at 2.6 million barrels a day, the reserve could last 182 years at the present rate of extraction.

Table 8. World Proven Reserves of Oil and Natural Gas of Major Countries, 2008[35]

	Oil	Natural Gas
	(Billion Barrels)	(Trillion Cubic Feet)
Country/Region	*Oil & Gas Journal*	*Oil & Gas Journal*
	January 1, 2008	January 1, 2008
Canada	178.592	58.200
Mexico	11.650	13.850
United States	20.972	211.085
Subtotal North America	211.214	283.135
Venezuela	87.035	166.260
Subtotal Central and South America	99.682	205.040
Subtotal Europe	10.565	143.680
Russia	60.000	1,680.000
Subtotal Eurasia	97.000	1,810.000
Saudi Arabia	266.751	253.107
Subtotal Middle East	742.658	2,518.962
Subtotal Africa	93.584	451.590
China	16.000	80.000
India	5.625	37.960
Subtotal Asia and Oceania	30.284	322.860
World Total	1,284.987	5,735.267

1 Proven reserves are estimated quantities that analysis of geologic and engineering data demonstrates with reasonable certainty under existing economic and operating conditions.

2 BP P.L.C., BP Statistical Review of World Energy June 2008, except United States liquids, and United States natural gas data are from the Energy Information.

However, over 95% of the oil reserves are found in the Tar Sands[36] which have to be mined in order to extract the oil. Most tar sands in Alberta are typically 40 to 60 metres deep, and sit on top of a limestone rock layer. The oil extraction system is an expensive process and can only compete against conventional oil if the price for a barrel of oil is high. In 2007, the average cost of producing a barrel of oil from the tar sands was $28[37] substantially higher than the cost of extracting conven-

---

35 Energy Information Administration, Official Energy Statistics of the US Government, http://www.eia.doe.gov/emeu/international/reserves.html, accessed September 17, 2008.

36 Wikipedia, "Tar Sands", http://en.wikipedia.org/wiki/Tar_sands#Canada, accessed February 6, 2008.

37 Guelph Greens, "The Tar Sands Fiasco", http://www.guelphgreens.ca/?page_id=268, accessed Sept. 17, 2008.

tional oil. Surface strip mining of the Tar Sands contributes to the high costs. Yet other forms of mining the Tar Sands are even more expensive. Pierre Fournier in an article in the Globe and Mail reported, "the production cost is close to $85 US"[38]. In addition to the heavy cost of extraction, which creates more jobs than the conventional oil extraction process, the process also produces much pollution, causing Green Peace Canada to call for a memorandum of a Tar Sand mining stoppage[39].

Of Canada's daily output of oil in 2007, nearly 50% came from the Tar Sands. Output is a heavy oil and the "market price for heavy oil is only about one-half of light crude oil"[40]. This means the Tar Sands extraction industry has a substantially higher production cost and receives a substantial lower price for its barrel of oil then the conventional oil industry. This consequently squeezes their profit margins. Another problem is its 'dirty oil' image. Because the mining requires a lot of water and produces a substantial amount of pollution, some political jurisdictions want to ban its purchase, even within Canada, as the recent discussion in Quebec suggests[41].

Natural gas is the other major energy reserve. Canadian production[42] at 17.1 billion cubic feet per day was the world's third-largest producer in 2006. Its proven reserves of 58.2 trillion cubic feet however rank it 19th in the world[43]. As can be seen from Table 8, the USA reserves are nearly four times Canada's and 13 other countries in the world have larger reserves than Canada[44].

How important are these two liquid hydrocarbon staples to the Canadian economy? The energy industry claims they contributed an "estimat-

---

38 Pierre Fournier, "One way or another, the oil sands will pay", Globe and Mail, October 20, 2008, Report on Business, B2. See also: http://www.theglobeandmail.com/servlet/story/RTGAM.20081017.wagendafournier1020/BNStory/GlobeSportsOther/, accessed Oct. 21, 2008.

39 http://www.greenpeace.org/canada/en/campaigns/tarsands/, Green Peace Canada, "Stop the Tar Sands", accessed Feb. 6, 2008.

40 Canadian Association of Petroleum Producers, "Crude Oil", http://www.capp.ca/default.asp?V_DOC_ID=689, accessed February 6, 2008.

41 Konrad Yakabuski, "Alberta's 'dirty' Oil a Sticky Problem for Charest", Globe and Mail, B2 Report on Business, September 19, 2008.

42 http://www.capp.ca/raw.asp?x=1&dt=NTV&e=PDF&dn=112818, Canadian Association of Petroleum Producers, Canada, "View the Canadian Statistics for the Past Eight Years", accessed February 6, 2008.

43 Infoplease Web Site, "Greatest Natural Gas Reserves by Country, 2006", http://www.infoplease.com/ipa/A0872966.html, accessed February 6, 2008.

44 Source: Energy Information Administration, Official Energy Statistics of the US Government, "World Proved Reserves of Oil and Natural Gas, Most Recent Estimates" http://www.eia.doe.gov/emeu/international/reserves.html, accessed September 29, 2008.

ed $27 billion to government revenues in the form of royalty payments, bonus payments and income taxes and the trade surplus contributed 80% of Canada's merchandise trade balance in 2006"[45]. In total, they contributed 3.25% to the Canadian GDP in 2010 (Table 3) but less than 2% of direct total employment even if one adds to their numbers employment in forestry, fishing, mining and quarrying (Table 2).

Like most resources, the oil and gas industry is uniquely distributed in Canada with 64% of oil output coming from Alberta, 13% from Saskatchewan and 12% from Newfoundland and Labrador[46]. In terms of natural gas, the figures are similar except that British Columbia replaces Newfoundland and Labrador for third spot. For Canada as a whole, the oil and gas conventional production sector value increased from $20.8 billion in 1995 to $55.9 in 2000 and $90.8 in 2005. Only about 45% of Canadian oil consumption comes from Canadian sources. The remaining 55% is imported mainly into eastern Canada as it is more economical then moving the oil east from the fields in the Prairies[47].

The more traditional hydrocarbon energy resource is coal and Canada has substantial production nearly half of which it exports mainly from western Canada. Yet at the same time it imports nearly as much to eastern Canada (Table 9). Most of the coal is used in generating electricity and some by the steel industry. Little is used in the heating industry now as was done in the first half of the 19th century. Liquefied natural gas is the fourth most important hydrocarbon based fuel in Canada. Hydro and nuclear contribute substantial energy to the economy, most of the former being located in Ontario (Table 9).

---

45 Canadian Association of Petroleum Producers, "Canada", http://www.capp.ca/default.asp?V_DOC_ID=603, accessed September 9, 2008.

46 http://www41.statcan.ca/2007/1741/ceb1741_000_e.htm, Statistics Canada, "Energy", accessed September 29, 2008.

47 http://www.statcan.ca/english/research/11-621-MIE/11-621-MIE2006047.pdf, Statistics Canada, Miles Ryan Rowat, "Boom Times: Canada's Crude Petroleum Industry", accessed, September 29, 2008.

Table 9. Canadian Major Energy Sources: Production, Exports and Imports, 2006[48]

	Coal	Crude oil	Natural gas (NG)	NG liquids	Hydro and nuclear	Refined petroleum
	Terajoules					
Production	1,339,754	5,935,706	7,219,725	701,362	1,599,479	4,676,361
Exports	656,986	3,915,262	3,906,417	218,763	153,849	932,872
Imports	565,243	1,899,155	369,342	17,547	85,046	626,023

Natural Gas Liquids (NGL) includes propane, butane and ethane produced by gas plant.

In part because Canada has great energy resources, Canadians are some of the highest energy consumers in the world. The fact that Canada has severe winters and continental hot summers, contributes substantially to considerable energy consumption, which is known to cause substantial pollution.

The Canadian industrial economy is still heavily oriented towards the first stage of the good production process, i.e. the processing of raw materials into raw but useable products. Mining and the formation of the first raw products for the manufacturing process require large amounts of energy per product produced. The extraction of these resources, at the first stage of their processing, also produces much pollution. Since Canada exports large amounts of semi-processed raw materials, some have argued that the pollution by-product blame should be shared between Canada and the nations that buy the semi raw material in order to produce finished products. If the final manufactured product in country X needs copper and buys it from Canada, then it should share in the pollution generating blame of mining the copper in Canada.

Although the Canadian government under liberal Prime Minister Jean Chrétien did sign the world pollution control Kyoto accord[49], Canada has missed its pollution reduction targets under the Chrétien government by a wide margin as well as by the following governments of Paul Martin and the conservative government Prime Minister Stephen Harper. The latter has abandoned the agreement altogether and has brought forth his own clean air proposal[50]. Some have speculated that

---

48 http://www40.statcan.ca/l01/cst01/prim72.htm?sdi=oil%20gas%20production, Statistics Canada, "Summary Tables, Coal", accessed September 29, 2008.

49 http://communities.canada.com/ottawacitizen/forums/post/54799.aspx, "Honesty about the Kyoto Accord (Ottawa Citizen editorial), Monday, February 26, 2007, accessed September 29, 2008.

50 http://www.cbc.ca/news/yourview/2006/10/tory_bill_aimed_at_cutting_gre.html, CBC News, "Tory bill aimed at cutting greenhouse gases in half by 2050", October 19, 2006, accessed September 29, 2008.

the Kyoto agreement was signed, in the first instance, just to be different from what the US did; they after all did not sign the agreement. However, another reason for Canada's failure to come even close to meeting the pollution reduction target under the Kyoto accord – they went up substantially – is that Canada is a loose confederation of provinces and most of the pollution activities take place under provincial and local government jurisdictions and not those of the federal government. Hence, what the federal government commits itself to do is frequently irrelevant at the provincial level.

Staple Summary

How have the historic different staples changed the export mix since the early 1980s? The staple model, as seen in Figure 1, suggests that collectively they still should contribute substantially to Canadian exports but their importance should have continues to decrease over the 25 year period. From Table 10 we can see that 'Agriculture and Fish Products' decreased from 12.7% to 8.1% between 1981 and 2006. Forestry products from 18.0% to 8.1% but energy increased from 11.1% to 14.6%. However, the latter was mainly due to high-energy prices in 2006 versus prices in 1981. By 2010, the total export values had increased 10% in comparison to 2006, yet agriculture and fish products had decreased by 4%, forestry products by 10% but energy products had increased by 3.6%[51].

Clearly, raw materials are even losing their significance in the export sector as they have in the total Canadian economy. Nevertheless, when compared to the absolute and relative growth of most other sectors of the Canadian economy as shown in Table 10, their decline is significant. To counter this image, resource industry proponents and analysts remind us that if their job multiplier in the total Canadian economy is considered, the actual importance of the resources sector is far greater than only their direct GDP or employment impact on the Canadian economy.

---

51 Statistics Canada, CANSIM II Tables by Subjects, "International trade/Merchandise exports", http://dc2.chass.utoronto.ca/chasscansim/, accessed January 10, 2012.

Table 10. Canadian Export of Goods; 1981-2006, in millions[52]

	year	1981	1986	1991	1996	2001	2006
Total Exports of Goods		116,257	162,127	197,308	297,134	406,195	444,776
	%	100	100	100	100	100	100
Agriculture and Fish Products		14,735	16,131	19,878	23,451	31,683	35,840
	%	12.7	10.0	10.1	7.9	7.8	8.1
Energy Products		12,945	19,170	28,708	40,568	45,106	50,639
	%	11.1	11.8	14.6	11.1	11.8	14.6
Forestry Products		20,956	24,288	27,043	35,792	37,051	36,141
	%	18.0	15.0	13.7	12.0	9.1	8.1
Industrial Goods and Materials		22,727	31,039	38,591	52,889	68,423	73,463
	%	19.6	19.1	19.6	17.8	16.8	16.5
Machinery and Equipment		14,922	23,086	29,378	62,587	99,758	105,808
	%	12.8	14.2	14.9	21.1	24.6	23.8
Automotive Products		24,450	42,333	47,985	65,403	93,173	105,623
	%	21.0	26.1	24.3	22.0	22.9	23.8
Other Consumer Goods		2404	3641	4184	10,310	16,452	19,272
	%	2.1	2.3	2.1	3.5	4.1	4.3
Special Transac-tions		944	782	2183	3480	8189	8572
	%	0.8	0.5	1.1	1.2	2.0	1.9
Other Balance of Payment A		5782	6184	4719	6442	5844	10452
	%	5.0	3.8	2.4	2.2	1.4	2.4

Note: This table uses 1997 chained dollars.

The Canadian Manufacturing Sector

As was mentioned earlier, Canada's manufacturing sector is proportionally one of the smallest of all the OECD countries. Still, as can be seen from Table 10, the 'Industrial Goods', 'Machinery and Equipment', 'Automotive Products' and 'Other Consumer Goods' together, dominated Canadian exports to the tune of 68.4% in 2006. In 1981, they contributed only 53.4%. Clearly, there has been a major increase over the years of the export of final/finished goods. However, in sharp contrast to the export picture, is the significance of the manufacturing sector in the total Canadian economy. Here it only contributes 15% to GDP (Table 1) and 12% of total employment (Table 2).

Where in Canada is manufacturing the strongest? As can be seen from Table 11, the heart of manufacturing in Canada is located in Ontario and Quebec. These two provinces accounted for 72.8% of all manufacturing sales in 2006. By 2010, only four years later, their proportion had declined to 70.8%. Most of the decline took place in Ontario

---

[52] Statistics Canada, CANSIM II Tables by Subjects, "International trade/Merchandise exports", http://dc2.chass.utoronto.ca/chasscansim/, accessed September 17, 2008.

where the auto industry, a major manufacturing player in this province, was in a slump over this period. Within Ontario the 'Golden Horseshoe' region of southern Ontario, stretching from Oshawa east of Toronto to St. Catharines on the western edge of Lake Ontario, is Canada's most important industrial region. As can be seen from Table 11, the smaller provinces, the most rural ones, tend to be the least industrialized, although in terms of Canadian manufacturing percentages, they have been increasing between 2006 and 2010. Yet if we compare their share of Canada's population with their share of the manufacturing sales, we see that all the smaller provinces, except New Brunswick, have a smaller share of total manufacturing than what their share is of total Canadian population (Table 11). The biggest two provinces, Ontario and Quebec, have a far greater share of total manufacturing then population.

What are some of the major specific manufacturing products that Canada produces? From Table 12 we can see that transport equipment with 19.03% is the biggest industrial sub sector, followed by food, petroleum and coal products, primary metals, chemicals, fabricated metal products, machinery, and paper, all of which were over 5% in 2007. Some specific manufactured products include cars, car parts, aircraft, railroad cars, Blackberry cell phones, processed foods, gasoline and heating oil, lumber, plywood and computers.

Table 11. Canadian Manufacturing Sales and Population by Provinces, 2007 and 2010

	Manufacturing in 000000 $ and %				Population in 000 and %			
year	*2006*	*%*	*2010*	*%*	*2007*	*%*	*2011*	*%*
Canada	605,527	100	529,847	100	32,976	100	34,483	100
N.L.	4,293	0.7	5,167	1.0	506	0.1	511	1.5
P.E.I.	1,333	0.2	1,207	0.2	139	0.0	146	0.4
N.S.	9,559	1.6	9,799	1.8	934	0.2	945	2.7
N.B.	14,730	2.4	17,257	3.3	750	0.1	756	2.2
Qué.	145,580	24.0	132,116	24.9	7,701	1.4	7980	23.1
Ont.	295,636	48.8	243,307	45.9	12,804	2.4	13373	38.8
Man.	14,862	2.5	14,422	2.7	1,187	0.2	1251	3.6
Sask.	9,866	1.6	10,912	2.1	997	0.2	1058	3.1
Alb.	65,091	10.7	60,074	11.3	3,474	0.7	3779	11.0
B.C.	44,480	7.3	35,542	6.7	4,380	0.8	4573	13.3
Y. T.	27	0.0	31	0.0	31	0.0	35	0.1
N. W. T.	64	0.0	9	0.0	43	0.0	44	0.1
Nvt	7	0.0	5	0.0	31	0.0	33	0.1

Source: Statistics Canada, "Manufacturing Sales, by Province and Territory", http://goo.gl/OXwF3, and http://goo.gl/OXwF3, accessed January 10, 2012.

Table 12. Canadian Manufacturing Sales, by Sub-sector, 2007 and 2010, ($ millions)

	2007	*%*	*2010*	*%*	*% diff.*
Total	612,879	100.0	486,666	100.0	0
Food	74,149	12.1	78,649	16.2	4.1
Beverage and tobacco products	10,949	1.8	10,550	2.2	0.4
Textile mills	2,074	0.3	1,503	0.3	0.0
Textile product mills	2,485	0.4	1,583	0.3	-0.1
Leather and allied products	522	0.1	366	0.1	0.0
Paper	31,035	5.1	24,938	5.1	0.1
Printing and related support activities	10,750	1.8	9,252	1.9	0.1
Petroleum and coal products	66,366	10.8	59,094	12.1	1.3
Chemicals	51,145	8.3	41,068	8.4	0.1
Plastics and rubber products	25,935	4.2	19,062	3.9	-0.3
Clothing	4,460	0.7	2,213	0.5	-0.3
Wood products	24,908	4.1	16,704	3.4	-0.6
Non-metallic mineral products	15,226	2.5	11,638	2.4	-0.1
Primary metals	53,818	8.8	33,902	7.0	-1.8
Fabricated metal products	36,858	6.0	29,292	6.0	0.0
Machinery	32,660	5.3	27,257	5.6	0.3
Computer and electronic products	18,738	3.1	15,510	3.2	0.1
Electrical equipment, app. and comp.	10,605	1.7	9,404	1.9	0.2
Transportation equipment	116,606	19.0	74,647	15.3	-3.7
Furniture and related products	13,936	2.3	10,428	2.1	-0.1
Miscellaneous manufacturing	9,654	1.6	9,606	2.0	0.4

Source: Statistics Canada, Summary Tables, "Manufacturing Sales, by Sub-sectors", http://www40.statcan.gc.ca/l01/cst01/manuf11-eng.htm, accessed January 10, 2012.

Like all sectors of an economy, components of the manufacturing sector are always changing and Canada is no exception. Between 2007 and 2010, sales in manufacturing decreased by nearly 20% (126 billion dollars). Of the 21 sectors listed in Table 12, eight also had their percentage of total sales decrease between these years. Transportation equipment was the hardest hit with -3.7% and it reflects the downturn in the North American auto sector, which has been in a recession for a few years. All other manufacturing sectors also experienced declines. Surprisingly, agricultural products increased their share the largest, 4.1%. This may be due to the agricultural marketing boards, especially in Ontario and Quebec[53], which are largely, price wise at least, insensitive to a recession. The 1.3% increase of petroleum and coal related products are due to the major increases in energy prices in the last few years. The

[53] Roslyn Kunin, "Time to Abandon Dairy Marketing Boards", http://cwf.ca/_blog/Canada_West_Foundation_Blog/post/Time_to_Abandon_Dairy_Marketing_Boards/, accessed January 10, 2012.

decrease in primary metals, -1.8%, reflects the high drop of metal prices worldwide in the last year due to the worldwide recession.

The Canadian Manufacturing Export Association claims on its web site, that 2.2 million people (see also Table 2) are employed directly in the manufacturing sector and another 2.5 indirectly. This makes the sector very desirable to every community in Canada. What is also attractive to communities is their pay; on average, it is 28% higher than the Canadian average. The fact that manufacturing output was exported at a rate of 70% (in 1980 it was only 25%), making up 80% of total Canadian good exports, means that the industry is a very 'basic' industry, one that communities love to have as they also then have good job multipliers[54]. Even though Canadian manufacturing jobs pay well per hour relative to other Canadian jobs, they only ranked fifth ($21) in US dollars per hour in 2007, after Germany ($32), UK ($24), USA ($23), and Japan ($22). China's hourly pay rate was only $0.54 and India's $0.38[55] per hour. However, one has to keep in mind that currencies fluctuate and close rankings can easily change from year to year.

Table 13. Private Foreign Canadian Manufacturing Firms, in top 40 Rank, 2008

Rank	*Name of Company*	*Sector*	*Revenue in 000,000*	*Ownership by Country (1)*
1	General Motors of Canada	Auto	31,675	USA
2	Chrysler Canada Ltd	Auto	15,226	USA
7	Honda Canada Inc.	Auto	12,200	Japan
8	Novelis Inc.	Metals	9,862	India
9	Ford Motor Company of Canada	Auto	9,363	USA
15	IBM Canada	Technol.	6,793	USA
24	ArcelorMittal Dofasco	Steel	4,139	Luxemburg
26	Ipsco Inc.	Steel	4,558	Sweden
30	Hewlett-Packard	Technol.	3,565	USA
35	Pratt & Whitney Canada	Industrial	3,300	USA

(1) This means the headquarter location of the firm. All are publicly traded companies and many Canadians also own their stock.

Source: Globe and Mail, "100 Biggest Private Companies", Report on Business, p. 94, July/August 2008.

Canada's history of having a 'branch plant' economy is well known. The first big trading company, the Hudson Bay Company was foreign

54 Canadian Manufacturing Exporter Association, http://www.cme-mec.ca/national/template_na.asp?p=4, accessed September 19, 2008.

55 The Hill Times, "Manufacturing Policy Briefing", July 17, 2007, p. 23, http://www.cme-mec.ca/pdf/HillTimes_CME.pdf, accessed September 19, 2008.

owned. When industrialization took hold after confederation in 1867, an attempt was made to protect local manufacturing by having high tariffs on imported goods. What it did however, was to encourage foreign firms to set up branch plants in Canada, i.e. behind the import barriers. The outcome of this process can still be seen today. For instance of the top 40 private firms in Canada in manufacturing, 10 have their headquarters in another country (Table 13). Similarly, if one looks at the next 60 firms, another 12 firms have headquarters outside Canada[56]. Not surprisingly, most headquarters are located in the USA. On the other hand, some Canadian publicly traded companies also have their headquarters abroad.

Table 14. Canadian and Provincial Manufacturing Sales, by Sub-sector, 2007

Provinces	CDN	N.L.	P.E.I.	N.S.	N.B.	Qué.	Ont.	Man.	Sask.	Alb.	B.C.
Total $ in Billions	613	5	1	10	16	150	293	16	10	67	43
Sub-Sectors	in %										
Food	12.2	23.7	x	21.0	x	11.2	10.1	22.3	22.9	15.0	13.3
Bev. and tob. prod.	1.8	4.1	X	X	1.5	2.3	1.6	1.3	0.4	1.2	2.7
Textile mills	0.4	X	X	X	X	0.7	0.3	X	X	X	X
Textile product mills	0.4	0.1	0.2	X	X	0.5	0.4	X	X	X	X
Leather and allied prod.	0.7	X	X	X	X	1.7	0.4	0.9	0.3	0.1	X
Paper	0.1	X	X	X	X	0.2	0.1	X	0.1	X	X
Print. and supp.	4.1	X	X	X	6.9	4.8	1.5	3.8	2.8	4.2	18.4
Petro and coal prod.	5.0	X	X	9.0	13.1	6.8	3.1	3.6	1.1	2.8	13.8
Chemicals	1.8	X	0.8	X	x	1.7	1.9	3.4	1.4	1.3	2.1
Plastics and rubber prod.	11.0	X	X	X	x	9.67	6.2	0.2	X	22.1	X
Clothing	8.1	X	7.8	X	x	7.6	7.5	X	X	20.9	3.2
Wood products	4.3	X	X	x	x	4.3	5.0	3.9	1.3	2.4	3.2
Non-metallic mineral prod.	2.4	X	X	x	x	2.5	2.1	1.7	1.3	3.6	4.7
Primary metals	8.7	X	X	x	x	15.2	7.3	X	X	X	7.4
Fabricated metal prod.	6.0	1.5	1.7	x	x	5.5	6.3	5.4	6.6	7.7	6.2
Machinery	5.3	0.4	X	x	x	4.0	5.3	6.9	9.4	9.3	5.5
Comp. and elec. prod.	3.1	X	X	x	x	3.1	3.9	X	X	1.1	3.4
Elec. equip., appl. and comp.	1.8	0.2	X	x	x	2.5	1.8	X	X	X	X
Transportation equipment	19.1	X	X	x	x	11.5	31.7	13.8	3.0	1.4	3.5
Furn. and related prod.	2.3	X	X	x	x	2.7	2.3	3.6	0.9	1.5	3.6
Misc. manufacturing	1.6	X	1.0	x	x	1.8	1.5	2.5	0.9	1.1	x

X means no information/confidentiality

Source: Statistics Canada, Summary Tables, "Manufacturing sales, by Sub-sector", http://goo.gl/NukQL, accessed 2012-01-20. The sub-classification is based on the three-digit NAICS (North American Industrial Classification System) classification, http://goo.gl/k30Cn, accessed 2012-01-20.

56 Source: Globe and Mail, "100 Biggest Private Companies", Report on Business, p. 94, July/August 2008.

A major component of the Canadian manufacturing scene is the auto industry[57]. Besides the four auto producers listed in Table 13, Canada has auto plants by Toyota, Hyundai, and Suzuki, and large parts producers like Magna International and Linamar. Nearly all auto manufacturing is taking place in southern Ontario and some 22% of all manufacturing in Ontario is in the automotive sector[58]. In fact, if one looks at the larger transportation sector as seen in Table 14, one can see that 31% of all manufacturing is in Ontario. The next closest province is Quebec at 13%. However, Quebec's output is more in the aircraft, railroad, skidoo, and seadoo areas. That Manitoba is ranked third is somewhat surprising but is due to bus and aircraft parts production. It is interesting to note that western and eastern provinces tend to have high percentages in food manufacturing. The western provinces also are above the Canadian average in machinery and fabricated metal production. This probably relates to the unique requirements in the resource extraction industries found there.

One of the manufacturing characteristics of the Canadian economy is its strong high technology sub-sector. Research in Motion, with its world famous mobile blackberry, has its headquarters in Waterloo, Ontario and leads this group in capitalization, and profits (Table 15) even though the number of the employees, 8,387, lags far behind those of Celestica (43,000) and Nortel (32,550). However, one has to remember that publicly traded companies can change in value substantially and quickly as did Nortel a few years ago. By 2010, Nortel was bankrupt and out of business. Although, Research in Motion, increased its numbers of employees to some 17,000 worldwide by 2011, its market capitalization, on January 10, 2012, was down to $8.13 billion. By January of 2012, MDA Ltd. of Richmond, BC was also down in its capitalization to $1.48 billion from a high of $1.7 billion in 2008. On the other hand, Open Text increased its market capitalization to $2.92 billion from $1.2 billion between 2008 and the start of 2012. Seven of the ten companies have headquarters in southern Ontario and two are in British Columbia. It is surprising that no large high tech company is found in Quebec. But it does have the headquarters of two large firms like Bombardier and BCE, Canada's largest communications company, which certainly have some high tech component in their operations.

---

57 http://en.wikipedia.org/wiki/List_of_countries_by_automobile_production, Wikipedia, "Canada in 2007 with 2,578,238 vehicles produced ranked ninth in the world", accessed September 29, 2008.

58 Canadian Manufacturing Exporter Association, 'Manufacturing in Ontario', http://www.cme-mec.ca/mfg2020/dates/ONReport.pdf, accessed September 22, 2008.

Table 15. Canada's 10 Largest, by Market Capitalization, Public High-Tech Companies

Name, date of reporting and headquarters in	Market capitalization (in $ million)	Profit (in $ 000's)	Employees
Research In Motion (Ma08) – Waterloo, ON	57,689	1,293,867	8,387
Nortel Networks (De07) – Brampton, ON	6,553	-957,000	32,550
MDA Ltd. (De07) Richmond, BC	1,705	94,967	2,975
Teranet Income Fund (De07) – Toronto, ON	1,581	136,649	680
Celestica Inc. (De07) – Toronto, ON	1,325	-13,700	43,000
Evertz Technologies (Ap07) – Burlington. ON	1,174	60,663	599
Open Text (Ju07) – Waterloo, ON	1,171	21,660	2,704
Sandvine Corp. (No07) – Waterloo, ON	730	18,962	280
Aastra Technologies (De07) – Concord, ON	546	35,767	1,690
Absolute Software (Ju07) – Vancouver	538	-5,876	169

Source: Globe and Mail, "1000 Biggest Public Traded Companies, Technology Firms", Report on Business, p. 78, July/August 2008.

Canadian Export and Imports

As was alluded to earlier, Canada is a major exporter and importer of goods. In 2007, the value of exports was about 38% of GDP (see Tables 1 and 16). Canadians clearly rely heavily upon exports for their high standard of living. With 77% of exports in 2007 going to the USA, Canada is heavily dependent on one trading partner. The value decreased to 73% by 2010 but still was very high. The cause of this high dependence is related to the high degree of interaction and ownership of Canadian firms with American ones. Secondly, geographic proximity is of great significance. It seems to substantiate what is frequently proposed as the first law of geography that says that nearer things are more related than further things. Another cause may be the fact that both societies have similar consumer value structures, which makes it easy for firms to cater to both. That many product technical requirements are also similar would further support the large flow of goods. The North American free trade agreement, NAFTA[59], which exists between Canada and the USA, expedites the flow of goods across the border. The small proportionate decline in exports to the USA in 2010 from 2007 is, in large part, due to the slump in the North American auto industry and not a sign for major changes on the long run.

---

59 Foreign Affairs and International Trade, "Canada and the North American Free Trade Agreement", http://goo.gl/kryHl, accessed 2012-01-20.

Table 16. Canadian Exports and Imports (in 000,000s) by Country or Country Grouping

	2007	*%*	*2010*	*%*
Exports	463,120	100.0	404,834	100.0
United States	355,732	76.8	296,672	73.3
Japan	10,027	2.2	9,717	2.4
United Kingdom	14,152	3.1	16,986	4.2
Other European Union	24,393	5.3	19,476	4.8
Other OECD	19,744	4.3	17,908	4.4
All other countries	39,074	8.4	44,076	10.9
Imports	415,683	100.0	413,833	100.0
United States	270,067	65.0	259,953	62.8
Japan	11,967	2.9	10,067	2.4
United Kingdom	9,963	2.4	9,561	2.3
Other European Union	32,404	7.8	30,788	7.4
Other OECD[3]	25,160	6.1	29,013	7.0
All other countries	66,123	15.9	74,451	18.0

Source: Statistics Canada: http://goo.gl/tzEW1, "Imports, exports and trade balance of goods on a balance-of-payments basis, by country or country grouping", accessed 2012-01-20.

All other countries and regions share the remaining 23% of Canadian exports. If one can assume that, 'other countries' represent the developing world, then the increase from 8.4% of exports in 2007 to 10.9% in 2010 is an encouraging sign. Part of this increase is probably due to more exports of raw materials to Asian countries like China by Canada.

Imports largely mirror exports in origin. The USA is again the dominant partner. At first glance, it would seem like an anomaly that a country would import and export so heavily to another country that has a very similar industrial economy. The imports from 'other countries' made up the 2.3% decrease in imports from the USA between 2007 and 2010. A substantial amount is probably due to the high price of imported oil, which Canada imports in the east, and it comes mostly from developing countries. Canada's balance in origin and destination of trade also means that it has substantial power in trade negotiations. Unless the trading partner cooperates, both imports and exports can decline. Many developing countries tend to sell to a different country or group of countries than from whom they import. This means less power in trade talks by smaller trading partners.

A quick glance at the main products Canada exported and imported (Table 17) for October of 2010 shows surprising similarities. Still, in all resource sectors Canada exported substantially more then it imported, while in the finished products sector it imported more then it exported. Clearly, this is still a remnant of its staple development days.

Table 17. Merchandise Trade of Canada, October 2011, in $ Millions

Principal commodity groupings	Exports	Imports
Agricultural and fishing products	3,605	2,755
Energy products	8,908	4,506
Forestry products	1,867	216
Industrial goods and materials	9,821	8,596
Machinery and equipment	6,943	10,676
Automotive products	5,034	6,054
Other consumer goods	1,400	5,046
Special transactions trade	301	597
Total	37,879	38,446

Source: Statistics Canada: "Merchandise trade of Canada (monthly)", http://www40.statcan.gc.ca/l01/cst01/trad45a-eng.htm, accessed January 2012.

The heavy dependence on the USA for imports and exports is quite different from the early period of Canadian development when staples dominated Canada's exports. At that time, most of the exports went to Europe. However, the emergence of the EU and its forerunners after World War II, created a 'fortress Europe' trade block and Canada had to look for markets closer at hand. The heavy presence of American automakers in Canada resulted in 1965 in a free trade pact in the auto industry[60]. NAFTA was just a continuation and expansion of this and brought Mexico into the agreement.

Canada's service sector

Like in all western countries, the broad group of service industries form the mainstay of Canada's employment economy. As can be seen from Tables 1 and 2, the service sector contributes some 69% to Canada's GDP and 76% to total employment. In a sense, much of it makes sure the goods produced in the primary and secondary sectors are taken to the final consumers. Furthermore, it supplies services to business both within Canada and abroad. In addition, Canada's large education, culture, health, and government employment numbers contribute immensely to the size of the services sector. The higher education demand for many of these services requires its own industrial classification category, which is labelled quaternary services. Since so many of the service jobs cater to the local population, they are frequently referred to as 'non-basic' jobs while primary and secondary industrial jobs are known as 'basic' jobs. Although the classification of a job into one of the different

[60] Government of Canada, "1965 – Canada – United States Auto Pact", http://www.canadianeconomy.gc.ca/English/economy/1965canada_us_auto_pact.html, accessed September 29, 2008.

categories in Table 2 is not precise, trade at 15.9% is the largest sector followed by health, education, professional, finance, accommodation, information and business services.

Over the seven-year period from 2003 to 2010, overall employment in the service sector grew by 13% led by professional, scientific and technical services (26%), educational services (19%) and public administration (17%). All four categories of the service sectors that require higher education, the quaternary sector, grew faster than the industry as a whole. Such growth has and will substantially change the nature of the economy in just a few years.

As mentioned above, public administration was one of the fastest growing service sectors in the 2003 to 2010 period. Neil Reynolds, in an opinion column in the Globe and Mail, compared it to a 'Dr. Seuss' story where the growth of parasites in the antlers of the moose finally becomes so dominant that the moose loses the freedom to move about[61].

He compares this to the provincial and territorial entities in Canada and their public sector employees. But the size of public sector employment varies tremendously between the provinces and territories. If one compares the number of employees to the total population of the political jurisdiction, as is done in Table 19, one can see it is very high for the territories, 23% for the North West Territories (NW), 21% for the Yukon Territories (YT) and 19% for Nunavut (Nv).

Table 18. Growth of Service Sector Employment, 2003 to 2010

Year	2003	2010	Growth
Employment	In '000		Fraction
Services-producing sector	11,747	13,301	1.13
Trade	2,468	2,678	1.09
Transportation and warehousing	791	806	1.02
Finance, insur., real estate and leasing	917	1,096	1.19
Professional, scientific and technical	1,004	1,267	1.26
Business, building and other support	609	672	1.10
Educational services	1,027	1,218	1.19
Health care and social assistance	1,679	2,031	1.21
Information, culture and recreation	715	766	1.07
Accommodation and food services	1,006	1,058	1.05
Other services	713	754	1.06
Public administration	819	956	1.17

Source: Statistics Canada, "Employment by Industry", http://goo.gl/rURD6, accessed 2012-01-20.

61 Neil Renolds, "The moose and the modern welfare state", Globe and Mail, December 28, 2012. p. B2. http://goo.gl/txl0D, accessed 2012-01-20.

In other words, around 20 out of 100 people living there work in the public sector in these territories. Since this includes children and the elderly, the proportion of people in the governmental labour force is probably closer to 50%. Of the public sector employment, by far the greatest portion works in the territorial governments themselves. Their percentage is far higher than their counterpart employees in provincial governments.

There is also substantial variation in the public sector service employment between the provinces, from a high of 14% in Nova Scotia, Manitoba and Saskatchewan to a low of 9% in Alberta and British Columbia. The difference of five percentage points comes close to a 33% difference between the lowest and highest provinces. The higher provinces tend to be the traditional 'have-not' provinces, the ones that can least afford a high number of government employees.

Table 19. Public sector employment as % of total, by province and territories, 2010

	CDN	NL	PE	NS	NB	Q	ON	MB	SK	AB	BC	YT	NW	Nv
Population (in 000)	34126	511	143	945	753	7906	13228	1235	1044	3721	4530	35	44	33
Public sector	11	13	13	14	13	11	10	14	14	9	9	21	23	19
Government	10	13	12	13	11	10	9	12	12	9	8	18	21	18
Federal general Gov.	1	1	3	3	2	1	1	1	1	1	1	2	3	1
Prov./Ter. Gen. Gov.	1	2	6	1	4	1	1	1	2	1	1	13	11	11
Health/Social Serv.	2	4		3	3	3	2	4	4	2	2	1	2	
Univer. and colleges	1	2	1	1	1	1	1	1	1	1	1			
Local general Gov.	2	1	1	2	1	1	2	2	2	2	1	2	4	6
Local school boards	2	2	2	2		2	2	3	3	2	2		1	
Gov. bus. ent.	1	1	1	1	1	1	1	2	2	0	1			
Fed. Gov. bus. ent.	0	0	1	1	0	0	0	1	0	0	0			
Prov/Ter. Bus. ent.	0	0	0	0	1	0	0	1	1	0	0			
Local Gov. bus. ent.	0			0	0	0	0	0	0	0	0			

Source: Statistics Canada: "Public sector employment as % of total population, by province and territories, 2010", http://www40.statcan.ca/l01/cst01/govt62a-eng.htm, accessed Jan. 3, 2012.

Table 20. Service sector employment, as fraction of total, by provinces, December 2011

	CDN	NL	PEI	NS	NB	Q	ON	MB	SK	AB	BC
Population (in 000)	34126	511	143	945	753	7906	13228	1235	1044	3721	4530
Services Emp. as % of Pop.	39.7	34.4	38.3	39.5	36.8	39.0	40.4	39.4	37.6	41.4	40.5
Services sector employment (in 000)	13,551	176	55	373	277	3,080	5,337	486	393	1,540	1,836
Trade	0.20	0.21	0.17	0.21	0.20	0.21	0.18	0.19	0.21	0.22	0.20
Transportation and warehousing	0.06	0.06	0.04	0.06	0.06	0.06	0.06	0.07	0.07	0.07	0.07
Finance, insurance, real estate, leasing	0.08	0.05	0.05	0.06	0.06	0.07	0.09	0.08	0.08	0.07	0.07
Professional, scientific and technical	0.10	0.05	0.05	0.06	0.06	0.10	0.11	0.05	0.07	0.11	0.11
Business, building and other support	0.05	0.03	0.04	0.05	0.06	0.05	0.05	0.04	0.03	0.05	0.04
Educational services	0.09	0.10	0.10	0.10	0.10	0.09	0.09	0.10	0.10	0.08	0.09
Health care and social assistance	0.16	0.19	0.18	0.18	0.19	0.16	0.14	0.20	0.18	0.15	0.15
Information, culture and recreation	0.06	0.04	0.05	0.06	0.05	0.05	0.06	0.05	0.04	0.05	0.06
Accommodation and food services	0.08	0.08	0.10	0.08	0.09	0.08	0.08	0.09	0.08	0.08	0.10
Other services	0.06	0.07	0.05	0.06	0.05	0.06	0.05	0.05	0.07	0.07	0.06
Public administration	0.07	0.11	0.15	0.08	0.10	0.08	0.07	0.08	0.07	0.06	0.06

1. Includes FT and PT employees both in and outside of Canada.

Source: Statistics Canada, http://www40.statcan.gc.ca/l01/cst01/labr67a-eng.htm, "Employment by major industry groups, seasonally adjusted, by province", accessed January 6, 2012.

The difference is further heightened by varying labour participation rates. This rate is based on the number of people 15 years of age and older in the labour force. In 2011[62], the younger group consisted of 16% of the total population. Labour participation rates generally go from low values in eastern Canada to higher ones in the west (60%-NFL, 68%-PEI, 64%-NS, 63%-NB, 65%-Q, 67%-ON, 69%-MB, 69%-SK, 73%-AB, and 65%-BC)[63]. If the labour force percentage is low and the public sector percentage is high, the overall proportion of people working for the government, compared to total employment, is high.

62 http://www40.statcan.gc.ca/l01/cst01/demo10a-eng.htm, Statistics Canada, "Population by sex and age group", accessed January 6, 2012.

63 Statistics Canada, "Labour force, employed and unemployed, numbers and rates, by provinces", http://www40.statcan.gc.ca/l01/cst01/LABOR07a-eng.htm, accessed January 6, 2012.

It is rather surprising how similar the provincial percentages are for health and education employment. In addition, government enterprise employments in the provinces all vary between 0 and 2%, the latter being in Manitoba and Saskatchewan. These two provinces have had a number of New Democratic Party governments in office over the years, which accounts for their high overall public sector employment.

If one examines the employment of the service sector as a percentage of a province's total population (Table 20) one can see substantial variation. Newfoundland has the lowest fraction at 34.4% while Alberta at 41.4% has the highest. In general, larger provinces have higher proportions.

An examination of the employment service subsectors of the provinces for December of 2011, as shown in Table 21, presents a rather strong overall similarity between the provinces. Except for the employment fraction in government and the professional, scientific and technical subsection, most proportions are similar in all provinces. It illustrates how ubiquitous these service employment demands are. Everyone, no matter where they live, needs the barber, sales clerks, waiters and other similar class services. Also needed everywhere are the trades, educators, finance/insurance and primary health care workers, which together make up about 15-16% of all service employees. At the top of the service sector are the highly trained lawyers, doctors, professors and other professional people. For this group we see some variation between the provinces, with the bigger provinces having around 11% of all service employees in this category while the smaller provinces have only around 5%.

Traditionally, the service sector has been seen as responding only to the local economy's needs and hence has been viewed as a 'non-basic'[64] sector of an economy. However, this perspective is changing. In a global economy, even services are "traded" or "exported" at the global level. Canada is no exception[65]. From 1981 to 2010, the export of services increased more than three fold. Commercial services, for instance, increased their share of the increase of the total export of services from 23% to 56% (Table 21) attesting to the globalization of commerce. The biggest proportionate decline occurred in financial intermediation, transportation, and travel. If one looks at the size of the Canadian economy in 2006, its GDP was 1,446,307 million. The service export of 70,122 million comes to 4.8% of total. In other words, the export of services contributes more to the Canadian economy than most of the resource sectors.

---

64 Answers.com, "Economic Base Theory", http://www.answers.com/topic/economic-base-theory, accessed September 29, 2008.

65 Foreign Affairs and International Trade Canada, "Canada and Trade in Services", http://goo.gl/xCBrI, accessed January 20, 2012.

Table 21. Canadian Export of Services, in Millions and Fraction by Sectors, 1981-2010

Year	Total Export	Travel	Transport.	Commercial	Government	Financial intermed.
1981	21,221	0.35	0.33	0.23	0.04	0.08
1986	25,286	0.35	0.23	0.32	0.05	0.06
1991	32,162	0.29	0.22	0.42	0.04	0.04
1996	46,928	0.29	0.2	0.47	0.02	0.03
2001	61,974	0.26	0.17	0.52	0.02	0.03
2006	70,122	0.22	0.16	0.52	0.02	0.02
2010	73,293	0.22	0.16	0.56	0.02	0.02

Source: Statistics Canada, "International trade/Merchandise exports", http://dc2.chass.utoronto.ca/.

Summary

The Canadian economy is diversified and dynamic. It has changed, in the last few decades, from a resource extraction oriented one to a high tech manufacturing and services one. The fact that Canada still has many resources makes it the envy of many other OECD countries.

From the above discussion of the Canadian economy one is tempted to argue there are as many economies as there are provinces and territories. This means that the economic well-being of individuals varies depending on where they live. Much of Canada's export and import trade goes south to the USA, 73% and 63% respectively in 2010[66], and not east and west between the provinces. However, much of the service sector is still locally based and has barriers for mobility between provinces. Provincial biased barriers exist in most professional accreditations, in governmental purchases and in subsidies to industries as well as to inter provincial trade. The political barriers, when combined with natural development differences between provinces and regions make for substantial differences in the economic well-being of Canadian people. Table 22 points some of these out. One can see that real GDP per person varied from the Canadian average, in 2003, from a high of 2.21 for the NWT, to a low of 0.74 for Prince Edward Island. Only Ontario, Alberta and the two territories Yukon and North West were above the average, the rest were below. By 2010, the same pattern still prevailed except that the extreme high for the North West Territories had decreased to 1.84 and the lowest had increased to 0.77. Overall, there was a coalescing of the differences during this period. Eight of the

66 2008 Statistics Canada, "Imports, exports and trade balance of goods on a balance-of-payments basis, by country or country grouping", http://goo.gl/Ol5Zs, accessed 2012-01-20.

13 regions moved closer to the Canadian average and only Ontario, Manitoba, Alberta and the North West Territories fell behind.

However, Real GDP per person, as an indicator of variation in wealth by province and territory, is hard to comprehend by the common person. Average hourly wage and salary is not. Table 22 also shows that in 2003 these ranged from 1.22 above the Canadian average for the North West Territories to 0.8 below for Prince Edward Island. By 2010, the difference had increased to 1.31 above the Canadian average for the North West Territories and 0.84 below the average for Prince Edward Island. All provinces except Nova Scotia, New Brunswick, and Ontario improved their status over these seven years.

That the financial differences between provinces and territories are not larger than they are is in part due to federal government transfer payments to the provinces, a long tradition in Canada. In 2009, the transfer payments to persons in the provinces alone came to 25% of total federal 'all government levels' outlays[67].

Table 22. Canadian GDP and Wages/Salaries by Province and Territories, 2003 and 2010

	GDP per person (in 2002 $)			Hourly Wages and Salaries		
	2003	2010	Change	2003	2010	Change
	$	$	$	$	$	$
CDN	37,081	38,826	1,745	17.24	20.44	3.20
	Proportion of CDN average			Proportion of CDN average		
NL	0.91	0.96	0.05	0.94	0.96	0.02
PEI	0.74	0.77	0.03	0.8	0.84	0.04
NS	0.79	0.82	0.03	0.91	0.90	-0.01
NB	0.78	0.83	0.05	0.92	0.90	-0.02
Qué	0.88	0.89	0.01	0.94	0.97	0.03
ON	1.07	1.03	-0.04	1.05	1.00	-0.05
MA	0.96	0.91	-0.05	0.9	0.93	0.03
SK	0.97	1.02	0.05	0.91	1.04	0.13
AB	1.33	1.27	-0.05	1.03	1.15	0.12
BC	0.92	0.95	0.03	1	1.00	0.00
YT	1.07	1.31	0.24	1.02	1.08	0.06
NWT	2.21	1.84	-0.37	1.22	1.31	0.09
Nvt	0.88	0.97	0.09	0.99	1.20	0.21

Source: Statistics Canada, "Real gross domestic product, expenditure-based, by province and territory", http://goo.gl/BzrEb accessed 2012-01-20, and "Earnings, average hourly for hourly paid employees, by province and territory", http://goo.gl/sEgKV, accessed 2012-01-20.

67 Statistics Canada, The Daily, "Government Finance: Revenue, Expenditure and Surplus", Thursday, June 16, 2005. http://goo.gl/ucrhk, accessed 2012-01-20.

Table 23 provides information on the source of transfer monies received by the provinces and territorial governments to run their affairs. The degree to which they depend on the federal government handouts is phenomenal. It varies from 10% for Alberta to 92% for Nunavut. Clearly, the territories and the eastern provinces receive amounts much higher than the average of the provinces and territories. Outside these regions, Manitoba's transfer payment is also high. It seems a dependency mentality must prevail in some parts of Canada. Although Canada is a loose federation in political terms, the high dependence of the provincial and territorial governments on the federal purse creates a major dependence. Yet, it seemingly is the only way for Canada to overcome regional differences in economic potential and historical momentum.

Table 23. Provincial and territorial transfer payments from the Federal government, 2009

	NL	PEI	NS	NB	QUÉ	ON	MB	SK	AB	BC	YT	NW	Nvt
Total Revenue	7,545	1,396	8,775	7,323	80,898	96,139	11,791	13,933	40,224	36,610	929	1,432	1,305
General transfers	3,280	367	2,387	1,824	9,333	4,207	2,458	359	1,224	1,663	575	820	979
Specific transfers	607	170	1,040	774	6,044	11,712	1,328	1,555	2,737	4,488	156	203	218
% of total	*0.52*	*0.38*	*0.39*	*0.35*	*0.19*	*0.17*	*0.32*	*0.14*	*0.10*	*0.17*	*0.79*	*0.71*	*0.92*

Source: Statistics Canada, "Provincial and territorial general government revenue and expenditures, by province and territory". http://goo.gl/Bns9Z, accessed 2012-01-20.

Review Questions

- Canadian development is associated with a particular development theory. Which theory is it and what are its salient characteristics?
- Canadian manufacturing employment, percent wise, is one of the lowest of all the OECD countries. Do Canadians not want to work in manufacturing places or is it because the demand for employees from the resource, services, health, etc. sectors are so great that people rather work in these sectors? Discuss.
- It is generally accepted that, in the past, Canada's wealth has come from the harvest of resources. Will this also be the case in the future? If not why not?
- One of the wealthier and bigger provinces in Canada is Ontario. Why?
- Canada's trade with the USA is huge. What are the major reasons for this?

Bibliographic Information

Unique references

Hecht, Alfred and Alfred Pletsch (eds.), *Virtual Geography Text on Canada and Germany*. http://www.v-g-t.de/, tri-lingual edition (English, German, French). Accessed January 20, 2012.

Major Canadian data source sites

Statistics Canada, "Today's News Release from The Daily", online: http://www.statcan.ca/menu-en.htm.

Statistics Canada, "Canadian Industry Statistics", online: http://strategis.ic.gc.ca/sc_ecnmy/sio/cis11-33defe.html.

Natural Resources of Canada, *National Atlas of Canada*, online: http://atlas.nrcan.gc.ca/site/english/index.html.

Historical Atlas of Canada, http://mercator.geog.utoronto.ca/hacddp/page1.htm.

A selective few classical references

Barnes, T. *et al.* (2000), "Canadian economic geography at the millennium", *The Canadian Geographer*, Vol. 44, No. 1, pp. 4-24.

Bone, R.M. (2005), *The Regional Geography of Canada*, 3rd edition, Don Mills: Oxford University Press.

Britton, J.N.H. (ed.) (1996), Canada and the Global Economy: The Geography of Structural and Technological Change, Montreal: McGill-Queen's University Press.

Innis, Harold A. and Contributor Arthur J. Ray (1999), *The Fur Trade in Canada*, Toronto: University of Toronto Press. (First published in 1930).

McCann, Larry and Angus Gunn, (eds.) (1998), *Heartland Hinterland, a Regional Geography of Canada*, Scarborough: Prentice Hall.

McGillivray, B. (2006), *Canada: a nation of regions*, Don Mills: Oxford University Press.

Watkins, Melville H. (1963), "A Staple Theory of Economic Growth", *The Canadian Journal of Economics and Political Science*, Vol. 29, No. 2, pp. 141-158.

Warkentin, John (ed.) (1968), *Canada: A Geographical Interpretation*, Toronto: Methuen.

Wallace I. (2002), *A Geography of the Canadian Economy*, Toronto: Oxford University Press.

CHAPTER 5

Canadian Political Geography for the 21st Century

Navigating the New Landscapes of Continental Division and Integration

Heather NICOL

Trent University

Canada, a Regional Nation

> "The interaction of physical geography and settlement history has made Canada one of the most highly regionalized countries in the developed world. This has both cultural and economic dimensions, and the two come together in the practical politics of making the federal system work. Harris (1987) has likened Canada to an 'archipelago' – a string of regionalized societies that differ significantly in ethnicity, political culture, and economic orientation" (Wallace, 2002, p. 21).

> "Though Canadian politics has always been refracted through a geographical prism, there is little consensus among sociologists, economists, or geographers, let alone political scientists, about how best to interpret this regional dimension" (Brodie, 1997, p. 261).

The modern nation of Canada was founded in 1867 (Figure 1), but even at this time there was much work to do to define the national territory claimed and governed by the new country. Canada had consisted of a collection of colonies and territories. It was several years later before British Columbia, Prince Edward Island and Rupert's Land, as well as the remaining northwestern territory of Canada joined the equation, and indeed several more years after that, before the territory of the young nation began to resemble that of the present day. For example, the Arctic islands were not ceded to Canada from Great Britain until the 1880s, and even by that time the northern territories and western provinces of Canada had yet to find their final shape. Moreover, it was not until 1949 that the final piece of the puzzle, Newfoundland, was added to Canada's political territory.

Figure 1. Canada's territory in 1867

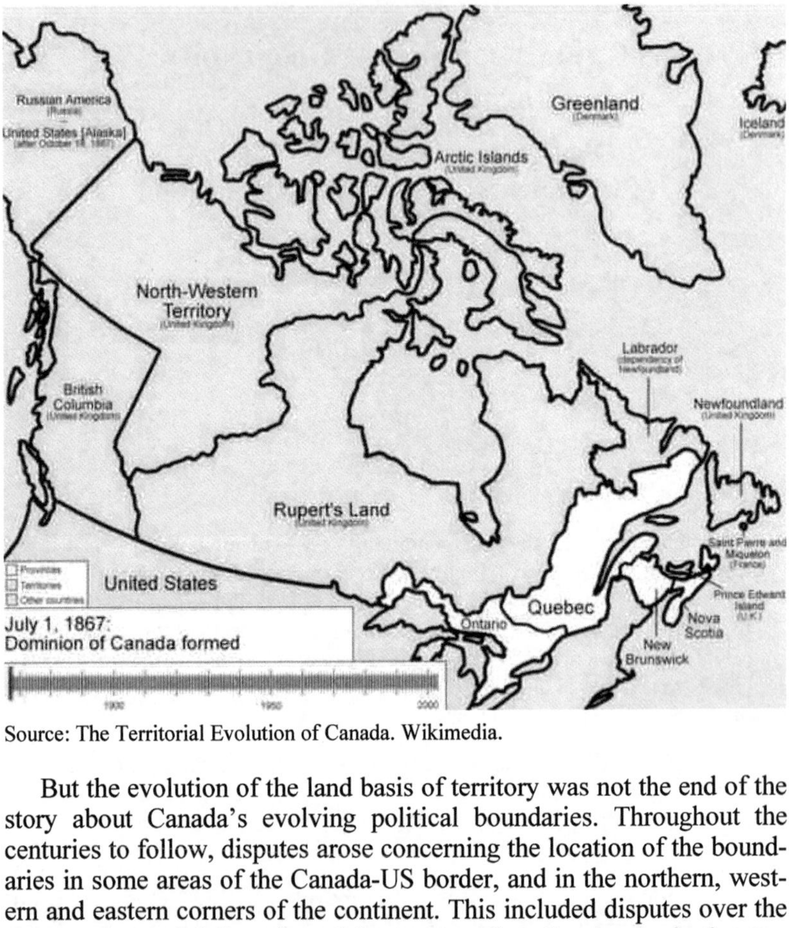

Source: The Territorial Evolution of Canada. Wikimedia.

But the evolution of the land basis of territory was not the end of the story about Canada's evolving political boundaries. Throughout the centuries to follow, disputes arose concerning the location of the boundaries in some areas of the Canada-US border, and in the northern, western and eastern corners of the continent. This included disputes over the demarcation and delineation of the western Canadian border in the area of the 'Alaskan Panhandle', which was not settled until the Hay-Herbert Treaty of 1903. This Alaskan dispute marked the end of a series of 19th century land boundary contestations between Canada and the US, a process which had begun even before Confederation. The Oregon Territory boundary as well as the Maine-New Brunswick boundary, for example, had provoked considerable disagreement between Canada and the United States much earlier in the century. In the 20th century, more disputes arouse concerning the location of maritime boundaries, particularly after the negotiation of the United Nation's Law of the Sea, in the last quarter of the 20th century. The Georgia Strait and Strait of San Juan de Fuca, for example, were contested areas in the late 20th century, as was the Internal Waters status of Canada's Northwest Passage. Some of these contestations remain ongoing in the 21st century. By the early 21st century, however, the evolving importance of the North in an international context, and the growing importance of Arctic energy and mineral resources, triggered a new round of potential boundary-making in the Arctic Ocean, not only between Canada and the United States, but between Canada and its European and Russian neighbours. Although these various claims have been or are now being articulated, they remain claims and are not yet translated into hard and fast maritime boundary lines.

The outlines of territorial evolution can be seen in animated form on Wikimedia[1]. In the beginning, Canada's territory appeared as in Figure 1 below. Graeme Wynne has observed that this territorial evolution was all the more remarkable because for the most part, the foundation of the new country was achieved in less than a century, creating a territorial foundation which amounted to six times the size of the original colonies (Wynne, 2001, 359). The political basis of the new country was a result of more than the shape of its external borders, however. An important component of the existing political geography of Canada was, and remains, the historical legacy of the evolution of regional and provincial boundaries, which were the outcome of an evolving relationship between federal government powers and delegated regional and provincial responsibilities and roles. Thus, the evolution of Canada's provincial and territorial map (Figure 2) paints a picture of the evolution of this complex political relationship at the regional scale, as the federal government responded not only to national agendas and policies, but to regional issues. It was this focus on regional boundaries and governance that came to dominate nation-building throughout much of the late 19th and early 20th centuries.

Why? Wynne (1987) suggests that these "internal boundary changes and shifting political jurisdictions reflected administrative requirements, the advance of settlement and competing territorial ambitions" (Wynne, 1987, p. 359). In this sense, the forging of the Canadian nation must be understood as a deliberate attempt to shape a political geography in support of political ideologies of the time: steeped in the idea of the imperative of the Westphalian Nation state, strongly tied to British imperatives of Empire, and responding to the American experiment to the south, Canadian government officials sought to build a transconti-

1 The outlines of territorial changes can be seen in animated maps on Wikimedia http://upload.wikimedia.org/wikipedia/commons/9/9d/Canada_provinces_evolution_2.gif.

nental nation which stood as a testament to the political and economic philosophies of the day.

Figure 2. Territorial evolution of Canada

Source: The Territorial Evolution of Canada. Wikimedia. Redrawn by Stephen Gardiner.

Since that time, we have talked about Canada as "a nation of regions" (McGillivray, 2006). But what exactly are the regions of Canada? The metanarratives that define the normative or traditional academic regional perspective are found in undergraduate Canadian geography texts, such as McCann and Gunn (1997), Bone (2003) and McGillivray (2006). To some extent, all of these narratives are informed by both the historical dynamics of staple thesis (Innis, 1956; MacIntosh, 1923), as well as by Robinson's seminal discussion of concepts and themes in the regional geography of Canada (Robinson, 1989), where the foundations for understanding regions as "building blocks" is developed within a framework of a hierarchy of regions. But where McCann and Gunn (1997) employ the concept of "heartland and hinterland", McGillivray is more explicit about what constitute core and periphery with reference to World System theory. Nonetheless, the overall regional divisions encountered in these texts are similar and indeed could be considered as "definitive": Canada is a nation of regions – between four and six – essentially defined by historical patterns in resource extraction and manufacturing.

Rather than being concerned with the small differences which each of these regional schemes exhibit, however, this chapter is more concerned with their similarities. Identifying the relationship between region and territorial map is really an exercise in understanding the political and economic forces which produced them. The aforementioned internal boundaries were, at first, not so much the lines defining region as they were political necessities marking empire, settlement frontier, or political expediency. But by the time of Confederation it was clear that regions were to be a hallmark of the Canadian political and economic system. However, even a quick glance at Figure 2, which defines the evolving boundaries of Canada as a nation-state will indicate that building Canadian territory has been a project which has taken almost two centuries. This figure suggests that there were essentially two important projects which contributed to the evolution of regions within the Canadian territory. One was the evolution of political control as the federal government as extended the impress of its power through the negotiation of new lands or through the definition of border. The other was through the delegation of territorial status and regional governance to Canadian territories to the north and west.

The construction of these territorial boundaries, both external and internal, was not completed until the late 20th century, and indeed is still underway. Indeed, the contemporary political map of Canada is only a decade old, as Nunavut came into its own. But even more subtle that the internal division of territory within Canada, have been changes to the rights and responsibilities of territorial governance and the division of territories into small and more empowered provincial governments. These changes have, in essence, captured the regional differentiation process and increasingly focused it upon sub-nation, provincial scales. The 20th century developments which increasingly equated French North America and Franco-Canadians with "Québec" (Waddell and Gunn, 1997), or new trends in which provinces represent themselves and their own interests in international economic negotiations (what Holroyd, 2009, has called "paradiplomacy") all stress the way in which provincial governments have increasingly operated as powerful agents in defining specific interests

Staples Theory and the Political Economy of Regionalism

Harold Innis' "staples theory" originally provided the framework for understanding national economic and political development in terms of natural resource extraction, and this economic imperative has been understood as the major influence which in large measure shaped Canada's regional character (Innis, 1956; Mackintosh, 1923; Easterbrook and Watkins, 1967). Fish, furs, lumber and wheat, connected by intricacies of forward and backward linkages, within a burgeoning world economy dominated by European powers, thus created a distinctive landscape of interlocked economic interdependencies, which, as the 19th century progressed, became increasingly oriented towards fuelling a Central Canadian industrial "growth engine".

The staples thesis has thus been the foundation for the historical understanding of regional development in Canada. Indeed, Clement (1996, p. 24) asks somewhat rhetorically "To what extent does the emphasis on staples capture the totality of the Canadian growth experience since the beginning" (1996, p. 24), and then responds as follows:

> The answer is that there has always been much by way of "local growth" that escaped the net, but staples such as fish and fur "dominated" the early economic history of Canada. The economy of what is today Canada has been from its Euro-Canadian beginnings an integral part of a larger North Atlantic economy and has tended to grow and fluctuate with it; the edge that has made the difference (good and bad) has been the varying fate of the staple trades and industries. Yesterday's "local growth" (home production for home markets) has become, in the modern era of free trade, export-dependent manufacturing and service industries ... in part because of an abundant endowment of hydro-electric potential and fossil fuels, plus the ability to shift from agricultural staples such as wheat to industrial staples such as minerals and newsprint (Clement, 1996, p. 24).

This enduring staples discourse, as Clement (1996) defines it in the paragraph above, suggests that Canada must be seen as a single and coherent nation, where regions connect like so many pieces of a jig-saw puzzle to produce a coherent whole, all linked by internal economic and political interdependencies, as well as by what McCann and Gunn (1997) describe as a broader series of hegemonic, heartland and hinterland relationships.

Within this ontological framework, regions were "valued" because of their relationship to the political and economic core of Central Canada. For example, the concept of a Canadian Shield Region, based upon resource extraction, meant that the near North in most Canadian regional geography texts (as well as the Arctic and sub-Arctic region lying above) was constructed as a peripheral region and an economic frontier. As a physical region, it cross-cuts almost every province in Canada, but was, because of this, politically powerless – remaining a geographical rather than political region. While ethnically and culturally diverse, and highly indigenous, the "Shield" and 'the North' remained categorized as a resource frontier – this was the regional logic which framed its very definition. Central Canada, on the other hand was situated in Southern Ontario and Québec, in the Saint Lawrence Lowlands, and as such the political economy region cross-cut provincial boundaries. Its position in the "heartland" made it de facto a powerful region. This process not only reflects the valuing of regions on the basis of physical characteristics, but also reflects, just as likely, the 'southern-oriented' regional perspective of the mapmaker and scholars involved in reproducing the categories of Canadian regionalism. By the late 20th century, the urban-industrial heartland, Central Canada, came to dominate the Canadian regional landscape, as a virtual 'Main Street' developed between Québec City and Windsor Ontario. Meanwhile, the surrounding regions became the "hinterland" of the central Ontario growth engine, itself strongly linked to US markets and economic growth (McCann and Gunn, 1997). All this, Paul Philips suggests in his volume *Regional Disparities*, resulted "in large measure from the fortuitous distribution for resource wealth" (Phillips, 1982, p. 4). It was also, however, a statement about the location of political power stemming from wealth.

Indeed, regionalism was understood as a manifestation of Canada's reliance on traditional resource industries, as they grew and increasingly contributed the raw materials of industrialisation for 19th and 20th century economic growth (Brodie, 1997; Wallace, 2002; McCann and Gunn, 1997). The spatialisation of resource extraction itself contributed to a pattern of uneven economic development: mining in the Canadian Shield and the Maritimes was essentially an activity located on the periphery. Logging occurred in the northern regions of Western Canada and British Columbia, the Canadian Shield, and the 'remote' river valleys of the Maritime Provinces, where a general failure of agriculture to sustain regional, export-oriented growth, occurred in the 19th century

(Phillips, 1982; Neill, 2002). In the latter region fishing, too, created a set of weak economically propulsive linkages, and ultimately condemned the Atlantic provinces to marginal economic growth and economic depression (Neill, 2002). Also a 'hinterland', the North stood intact as a resource frontier under the aegis of internal colonialism (Bone, 2003: Able and Coates, 1999). Thus, according to this understanding of development, staples industries sustained the linkages and defined the nature of regionalism in relation to the Central Canadian nexus (Mackintosh, 1923; Easterbrook and Watkins, 1967). At the same time, a lack of innovation, technological up-dates, economic dependency on both European and then US capital, markets and branch-plants, and a host of other 'made-in-Canada' difficulties, ultimately worked against technological and economical diversification maturation, so that the economic profile of this modern nation-state would retain the profile of an economically developing nation – dominated by raw material exports and highly regionalized as a result of uneven development processes. Indeed, in the opening to his seminal 1986 study "*Regional Economic Development*", Donald J. Savoie argued that "Among the countries currently classified by the United Nations as 'industrialised market economies', Canada is surely one of the most highly regionalized, and its economy is accordingly one of the most badly fragmented" (Savoie, 1986, p. 3).

This understanding of Canadian regionalism is a political as well as an economic one, and it is an important factor contributing to the calculation of Canadian national identity. As an historical argument, it suggests that economy and politics combined to allow the Central Canadian "core" to dominate a series of resource-oriented hinterlands. Unlike the American West, where wagon trains created a moving frontier, in Canada disparate regions were sewn together by a political ideal and the desire to create a nation (Harris, 1997). Indeed, as late as the 1980s and 1990s, Canadian geographers still spoke about the functional, yet delicate political and economic balance inherent in "heartland and hinterland" regionalism (McCann and Gunn, 1997). This regional system explained how Canada developed as an "east-west nation" against all odds, and certainly against the continental "grain" and pull of American markets to the south. It was, in effect, a geography which recognized the regionalized nature of Canada's economic, demographic, social and political structures, and recognized the degree of confluence between provincial boundaries and regional borders, since both evolved somewhat simultaneously. Moreover, it was based upon a specific understanding of Canadian economic development. This staples-oriented overview constitutes the metanarrative defining the geography of Canada, although over time it has been increasingly challenged by scholars who suggest that domestic markets may have been a much stronger influence upon the development of the 19th century national economy than staple theory would have us believe. That is to say, that domestic-led rather than export led regional development has perhaps been underrated. Indeed, Robert Neill, explains that

> The staple export paradigm associated with Innis and Mackintosh disposed viewers toward unwarranted assumptions about the roles of agriculture, commerce, and manufacturing in Canadian economic development; assumptions that were the subject of a question prompted by O.J. Firestone (1958), and raised by Ken Buckley (1958). "How much investment took place, in what sectors and regions, and what interconnections were involved?" The question challenged the extension of the Imperialist-staples view of late nineteenth century Canada to the whole range of Canadian history, but it suggested nothing to replace that view (Neill, 2002, p. 4).

Despite such criticisms, the end result of these "unwarranted assumptions", as Neill dubs them, was a Canadian political geography, if indeed we can call it such, which reflected the themes of economic geography from a political economy perspective – a hierarchy of power which found its nexus in Central Canada, and which faded towards the north, east, and west. Political issues essentially correlated with the issue of Québec separatism, and to a slightly lesser extent, the politics of aboriginal land claims. Because the latter were political issues delegated to the physical and economic peripheries, they remained a sidebar in the political discourses of regionalization, outside of their application to regional descriptions of "the North".

Regions and State: A Complex Relationship

Although Canada has been, since Confederation, involved in an ongoing process of internal territorial differentiation and regional construction, largely based upon economic imperatives, it is also true that for much of its history since the late 19th century, Canada has also been involved in a political project of federal consolidation. The federal government has attempted to cobble together a nation of regions and cities within a country clearly 'fighting' its natural physiographic orientation – which runs north south along the 'continental grain'. Politically, however, Canada is an 'east-west' nation (Brodie, 1997), originally cobbled together by 19th century National Policies and Confederation, against the 'continental grain', an attempt perhaps to ward off the threat of 'American expansionism' (Aitkin, 1959). Thus the tension between continentalism and nationalism, along with the pull of US markets and Canadian market protection, as well as the necessity for the political "glue" provided by Confederation to hold the disparate regions together, are facts of life for the young country of Canada.

These facts have been understood since the time of its founding, and have informed the way in which the country has understood politics and regionalism since that time. This means that academic focus has, at least until quite recently, been squarely focused upon examining the considerable effort by federal agencies to build the east-west linkages which "glued" the country together. Early on, during the 19th century, this government effort was invested in railways, and later, it was invested in federal projects which attempted to stimulate regional development through incentives, transfer payments and other means (Wallace, 2002; Savoie, 1986; Albo and Jenson, 1997). Indeed, these policies are what Holroyd (2009, p. 61) calls "the staples of Canadian politics – revenue sharing, education, health, infrastructure and regional economic development". All this means that until the late 20th century, even the existence of the international boundaries which divided Canadian territory from the US were understood in normative concepts of an east-west oriented regionalism and associated also with geopolitical concepts like defensive expansionism and the Laurentian thesis (Creighton, 1937). The latter implied that an east-west line or a boundary was inherent in the structure not only of continental geography, but that of the political division between Canada and the US boundary lines that were to reinforce what history, culture and geography had divided (Laxer, 2003).

Thus, for much of its history, the study of political geography in Canada existed as a defined field which relied upon an understanding of regionalism as a "19th century" phenomenon which extended into the 20th century. Regions were seen as the historical outcome of the settlement process and the impact of resource-oriented economies. Not only is this understanding somewhat problematic from the perspective of the 21st century, it has also 'naturalised' the regional structure of 19th century Canada, as nation-building political processes privileged Central Canada. Sanford MacDonald, for example, Ontario's Premier in 1867, feared that Confederation would shatter the political and economic unity of the Saint Lawrence and Ottawa River valleys because it would deflect the political and economic focus from what was to become Eastern Ontario "from its natural entrepot of Montréal in favour of the Grit and Orange dominated west" (Hodgins, 1971), But others worried that Confederation would give "at once a preponderance to the west of Toronto" (for example, see the newspaper *The Cornwall Freeholder*, March 22, 1872). Thus, if no one was certain of what the regions would be, they were certain that regionalization would result. Indeed, it was the idea of a Canada of regions – divided along broadly physiographic lines which later correlated with territorial and provincial boundaries, which became the dominating paradigm used to assess the effects of Confederation.

Thus while Goldwyn Smith and his contemporaries described Canada as a series of North-South regions held together by the glue of Confederation, in doing so not only did they build upon popular understanding of the impact of Confederation and National policies upon 19th century Canada and the improbability of Canada, they also perpetuated such ideas. To some extent, it created a juxtaposition about the "unnatural" relationship between Canada and the US, in the face of what Martin W. Lewis and Kären Wigen (1997) have called, "the myth of continents": whereby meta-geographical assumptions about national and regional affinities grow out of cultural concepts, like the naturalised context of continentalism. Unlike some scholars of Canadian political and economic geography, Wynne (1987) suggests that the idea of Canada was very much a 20th century construction (forged in Canada's power houses and central places), and that this 20th century idea, created centrality with respect to the role of Central Canada's urban places in regional context. Canada today incorporates the "idea" of a 19th century nation forged by 20th century urban industrialisation.

All of this suggests that for much of the past 150 years since Confederation – even as Canadian policy-making invested in carving out the country, organizing it into the east-west framework supported by National Policies, and rationalising a border with the US – the regional map did not remain static. The regions of Canada emerged over time out of the fabric of clusters of settlements in the south and east, and out of a broad undifferentiated frontier in western and northern Canada. These regions were often seen as separate places with more in common with each other than with the rest of the country. They existed because of the potential of staples industries, like fishing, logging, or wheat production, and the nature of the economic development base determined the role of region in the Canadian federation, and the boundaries of regional distinction.

In all of this, the role of a federal government was essential to glue the bits of the fabric together, somewhat like a patchwork quilt. But within this policy framework, it was the economic incentives which most interested regions – the freight rates and tariff structures, public subsidies for industrialisation, colonial allegiances, and even free trade with the US were all discussed and acted upon, depending upon regional demands and political imperatives. Thus, the *raison d'être* for regional politics remained rooted in a staples-based economy within a federal framework, so much so that Clements could suggest that staples theory really 'captures' the essence of the nature and pattern of Canadian economic development since its inception.

And while not everyone would agree that the staples thesis captures the economic development experience since Canada's inception, it is important to note that the *belief* in the centrality of export-led resource development was essential in the process of building a political narrative in late 19th century Canada (Brodie, 1997). It was the idea that regional development was a key role of the state, which really institutionalised practices concerning regional definition and development policies, and which launched the federal government upon "an elaborate set of balancing acts" (Wallace, 2002). Eventually, as Brodie (1997) observes, the effect of this belief was to reduce "regions" to institutional boundaries within federalism", where "regional differences" were increasingly equated with provincial institutional focuses.

This notion of regionalism as an outcome of a national political economy largely focused on staples production had implications for regionalism in the 20th and perhaps even the 21st centuries. Whereas National Policies along with railroads served as the premier policy solutions in 19th century Canada, a new relationship between regions and the state developed as the 20th century progressed. The idea was not to recruit settlers to what has been dubbed the 'last best west' to drive the industrial engines of Central Canada, but rather to create a policy framework which underscored the role of state (i.e. the federal government) in new ways. As Wallace notes,

> The federal political arrangements adopted in 1867 reflected the experience of a society more geographically compact and culturally homogeneous ...than exists today. It was expected that 'national' political parties would find nationwide support, leading to a regionally representative federal cabinet that would ensure that the interest of each part of the country had a voice. But by the last decades of the twentieth century it became clear that such an assumption is no longer tenable (Wallace, 2002).

Indeed, Donald Savoie argues that the strength of regional political clout is more than the politics of the economy:

> political power in Ottawa is increasingly concentrated in the hands of the prime ministers and a few advisers. One of the reasons for this, ironically, has been the regional factor, combined with national unity concerns. But in their attempts to manage developments on this front, they have made matters worse. That is, concentrating political power in the hands of a few people has greatly inhibited the ability of national political institutions to understand regional forces, let alone accommodate them in policy- and decision-making (Savoie, 2000, p. 5).

But along with a changing role for sub-national governance, and the rising star of provincial powers, other developments began to reshape the structure and nature of the Canadian state, the political economy of its regions, and the structure of provincial and federal relations. Two things contributed to this. First, in the post-World War II years, Canada saw large-scale exports of resources and large scale imports of American capital which provided funds to expand the staple sectors, including such new staples as oil and gas, iron ore, uranium, and potash. Such revenues stimulated development of industry as well as natural resources. Secondly, the federal government launched a new series of policies designed to address issues of growing regional disparity. The 1960s for example, stand out as the era of "regional policy" (Wallace, 2002, p. 22) whereby Federal regional development policies directly attacked issues of regional differentiation and economic inequality in Canada. Indeed, Wallace argues that the reason for such policies were inherent in the changes to the Canadian economy which had occurred throughout the first half of the 20th century, and indeed which accelerated after World War II. Yet, Savoie argues, such policies were remarkable failures

> the United States has looked mainly to its constitution, political institutions (i.e. the Senate), and the market to address regional differences in economic modernization and only to a very limited extent, regional programmes. Canada, meanwhile, has mainly looked to regional development programmes to address the problem. The Canadian constitution (section 36) commits federal governments to "furthering economic development to reduce disparity in opportunities and providing essential public services of reasonable quality to all Canadians." From the Diefenbaker era to the end of the Mulroney years, federal regional development programmes were a sacred cow in Canada, and successive governments rivalled each other in seeking new forms of organization to carry it out. But in all its guises, the sacred cow limped badly, and never achieved significant reductions of regional gaps in economic development (Savoie, 2000, p. 3).

Structurally, Canada's 20th century industrialisation had always been reliant on foreign direct investment, mainly from the US, which had contributed to both the centrality of the resource extraction sector in the Canadian economy, as well as its "branch-plant structure". It had also contributed to the structure of labour and encouraged an export led process of development (staples) rather than domestic investment led development. While staples theory played a contributing role, in the sense that the Canadian political economy concentrated industrialisation, industrial capital and the industrial labour force within Central Canada. Thus regional development policies initiated at the federal level, and designed to counter regional inequalities, actually reinforced them (Wallace, 2002). This is because many of these regional policies actually reinforced the historical pattern of economic and political regionalism with an industrialisation core located in Central Canada: "The geographical extent and regional diversity of the Canadian economy had, ...prompted the adoption of de facto regional policies from a much earlier date" (Wallace, 2002, p. 22). This included rail freight agreement rates, along with other transportation provisions, the Prairie Farm Rehabilitation Act, the Maritime Marshland Rehabilitation Act, regional policies which sought to assist poorer regions in achieve greater levels of economic growth, and policies to finance health, education and infrastructure inequalities (*ibid.*, p. 23). Also included were industrial subsidy programs designed to enhance regional development and which designated regions in terms of their incentive rating (*ibid.*, p. 24-25). And, in 1957, the Government of Canada launched a program of transfer payments, to provide "significant financial support to provincial and territorial governments on an ongoing basis to assist them in the provision of programs and services" (Government of Canada, 2009).

Speaking from the perspective of the experience of the Maritimes, Donald Savoie underscores the importance of political economy and resulting policies in regional entrenchment:

> a good number of Atlantic Canadians believe that a key, if not the most important reason, why their region trails others economically is misguided federal policies that have, over the years, strongly favoured – and continue to favour – central Canada. They look back to economic protectionism and the National Policy, which, soon after Confederation, forced producers in Atlantic Canada to ship their goods on expensive routes to central Canada rather than to their traditional markets in the New England states. Moreover, the National Policy forced the United States producers to establish branch plants in Canada, nearly always in central Canada, not the Maritimes or the West. Lastly, Ottawa's decision to concentrate the bulk of its activities in support of the war effort during the early 1940s served to strengthen considerably Ontario's industrial capacity in relation to other regions (Savoie, 2000, p. 4).

Flash forward to the 1960s. By then, Canada was in what some had called an economic Golden Age (Urmetzer, 2005; Wallace, 2002), a result of high levels of US investment. Ironically, while the political economy revolved around such investment, a corresponding national discourse increasingly revolving around Canada's need for independence from US culture, economy and politics (Clement, 1996; Wallace, 2002; Britton, 1996; Brodie, 1997). And indeed, it was clear by this time that a different focus, had emerged as the basis of the Canadian economic development. This is when natural resources, automobile and auto part manufacturing, and machinery and equipment became the dominant sectors in Canada's merchandise trade (Wallace, 2002, p. 14), and when Ontario reached its zenith as manufacturing capital of Canada and its wealthiest province.

Thus, manufacturing and resource extraction rooted in the modern industrialisation era, and not just resource extraction industries, contributed both to the economic development of Canada and to a series of regional disparities which ran deep, and which reinforced the idea that there were immutable "regional building blocks" of Canadian geographical landscapes. Transfer payments, freight rates, development programs and other forms of incentive and equalisation payments did not resolve the deeply ingrained political and economic divisions. As the century drew to a close, although faith in the politics of the Canadian liberal democratic tradition and associated federal policies for equalisation and regional development remained strong, the context in which such policies were implemented was accompanied by the increasing domination of Canadian culture and markets by American products and resilient regional economic disparities (Brym, 1986, p. 12).

A Changing Political Economy

While not wanting to over-state the case, it must be recognized that the influence of 20th century, as well as 19th century structures on Canada's political geography have been immense. While the automobile industry survived after restructuring, it was the realisation of the energy sector in Western Canada, and the rise of the service industries elsewhere in Canada which began to alter the old heartland-hinterland economic and political structure. Power vested for so long in Central Canada was challenged by a rising west coast political economy and the emergence of a new Alberta. Under a booming late 20th century Southeast Asian economy, and exodus from Hong Kong, British Columbia became integrated into the Pacific Rim – a world away from the old economy oriented towards the North Atlantic, or even the "new economy" with its immense north-south orientation under the NAFTA. Indeed, the "new economy" as it was dubbed did not find its foothold in the old industrial economies *per se*, but within metropolitan regions and new hinterland relationships. The rise of north as a new resource frontier, the role of energy and service sectors had begun to see a creeping reorganisation of political power, alignments and understandings about the basis of Canada's political geography.

At the same time, it is important to note, that if indeed the mechanisms of staples production are viable in the current Canadian political economy, and there are many who suggest we are in a "post-staples" phase (Wellstead, 2007), then the impact of staples theory on contemporary regionalism has to do with the way in which the existing "industrial mix" of each region is structured. Each region will remain different, and each will respond differently to outside change. This means that regional economies will continue to be volatile as conditions change in external trade of staples commodities, and that these different reactions reinforce regional diversity. Thus political regionalism is likely to remain invested in economic regionalism. This increased regionalization is more than an economic detail, therefore, and it is strongly coupled to political processes which impact upon the strength and voice of regions. Québec's economic successes in the late 20th and early 21st century, for example, have contributed to its growing political voice and provincial/national agenda within the Canadian confederation.

Moreover, while staples theory, "metropolitan" modes of European colonialism, and reification of this dependency relationship in space and place informed earlier studies in the 20th century, the contemporary referent for regionalism today is increasingly "globalisation" on the one hand, and on the other, "localisation". The study of the relationship of regions to the state has become reinforced by two new framing processes. One, as Nurse has described earlier in this chapter, has a clearer focus on local diversity within regional constructs. The other is a more spatially ambitious project aimed at reconstructing regions and repositioning them with respect to the relationship of the region to the continent, and to the broader world economy (Garreau, 1981; Courchene, 2005; Florida, 2009; Gertler, 2001). Or, more accurately stated, a variety of scholars are today attempting to identify the spatial frames for new types of regions which have emerged as globalisation and continental integration proceed to have an effect on the political economy of Canada. In order to understand the effect of contemporary political and economic forces on regionalism within Canada, and specifically regionalism in the early 21st century, we need to understand what has changed and how it has changed. The next section of this chapter looks at the contemporary face of regionalism in Canada and the global pressures which are causing a significant reconfiguration.

A New Legacy

> A new economic landscape points to the need for Canadian regions to integrate themselves differently in the emerging economic order. It is becoming clear that East-West economic ties will matter less in the future and thus political ties will also matter less (Savoie, 2000, p. 10).

As we have seen, the legacy of staples theory led to regional frameworks which recognized Atlantic, Western, Central and Northern Canada, and in some cases, the Canadian Shield as major regions. This was a framework which structured the relationship between economic forces and regionalization in 19th century terms. By the mid-20th century it was clear that the driving force which consolidated Central Canada's economic lead was automobile manufacturing, large-scale resource extraction, and the production of machinery and equipment. Nonetheless, most investment came from outside the country, making the economy a very open one, and tying it closely to the vicissitudes of the US market. In late 20th century, however, another compelling debate was taking place about Canadian regionalism which was not unrelated to the staples-based paradigm, and to the positioning of Canada with respect to the American economy. This debate suggested that human agency, including policy and business practices (see Acheson, 1972; Naylor, 1975) not just basic and impersonal economic processes had created regional inequality and had influenced the relationship between economic process and regionalism (Brym, 1986). The question was which policies to pursue – nationalistic or integrationist.

Discussion focused upon the negotiation of a Free Trade Agreement with the United States, and the outcome of these negotiations shook regionalism at its roots – Canada shifted its political gaze from across the Atlantic towards North America. The propelling influence was globalisation, or rather North American continentalism, under an emerging neoliberal political and economic ideology. Neoliberalism, unlike previous Keynesian welfare state policies, found little role for strong federal visions or political interventions (Albo and Jenson, 1997). As such, what Neill has described as 19th century continental forces which played upon staples and non-staples economies, and regional integration within 19th century Canada, was reinvented to incorporate the modern and indeed post-modern regional political economy.

Indeed, the recent modernisation of staples industries, new imperatives for innovative practices, the trade effects of the NAFTA, the rise of the service industries and especially business services in large metropolitan areas like Toronto (See Wallace, 2000) all point to the fact that the Canadian economy has changed in the past two decades. Not only did it survive the 'crisis of Fordism', but it made the transition to a post-industrial economy in many areas of the heartland. Wellstead (2009) would go so far as to suggest that Canada has entered a post-staples phase, and since the 1960s new regional and national influences have reshaped the economy. Citing a number of scholars, Wellstead observes that "substantial natural resource depletion, increasingly capital and technological intensive resource extraction from lower-cost staple regions, and the transformation from pure extraction to increased refining and secondary processing of resource commodities are characteristic of a mature staples economy, and that "a sectoral shift from natural resources to the service sector, and from periphery to centre, as the focus of economic growth could occur, ushering a new 'post-staples' political economy" would result if these trends persist (Wellstead, 2009, p. 8).

This economic transition has been felt in all regions in the periphery and in the core areas. The latter are the urban-industrial complexes which in a sense have propelled Canada's economic growth and insertion into the global economy. They have also made the economy more open and globalised, or rather reinforced openness and globalisation in new ways. Canadian transnational corporations (TNCs) are among the most globalised in the world (although not necessarily the largest and most powerful of TNCs). One of the instruments of transition in the late 20th century was clearly the Auto Pact, which led the charge to what was eventually transformed into an integrated North American automobile production industry, first under the Canada-US Free Trade Agreement (FTA), and then under the NAFTA. The influence of the Auto Pact and then the NAFTA was stunning to Canada's political geography because it essentially changed the orientation of Canada's commitment to North America. Canada invested in regionalism – but at the continental level. This is because NAFTA reorganized Canada's branch plant landscape and radically influenced its late 20th century trade relationship with the US. And yet, as Konrad suggests, the invoking of the fast-track option to the NAFTA in the 1990s served to accelerate a continentalist process within North America which, ironically perhaps, had its antecedents in the late 19th century (Konrad, 1992).

The NAFTA has spatially translated into border regions surrounding the points of transnational crossings between Canada and the US. The 9/11 co-operation impetus has resulted in restructuring of border relations, and new technological applications to keep the trade moving through the checkpoints. The largest of these checkpoints are well integrated into an emerging continental corridor network, linking the old east-west regions of Canada to the interior and southern US, as well as the north (Konrad and Nicol, 2008). Corridors linked Canada to the Deep South, as NAFTA moved through the now deregulated trucking industry, challenging the old east-west orientation of Canada's major transportation networks. So strong was the pull of trade that after 9/11 there remains serious discussion and advocacy by many in Canada's business community, for a North American 'security perimeter' to protect the two-way trade relationship, and indeed Canada's current political position with respect to border management stresses the importance of unimpeded flows – successfully or not (Nicol, 2005).

These economic developments create their own sets of linkages, both political, economic and regional. In some cases, for example, in northern Ontario, Québec, Nunavut, the Yukon and the Northwest Territories, there are political and economic convergences that suggest today's region transcends provincial or internal boundaries – as Keskitalo's (2004) study of the regionalism and negotiating the "international North" indicates. On the other hand, the rapid development of Alberta's energy sector has created a regionalism somewhat specific to that province, prising it loose from its transitional location in the normative regional framework as one of the three "Prairie Provinces" along with Saskatchewan and Manitoba. Overall, the late 20th and early 21st centuries have seen substantial shifts in the understanding of regionalism in Canada, including the shifting geographical demarcation of (Nunavut), the positioning of territory within federal frameworks of transfer payments (Ontario, Newfoundland), and devolution of the power and role of regions, particularly in relation to the globalisation of regional cities

(Florida, 2009; Courchene, 2005). Indeed, as the 20th century drew to a close, the power of the provinces as "regional advocates" was drastically reduced by the new paradigms of governance – devolution, downsizing and reduction in budgets has taken its toll. At the same time, the rise of "Global City Regions" (GCR) has come to represent "a range of new roles and rationales that are catapulting cities onto the policy and jurisdictional centre-stage. Included under this rubric will be brief discussions of why cities are now the key players in both the old geography (the space of places) as well as in the new (the space of flows)" (Courchene, 2005, p. 1).

In the late 20th century too, the special problems of geographical scale, regional disparities, and proximity to the economic shadow of US markets combined to create a narrative about Canada which suggests that rather than a nation structured by a core-periphery framework (made up as much of federated regions as provinces), and indeed a economically "problematic" nation whose uneven development experience is tenable only because of federal policies and programs, Canada must be seen in globalisation terms. That is to say rather than existing as simply an "edge" in the American market, a linked economic periphery within a continental economy, Canada became understood as a series of interconnected "regions" of North America. Garreau (1981) for example, identified a North America of interconnected regions, divided not by nationality but rather by economic and socio-cultural similarities. And while Garreau's regionalism was decidedly US-centric (Detroit is seen as a more important city than Toronto, for example), it did bring home the point that globalisation in the form of continentalization, would have an impact on Canadian regionalism. Indeed, a series of north-south regional corridors were envisioned as a result of the impact of first the Canada-US Free Trade Agreement, and then the NAFTA, while regional borderlands assumed a larger and larger role as conduits for the free trade traffic (Konrad and Nicol, 2008). It is therefore not surprising that Neill (2002) has argued that if the first half of the 20th century was devoted to building the staples-based concept of a Canadian nation, the late 20th century saw a concerted effort to understand regionalism within new and more globally-framed networks. This is a trend which has carried into the 21st century.

Indeed, Savoie argues that today, globalisation and neo-liberal policies are redefining how Canadian regions relate to one another, Canada's national political institutions are "no longer in a position to promote national political integration". The outcome of this lack is potentially a weakened federal government, no longer be able to rely for political support, upon its traditional supporters. Savoie indicates that the real question for Canadians will be "how Canada can continue to function as a national union, as we come to terms with the fact that all things Canadian will increasingly become regional" (Savoie, 2000). As we see below, this regionalism may not be as traditional as Savoie's structure would suggest. In addition to the strengthening of provincial powers as part of a new regionalization process, are new spatial dynamics which may lead to a significant reconfiguration of the conceptual basis of Canada's geographical regions.

So what does all this mean for the political geography of Canada? Are the new regional shapes which are beginning to emerge indicative of more permanent structures to come? Is the geography textbook classification of regionalism identified earlier in this chapter beginning to be challenged by events which may demand a much more complex system of regional conceptualisation? In some ways the answer to this question can only be "yes". For example, even if British Columbia is considered as a region quite distinct from Western Canada, the rising star of Alberta's political economy suggests that there is a "new western region" based upon the energy sector, and that this regionalism divides the western provinces of Alberta, Saskatchewan, Manitoba and British Columbia into three regions, rather than two, while strongly orienting British Columbia towards a more integrated north-south relationship with the US through the framework of various "Cascadia" agreements. The new North and Northwest, for example, and their mineral and energy sectors, are the source of a new economic emphasis among the western provinces, particularly Alberta and Saskatchewan, prompting Wallace (2002) to claim that there are significant regional differences between the old agricultural base of the prairies and its new energy sector. The rise of Calgary, the new financial clout of the province, and the position of Saskatchewan as the fastest growing province in Canada suggest a fundamental re-orientation of this jig-saw piece in the east-west political regional landscape.

Similarly, the rise of communications and knowledge infrastructure in the Atlantic provinces have helped to stem the tide of westward migration, and bring a new element of "lifestyle region" to formerly depressed areas, while the energy sector and tourism, the world's largest industries, have discovered Newfoundland, while Newfoundland has developed its own strong voice in federal politics. Meanwhile, economic slowdown and structural changes to the economy have challenged the "heartland", and Ontario became, in 2010, the recipient of federal transfer (equalisation) payments for the first time.

An important question is whether this shift will see a major challenge to power and economic influence in Central Canada? Courchene (2000) suggests not, arguing that Central Canada is better positioned than anywhere else in Canada to take advantage of the global economy, but

Ontario more so than Québec. He notes that Ontario stands as it own "region state" within North America, a dominant region which is remains the national leader. While historically, Canada's urban centres were seen as nodes connecting the resource-based economy to a distant metropolis in Britain and Europe, and as such they organized the exploitation of Canada's remote regions or hinterlands (funnelling and concentrating wealth overseas), by the 20th century they were implicated in a national economy which was firmly divided into heartland and hinterland. As such, these urban centres both contributed to, as well as mediated, Canadian regionalism.

In the 21st century, however, the creation of powerful regional linkages surrounding Canada's major cities, and in particular Toronto, has challenged traditional thinking about Canadian regions. Urban and metropolitan growth has been rapid, propelling Central Canada into the global spotlight and presenting it as a region-state to the broader world economy. No longer the 'middle-man', cities like Toronto, and the Toronto-centred region, are directly engaged with the world economy, creating their own synergies.

While the classification of Ontario as a region-state focused on Toronto may indeed be debatable (and may indeed be under reversal from economic problems experienced by the Canadian economy in the early 21st century) its speaks to the fact that first the Auto Pact, and then the NAFTA, created disproportionate benefits for this province, even in comparison to Québec and in comparison to other regions in Canada. A regionalism which ignores the political and cultural situation to pronounce Ontario and Québec one and the same region, is thus problematic, and may obscure the forces which are increasingly differentiating the region.

It also speaks to the fact that Canada's major cities have become important growth engines and political centres for the nation, replacing regions to some extent. Winnipeg, Calgary, Vancouver, for example, have all achieved an importance which reflects much more than simply their regional importance. Thus, today, fresh economic forces accentuate political divisions within Canada's national landscape. The question is whether such divisions will be entirely new, or whether they will fall along pre-existing regional fault-lines. Are they consistent with a staples-based regionalism, or are new contours reworking the landscape? In answer to this question, Howlett and Brownsey go so far as to say:

> No longer tied to the original staples industries, Canada has become an advanced industrial economy but one which remains different from the typical model of advanced manufacturing and services found in Europe, the US and Japan. The new base of the Canadian economy retains its origins in early staples industries with many new activities grafted onto those traditional

> sectors. This transformation of the old staples political economy has ushered in some elements of a new political and social order at the same time that it has exacerbated or worsened many elements of the old. Over the last several decades the transformation of significant components of Canada's staples economy has led to demands for many institutional, legal and political reforms which would better represent Canada's new globalized and regionalized configuration of business and social life (Howlett and Brownsey, 2005).

Thus, we might want to understand Canadian regionalism as a product of ongoing policies and practices as much as reflecting the recalcitrance of historical development, and not so much in terms of a direct comparison with other industrialised countries like those of Europe and the U.S. This means that understanding the structure of regionalism as a process linked to developments both within and without Canada raises the possibility of understanding regionalism as both dynamic and sensitive to broader flows of globalisation. Indeed, in the 21st century, Canadian regionalism needs to be considered in the context of forces which are clearly transforming the economic and political weight and nature of Canadian regionalism. Central to this are the impacts of urban regionalism, continental integration and even the post 9/11 securitization of borders and borderlands. All of these new models of regionalism have two things in common. The first is that they challenge the understanding of Canadian regionalism as building blocks in an east-west economy, and the second is that they are intimately linked to an understanding of regions as products of uneven development in an era of globalisation.

This leaves us with the question as to where we go from here. The future remains uncharted. Wellstead suggests that we are at a crossroads, and argues that a shift from dependence on staples production has been underway since the 1960s, leading to the emergence of a new form of political economy. "Exactly what this political economy is, however, remains uncertain and controversial" he suggests, but there is evidence "in some regions of the emergence of a "post-staple" based economy and with it, new forms of governance" (Wellstead, 2007, p. 8).

The North is a case in point. Recent academic discussions have focused on the way in which Canada's North has been incorporated into a broader circumpolar region, or even an international North which transcends national boundaries (Heininen, 2005). To many Canadians the concept of the North as a region has been correlated with the idea of a resource rich frontier, a virtual untapped resource rich region on the margins of the civilised world – "there for the taking" – even the popular dichotomy "Northern Frontier Northern Homeland" (Berger, 1970) implies that the North is both of these things. These are extremely important conceptual frameworks, and are significant devices in orienting support if the Arctic is to be portrayed as an abstract space for resource exploitation.

But the late 1980s and early 1990s saw the emergence of a new understanding about the north on the part of Canadians and their policymakers. Canada signed on to the Arctic Environmental Protection Strategy Arctic (AEPS), and successfully campaigned for an international institutional presence in the North which was to become known as the Arctic Council. A comprehensive land claim process took place, and new attitudes and sensitivities towards indigenous cultures were incorporated into both Canada's and the broader international circumpolar approach. This culminated in the development of a 'northern dimension' for Canadian foreign policy – as an explicit set of ideas and approaches to northern Canada and its neighbours, which were to differ from those of the south. But even within the circumpolar North there have been significant developments in devolution which have led to the emergence of new political regions within the former "territories". Globalisation is driving the reorganisation and reconceptualization of Canada's North in many ways, just as it is having an impact upon energy and new resource exploitation in the west and northwest (Abel and Coates, 2001).

It would be true to say that in the past few years this realisation of the link between fragile Arctic environments and the survival of human societies, ecosystems and the national borders which define them has assumed larger and larger proportions in a more comprehensive list of globalised threats. The Canadian Government has assumed a more aggressive posture with respect to the politics of northern sovereignty as a result of this new realisation, while other countries like Russia, Denmark, the U.S. and indeed all the circumpolar or Arctic states have weighed in with their concerns and imperatives. Never, for example, since the Cold War, have there been so many articles in Canadian newspapers and newsmagazines, which focus upon the territorial definition of Canada in the North, and the impending threats it faces from the broader international order. Indeed, this is a new kind of "regionalism".

Conclusions

The argument has been made that there is a new political landscape emerging within Canada, a new political geography, which reflects a new dynamic of power, regionalization and globalisation. Globalisation promotes regional responses, often within new regulatory frameworks, which are consistently different in scope and scale than in earlier eras. In Canada, in the future, this might mean that western Canada is no longer dependent upon financial and business structures of Central Canada, and its manufactured goods, or that the North realises the rewards of self-government in terms of royalties and mega-project development –

generated wealth. It might also mean a Pacific Rim which is highly oriented towards broader circumnavigational patterns, and a Central Canada which does not shrink from such developments elsewhere, but which turns towards global connections. All this may be overly optimistic but it still serves to stress the point that heartland and hinterland may no longer be appropriate models to understand the political and geo-economic regionalization of Canada. Moreover, it may well be that the staples thesis, which has been evoked to explain the configuration of wealth, space, power and regional interests in Canada for over three centuries, is no longer a useful explanatory model.

But it also suggests that Canadian regionalism is sensitive to the way in which we understand it, just as much as to the way in which global, continental and national political, economic and socio-cultural forces work to define specific places and spaces which are imbued with regional character or similarities. Indeed, the rationale for, as well as the framework and even nomenclature of Canadian regionalism is by no means agreed upon within the framework of political, academic and popular thought. Initially, academic texts on Canadian defined the country in terms of its physical geography and corresponding human geography, in ways which were categorical and descriptive. Today, however, regionalism is understood in less reified and more nuanced ways, based upon the understanding that regions are often recursive constructions – filtered assemblages of facts seen through lenses which value or prioritise certain details and discard others. Thus, regions are not just physically verifiable territories, or even artefacts of "the interaction of physical geography and settlement history" (Wallace, 2002) which took place long ago.

Wallace (2002), for example, notes that the Canada which took shape in the 19th and early 20th centuries was quite different than the Canada of today. Indeed, the broad contours of regionalism which defined Canada one hundred years ago, or which challenged the Canadian government at Confederation, appear quite different in the 21st century. Alberta can no longer be considered as part of an agricultural frontier lying distant from the powerful elites of the Canadian state. Newfoundland has asserted its economic and political independence in ways which are not reflected among the other Atlantic provinces. British Columbia can, by no stretch of the imagination, be correlated with a "Western Canada" which stretches as a frontier from the western edge of Ontario to the Pacific Ocean. Yet it would be wrong to suggest that there are no traces of the imprint of this past in Canada's contemporary the regional landscapes.

Review Questions

- What is meant by a staples-based economy?
- How have broad economic developments contributed to the basis of Canadian political regionalism?
- When was the nation of Canada founded, and what has happened to its territorial base since then?
- What are the geographical regions of Canada?
- What kinds of criteria are used to define Canada's geographical regions?
- Why has regionalism led to unequal development?
- Why has the Canadian government been interested in managing regional disparities?
- Why are new regions emerging in Canada?
- What are some of the new regions?
- In your opinion, will Canada's regional geography remain important in the future?

References cited

Abel, Kerry and Ken S. Coates (1999), *Northern Visions, New Perspectives on the North in Canadian History*, Toronto: Broadview Press.

Aitken, Hugh G.J. (1959), "Defensive Expansionism: The State and Economic Growth in Canada", in *The State and Economic Growth*, Hugh G.J. Aitken (ed.), New York, Social Science Research Council, pp. 79-114.

Gregory Albo and Jane Jenson (1997), "Remapping Canada; The State in the Era of Globalization", in *Understanding Canada, Building on the New Canadian Political Economy*, edited by Wallace Clement, Montréal and Kingston, McGill Queen's Press. pp. 215-239.

Hayter, Roger and Trevor Barnes (2001), "Canada's resource economy", *The Canadian Geographer/Le Geographe canadien*, 45, No.1, pp. 36-41.

Berger, Thomas R. (1988), *Northern frontier, northern homeland: the report of the Mackenzie Valley Pipeline Inquiry*, Vancouver: Douglas and McIntyre.

Bone, Robert M. (2003), *The Geography of the Canadian North*, Toronto: Oxford Press.

Britton, John N.H. (1996), "Canada and the Global Economy. The Geography of Structural and Technological Change", Canadian Association of Geographers Series in *Canadian Geography*, No. 3.

Brodie, Janine (1997), "The New Political Economy of Region", in *Understanding Canada, Building on the New Canadian Political Economy*, edited by Wallace Clement, Montréal and Kingston, McGill-Queens Press, pp. 240-261.

Brym, Robert J. (2009), *Regionalism in Canada*, Richmond Hill: Irwin Publisher.

Clement, Wallace (1996), *Understanding Canada: Building on the New Canadian Political Economy*, Montréal, PQ, CAN: Montréal: McGill-Queen's University Press, p. 24.

Courchene, Thomas J. (2005), "City States and the State of Cities. Political-Economy", *IRPP Working Paper Series*, No. 2005-03, Dimensions and Fiscal-Federalism, June.

Creighton, Donald Grant (1937), *The Commercial Empire of the Saint Lawrence*, Toronto, New Haven and London.

Easterbrook, W.T. and M.H. Watkins (1967), "Approaches to Canadian Economic History", *The Carleton Library*, No. 31, Toronto, McClelland and Stewart Ltd.

Finance Canada (2009), *Federal Transfers to Provinces and Territories*, Ottawa. Feb. 1, 2009, http://www.fin.gc.ca/access/fedprov-eng.asp.

Florida, Richard (2009), *Ontario in the Age of Creativity*, Martin Propserity Institute, Toronto.

Garreau, Joel (1981), *The Nine Nations of North America*, Houghton Mifflin.

Gertler, Meric S. (2001), "Urban Economy and Society in Canada: Flows of People, Capital and Ideas", Isuma: *The Canadian Journal of Policy Research*, 2: 3 (Autumn), 119-130.

Harris, Richard Cole (1997), "Regionalism and the Canadian Archipelago", in *Heartland and Hinterland: A Regional Geography of Canada*. L.D. McCann and Angus Gunn (eds.), Scarborough: Prentice Hall Canada, pp. 395-421.

Heininen, Lassi (2004), "Circumpolar International Relations and Geopolitics", in *Arctic Human Development Report*, Akureyri: Stefansson Arctic Institute, pp. 207-226.

Hodgins, Bruce (1971), *John Sandfield Macdonald 1812-1872*, Toronto: University of Toronto Press.

Holryod, Carin (2009), "Deconstructing Canada in an Age of Globalization", *Policy Options*, May 2009, pp. 57-62.

Howlett, Michael and Keith Brownsey (2005), *The Post-Staples State: The Political Economy of Canada's Primary Industries*, Vancouver: University of British Columbia Press, www.sfu.ca/~howlett/NR06/NR06rev2.doc. Accessed April 7, 2008.

Innis, Harold A. (1956), *The Fur Trade in Canada*, Toronto: University of Toronto Press.

Keskitalo, E.C. (2004), *Negotiating The Arctic: The Construction of an International Region*, New York and London: Routledge.

Konrad, Herman W. (1992), "North American Continental Relationships: Historical Trends and Antecedents", in *North America without Borders, Integrating Canada the United States and Mexico*, Stephen J. Randall (ed.), Calgary: University of Calgary Press, pp. 83-104.

Konrad, Victor and Heather N. Nicol (2008), *Beyond Walls: Reinventing the Canada-US Borderlands*, Surrey UK: Ashgate Press.

Laxer, James (2003), *The Border: Canada, the US and Dispatches from the 49th Parallel*, Toronto: Doubleday.

Lewis, Martin W. and Kären E. Wigen (1997), *The Myth of Continents: A Critique of Metageography*, Berkeley and Los Angeles: University of California Press.

Mackintosh, W.A. (1923), "Economic Factors in Canadian History", *The Canadian Historical Review*, Vol. IV, No. 1, March 1923. pp. 12-25.

McCann, L. and Angus Gunn (1997), *Heartland and Hinterland*, Scarborough: Prentice Hall.

McGillivray, Brett (2006), *Canada: A Nation of Regions*, Don Mills: Oxford University Press.

Naylor, R.T. (1975), *The History of Canadian Business, 1867-1914*, Toronto: J. Lorimer.

Neill, Robin (2002), The continentalization paradigm of Canadian economic development: with special reference to economic growth in the Maritimes, http://www.unb.ca/econ/acea/ACEA2002-Neill.pdf. Accessed April 7, 2008.

Nicol, Heather (2005), "Resiliency or Change? The Contemporary Canada-US Border", *Geopolitics*, Volume 10, Number 4, Winter 2005. pp. 767-790(24).

Nurse, Andrew (2002), *Rethinking the Canadian Archipelago: Regionalism and Diversity in Canada*, Report Prepared at the Request of Canadian Heritage, 2002, Mount Allison University.

Philips, Paul (1982), *Regional Disparities*, Toronto, James Lorimer and Co.

J. Lewis Robinson (1989), *Concepts and Themes in the Regional Geography of Canada*, Vancouver: Talon Books.

Savoie, Donald J. (1986), *Regional Economic Development Canada's Search for Solutions*, Toronto: University of Toronto Press.

Savoie, Donald J. (2000), "All things Canadian are now regional", *Journal of Canadian Studies*.

Scott, James (2009), *Decoding New Regionalism: Shifting Socio-political Contexts in Central Europe and Latin America*, Surrey UK: Ashhgate Press.

Urmetzer, Peter (2005), *Globalization Unplugged: Sovereignty and the Canadian State in the Twenty-First Century*, Toronto: University of Toronto Press.

Waddell and Angus Gunn (1997), "Québec: A people and Place", in *Heartland and Hinterland*, Scarborough: Prentice Hall.

Wallace, Iain (2002), *A Geography of the Canadian Economy*, Toronto, Oxford University Press.

Watkins, M.H. (1963), "Staple theory of Economic Growth", *Canadian Journal of Economics and Political Science*, Vol. XXIX, No. 2, pp. 141-158.

Wellstead, Adam, "The (Post) Staples Economy and the (Post) Staples State in Historical Perspective", *Canadian Political Science Review*, Vol. 1(1), June 2007, pp. 8-25.

Wilson, Elana, "Arctic unity, Arctic difference: mapping the reach of northern discourses", *Polar Record*, 43(225), p. 125-133 (2007), p. 125.

Wynn, Graeme (1987), "Forging a Canadian Nation in North America", in *The Historical Geography a Changing Continent*, Michell and Paul Groves (ed.), Totowa, N.J.: Rowan and Littlefield, pp. 373-409.

CHAPTER 6

Canada

An Urban Nation

Hélène BÉLANGER and Yona JÉBRAK

Université du Québec à Montréal

Introduction: Why Study Cities

With more than half of the world's population living in cities, we inhabit an urban world (UN DESA, 2012). But this figure hides a discrepancy between countries and regions. Africa, for example, with less than 40% of its population residing in cities, is the least urbanized region of the world. At the other end of the spectrum, North America is the world's most urbanized region. Canada has been an urban nation for over 90 years, with more than 80% of its population living in cities and towns (figure 1).

Figure 1. Urban and rural population in Canada, 1851-2006

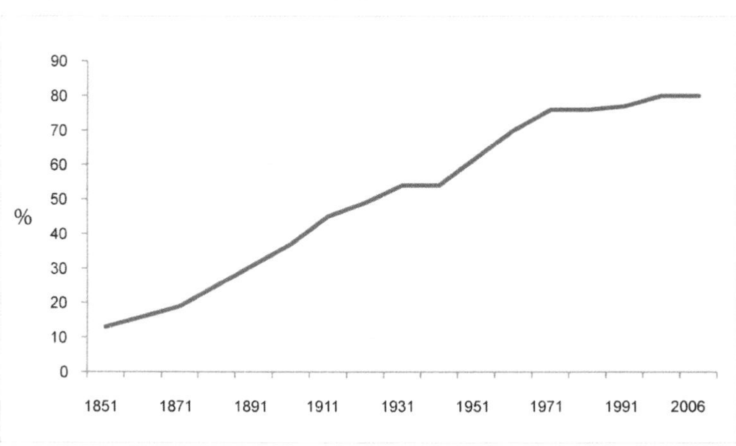

Source: Statistics Canada, Census of Population, 1851 to 2006. For more statistical information, refer to the 2006 Census.

Finding a suitable definition for "city" or "urban area" can be a challenge. According to Statistics Canada, a *city* is an area concentrating a minimum of 1,000 residents with a density of at least 400 people per square kilometre. Beyond this technical definition, the difficulty in defining the concept "city" can be found both across and within disciplines, such as in the case of the discipline of urban geography. In one way, the definition depends on perspective. For example, if the concern is the built environment, characteristics such as density and continuity of the built fabric will distinguish urban from non-urban areas. Where the interest is on stakeholders involved in an area, the distinction between urban and non-urban areas will be based on characteristics such as political or administrative boundaries of the area under study. These two examples give insight to the complexity of the concept "city" and (by extension) "urban". It is beyond the scope of this book to contribute to the debate concerned with these definitions, however. Nevertheless, scholars generally acknowledge that "cities are places of concentrated settlements whose occupants engage in relatively specialized kinds of non-agricultural economic activities" (Filion *et al.*, 2000: 1).

Cities are the product of civilization (Allain, 2004; Olson, 2000) and therefore are studied by scholars from many disciplines. Geographers, morphologists, planners and designers (among others) tend to study the city as the container, the décor or the habitat of the society and its activities. Sociologists, anthropologists, political scientists and economists tend to be more interested in the activities and social composition of different groups. As for environmental psychologists, they tend to be interested in the reciprocal relationship between individuals or groups (the society) and the built environment (the container). Studying forms, functions, and social characteristics of cities informs us about the society that produced, and is still producing, the city. Thus, by studying Canadian cities, we are studying Canadian society.

This chapter presents an overview of Canadian urban geography and is divided into three sections. The first section will present the Canadian urban system and how it is related to Canada's discovery and colonization. The second section will present urban forms, by briefly introducing the dimensions and dynamics shaping Canadian cities, followed by a focus on historical differences and general common tendencies, including an exploration of the internal structure of the city. The third section, entitled Building places, will propose a closer look at the Canadian planning system itself and how cities are managed on a day-to-day basis by the various actors that shape them.

Various concepts will be used in this chapter that may be new to the reader. These concepts are identified in *italic*, with the definition presented directly in the text.

The Canadian urban system

Canada is a large country of almost 10 million square kilometres, with a population of more than 33 million people (according to the 2011 Census). The national population density is very low with only 3.69 people per square kilometre. However, this figure obscures important density variation throughout the country. The majority of the Canadian population is concentrated in four provinces: Ontario, Québec, British Columbia and Alberta. And, due to its physical environment and climate, 83% of the population lives on only 13% of the territory (see figure 1 in Chapter 3). This area, in the southern part of the country and in close proximity to the international border with the United States of America, includes Canada's three largest cities: Toronto (population of 2.6 million), Montréal (population of 1.6 million) and Vancouver (population of 0.6 million).

The Canadian *urban system* – how cities are distributed in the national territory, forming a network – is rooted in the 17th century discovery and exploration of what we know as Canada today (see McCann and Simmons, 2000). The sites selected for the first permanent settlements were found mainly along the coasts or waterways (which at the time served as communication infrastructure). Permanent settlements were slowly built by Europeans, some on sites initially occupied or seasonally inhabited (or visited) by Amerindians. This was the case for the sites now known as Québec City (then called Stadaconé) and Trois-Rivières. Key places emerged from these early settlements that were first dedicated to fur trade or the evangelization of Amerindians. These activities spread to the Atlantic Provinces in the 18th century and later made their way to western Canada (figure 2). These centres are important in Canada's urban system today.

From Colonization to Urbanization

In 1666, the first Canadian census registered 3,215 inhabitants in the colony that was then called New France (see Statistics Canada, 2008). Except for a handful of settlers, most of the inhabitants were concentrated in three major permanent settlements: Québec City, Montréal and Trois-Rivières. Since the beginning, transport infrastructure has had great impact on the urban system and the urban form. At first, it was the hydraulic system that favoured the localization and development of the first permanent settlements as well as the economic development of "what is now Canada" (Hiller, 2010). In the back country, transport was done in very difficult conditions by horses, leaving settlements isolated during the long winter months prior to the development of the railway.

Figure 2. Foundation of main cities throughout Canada

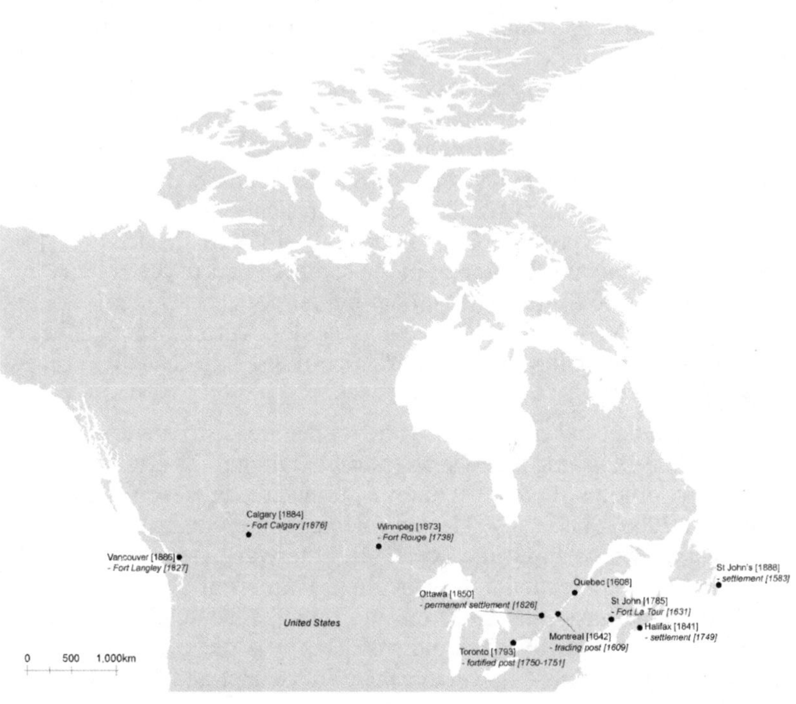

Source: Based on *The Canadian Encyclopedia* (2012) [available online]; *City of Ottawa* (2012) [available online].

Transport infrastructures were similar in the United States, however the level of territory occupancy was quite different. In the United-States, the urban system covers the territory in all directions while in Canada, only a small part of the country holds the majority of the population. Topography and weather explain this distinction, at least in part. Yet Canada's urban system also has roots in the exploitation of natural resources. Urban centres developed in the vicinity of natural resources at the foundation of major industries: fisheries along the East Coast, agriculture in the St. Lawrence Valley, energy in the Prairies, and wood extraction on the West Coast. But the railway also had a significant impact on the development of new permanent settlements and urbanization. The railway system meant greater distances could be covered in shorter periods of time. At first, the objective was economic: to transport goods between Canada and the United-States. Later, the network expanded within Canada, allowing a rapid expansion to the West (see Hodge, 1998 among others). Where the train went, villages emerged (figure 3).

Figure 3. Railway network in Canada [circa 1957]

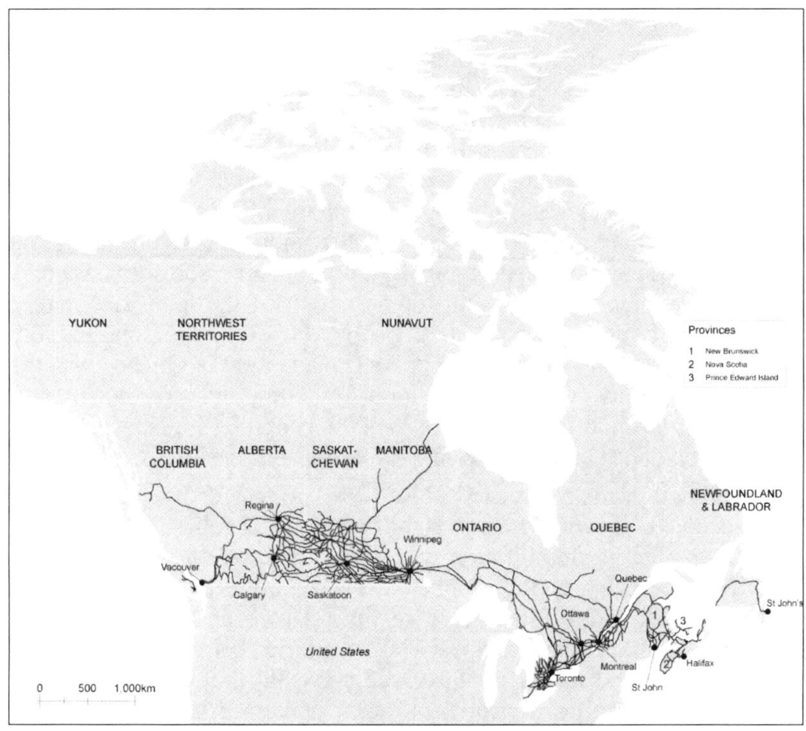

Source: Based on Geogratis, *Atlas of Canada: Railways* [available online].

The rail did not only have an influence on the urban structure but also on the urban form and urban growth. Since trains need to be built and maintained, this industry requires significant manpower favouring *urbanization* – the transition from a rural population to an urban population, or the movement of individuals and groups from the rural to urban environments – and *urban growth* – the increasing size of cities. *Industrialization* – the passage of an economy mainly based on the primary sector (agriculture, extraction, etc.) to an economy based on factory production and the division of labour required – in Canada, as in other countries in the world, was the catalyst for the acceleration of the urbanisation process. It is true that cities existed long before industrialization but industrialization requires infrastructures, activities concentration and an important flow of labour to operate the machinery which supports the concentration of population in the vicinity of industries. This is why the urbanization process accelerates greatly with industrialization.

In the case of Canada's industrialization, ship building and the railway system were important because they occupied two different roles: 1) ships and trains were facilitated the transport of goods, essential for the development of new markets, and; 2) ships and trains were an important industrial product, attracting population due to the availability of jobs. Thus, it is not surprising that the cradle of Canada's industrialization was Montréal, specifically an area along transport infrastructures around the Lachine canal (which opened in 1825 and was later widened in 1848). The canal was not only a transport infrastructure but also a water resource and a power source. Rapidly, industries settled along the canal. New populations settled rapidly in the vicinity and formed working class neighbourhoods. With the attraction of the industries, and a population of more than 107,000 people in 1871, Montréal rapidly dominated the urban system (see Parks Canada, 2007). Québec was the second largest city (with a population of 60,000) while Toronto was in third place (with a population of 56,000 people) (Statistics Canada, 2008). Canadian society remained a rural one until the 1920s.

In only 70 years, between 1891 and 1961, Canada's population nearly quadrupled, growing from 4.8 million to 18.2 million people. This growth was fed by immigration waves and important *natural growth* – the difference between the number of people born and the number of people who died during a period of time. However, growth occurred mostly in rural areas, most rapidly in the prairies before the 1920s (the period of time that marks the threshold between rural and urban Canada). With this significant growth, Canada witnessed internal migration movement – to the West for English Canadians and cross-border movement to New England for French-Canadians. It is estimated that 900,000 French-Canadians moved to New England between 1840 and 1940, attracted by job opportunities in factories (Lavoie, 1981).

Canada became an urban society in the 1920s. Since that period, towns and cities have continued to grow, expand and eventually *sprawl* – a low-density urban growth at the outskirts of towns and cities. With time, technical advances and increases in productivity have promoted the growth of the service sector economy. This change in the economic base favors larger cities over smaller ones. As Bourne (2000: 26) puts it, "Canada became a metropolitan nation (with settlements of 100,000 and more) in the mid-1960s and a suburban nation (with 50% of the population living outside the central city boundaries) in the early 1970s". It is during this period that Montréal lost its position as the Canadian economic metropolis to its neighbour to the west, Toronto. As Polèse (2008) explains, this peculiar change in the rank order is the result of the Québec Quiet Revolution and the rise of nationalism. Historically, Montréal's wealth was linked to the prosperity of the Anglophone elite. With political tensions, Anglophone elite re-located to Toronto (with their money and businesses) leaving Montréal in a relative decline while giving Toronto growth stimuli. Despite this rank-order change, the structure of the urban system remained, albeit with a new dominant metropolis (Toronto) and "an integrated hierarchy of cities" (Simmons and McCann, 2000: 97). Montréal is now considered to be the regional metropolis of the province of Québec. However, as Fillion highlighted (2010: 537) "[i]t rather appears that changes affecting the economy, policy-making and demography will result in a highly polarized pattern consisting of a few large urban areas experiencing high rates of growth and numerous smaller centres caught in a cycle of decline".

Urban Forms

The urban form is composed of different elements. Several typologies exist to classify these elements (see for example Allain, 2004; Kostof, 1992) and can be synthesised using these four dimensions (or categories of elements): the site, the plan, the plot and the built environment. Even if cities are often defined as *palimpsests* – (re)constructing themselves over previous versions – unless a city is faced complete destruction, these elements do not evolve at the same pace. Some of these elements are more stable than others, such as the site. The geomorphological environment (the site) and topography gives insight into the location and (often) the design of most cities. Great variation in the topography creates particular landscapes and more or less desirable areas in the city. Beyond the site, there is the way in which the city was (and is) designed (is the plan), namely the cadastre, street layout and land uses. The plot – is how the city blocks (which are part of the plan) are divided into public and private lands – will impact what is built. The built environment – consisting of volumes and open spaces – is the final element. The legacy of these dimensions can sometimes be considered as a burden in the development of cities. It was the case in many old centers during the 1960s when authorities were trying to modernize the city with various urban renewal programs.

These dimensions of the urban form are forged by a complex mix of forces (see Hall and Barrett, 2012): the economy, laws and rules, the 'population factor', technical innovation, the cultural context and dominant ideologies (see Allain, 2004).

1) The *economy* includes the local economic base but also macroeconomic forces (such as globalization) and their local effects, and building cycles and their role on economic growth and urban development. For example, the tertiarization of the economy in Canada, as in other regions of the world, favors the growth of the largest me-

tropolis (Toronto, Montréal, Vancouver and, recently, Calgary) at the expense of smaller urban centres.

2) The rigidity and extension of *laws and rules* impact the urban form. In the 19th century, new sets of rules were adopted in the building codes of Canadian cities following major fires. During the 20th century, *laws and rules* explained, in large part, the distinction between urban growth containment or sprawl in the United States, Canada and Europe.

3) '*Population factor'* is an expression, borrowed from the economist William Alonso (1980). The 'population factor' refers to not only demographic characteristics but also the demographic transition which led to changes in lifestyle and consumption preferences. For example, demographics explains the demand in housing, which constitutes 30% of land-use in towns and cities across Canada (Harris, 2000: 380) but the 'population factor' explains the household preference for a bungalow at the periphery versus a preference for an apartment downtown (closer to work and cultural infrastructures) (see Ley, 1996; Rose, 2009 among others).

4) Beyond the innovation that led to industrialization and urbanization, and to the increase in productivity and the transition to a service economy, *technical innovations* helped shaped Canadian cities directly. The invention of the escalator preceded an increase in density by means of vertical construction. The bridging of the St. Lawrence River accelerated the development of suburbs in Québec, Trois-Rivières, Montréal, Toronto, etc. (Olson, 2000: 227)

5) *Cultural context and dominant ideologies* encompasses two types of forces. First, culture influences the way physical and social environments are perceived in addition to the way individuals and groups act/react to these environments. Second, dominant ideologies impact the way cities are planned.

In sum, urban forms resulting from a complex mix of these components range from *geomorphic* (Moholy-Nagry, 1968, in Cuthbert, 2006) – forms that result from geomorphological elements such as hills, plains or water – to planned forms. Urban forms along this spectrum can be found throughout Canada (see figure 4).

In the Beginning… Cities with Different Shapes and Forms

The planning of the first cities was influenced by the founders' country of origin. For example, Québec City had a French medieval character based on a radio concentric plan around a central place (the *place d'armes* in French) (see Bourget, 1995); Montréal was inspired by the French bastide with a central street, while St. John's, Newfoundland and

Labrador had an orthogonal grid (see figure 4). In these three cases, the wall to defend citizens from threats was an important part of the design. These models of early settlements were, according to Stelter (1990), imposed by European rulers as a way to extend their kingdoms.

Canada's human settlements during colonization were first based on the cadastral system of "cotes" and "rangs". This seigneurial cadastral system divided the plot into long and narrow strips of land. This particular division gave settlers greater access to the main transport infrastructure: the river (see figure 5). The legacy of the orthogonal grid has greatly influenced Montréal's urban form (see figure 6). But by the end of the 18th century, the system of townships was put in place. A township is a square area of about 16 square kilometres dedicated to colonization. The basic design of Canadian cities seems very flexible compared to the strict set of rules imposed in Spanish colonial towns and cities. The Laws of the Indies, the first planning legislation in the colonial Americas, specified criteria for the localisation of the city to be founded, the size and orientation of the blocks, the localisation of the institutions, the distribution of open spaces, and the distribution of the population according to their ethnic and economic status.

Before industrialization, urbanization and urban growth was slow in Canada. Starting in the 19th century in the province of Québec, industrialization rapidly expanded into the province of Ontario, and later into the central provinces, where it lead to the same effect in all cases: urbanization and a tendency to urban growth. With industrialization, smaller towns and cities tended to specialize while larger areas developed a more diversified economy. This was the case for Montréal, with naval and railway industries, garment factories, mills etc. In smaller centers, specialization was at first beneficial, but with the evolution of the economy and de-industrialization, specialization sometimes led to impoverishment, especially in the Maritimes, "now Canada's least urbanized region" (McCann and Simmons, 2000: 92).

Figure 4. The urban grid of Québec City, Montréal and St. John

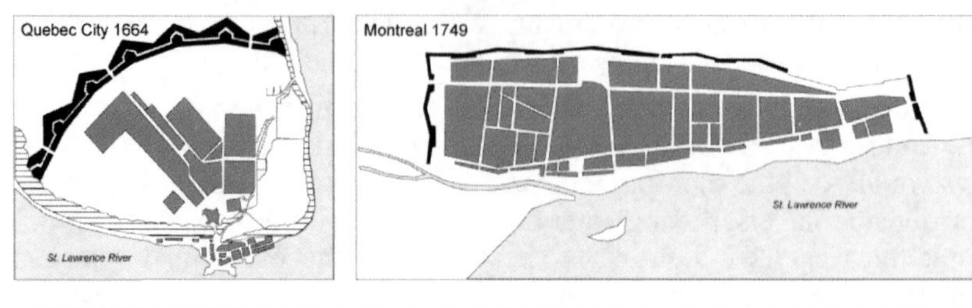

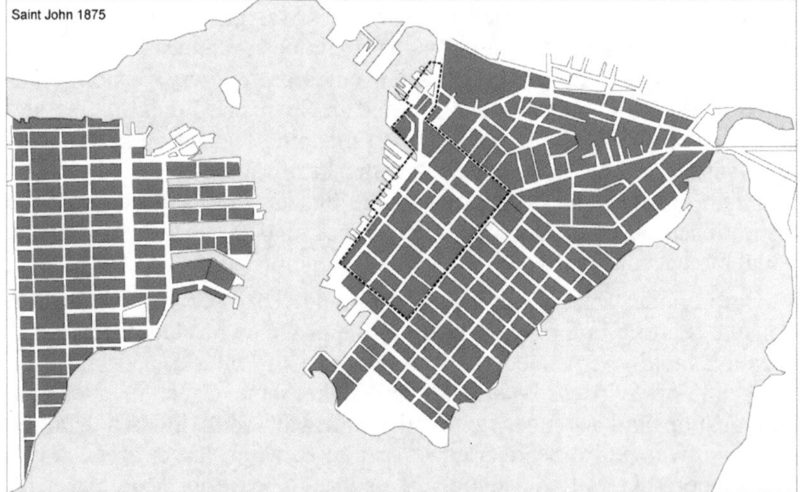

Sources: Based on historical maps for Montréal and Saint-John; based on a sketch by Noppen *et al.* (1979) in Bourget 1995 for Québec City.

Figure 5. Cadastral systems of 'rangs' and 'cotes' and townships

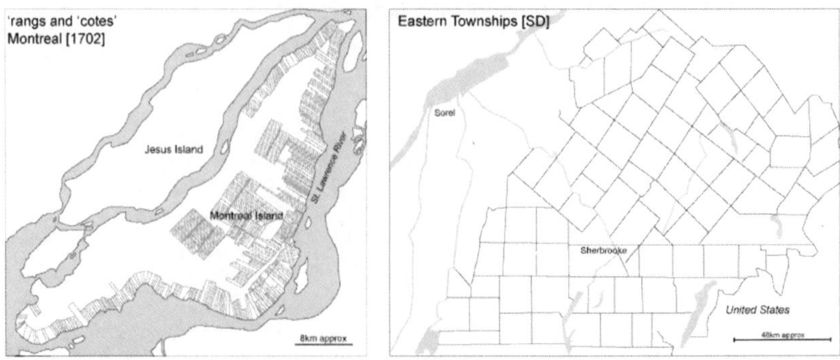

Sources: Based on Beauregard (1984) for Montréal and SD for townships.

Evolution of Urban Forms... Some General Tendencies

Despite historical differences and local disparities resulting from the complex dynamics affecting urban forms, a general pattern emerges in the evolution of Canadian cities. This evolution is similar to other cities in the United States and in Europe. The first general tendency is, during the industrialization of cities in the 19th century, towns and cities witnessed a decentralization process of the population, leading to a socio-economic decline of central neighbourhoods (Jackson, 1985 among others). The rapid industrialization, along with the concentration of factories and the massive arrival of the population, caused increased pollution, noise, and degrading health conditions, pushing wealthier populations to the periphery. This movement to the periphery was facilitated by technological progress: first the railway, followed by the tram (Ghorra-Gobin, 1994; Knox, 1994; Macionis and Parrillo, 1998). The decentralisation movement liberated space at the center which allowed the development of the central business district housing the highest concentration of offices and retail stores. With technical innovations such as the elevator and the steel structure, the central business district witnessed vertical densification. It was during this period that finance centers, anchored along streets, emerged in major cities: St-James Street in Montréal, King Street in Toronto and West Hastings Street in Vancouver. This phase of development was well illustrated for American cities by sociologists from the Chicago School using concentric circles (see Burgess, 1925). Later, industries followed the decentralisation movement for reasons related to production techniques and easier access to transport infrastructures. But it took time for the road network to develop substantially. The invention of the combustion motor, and its arrival in Canada at the beginning of the 1920s, marked the beginning of a new era. As early as 1907, more than 2,000 cars were registered in Canada (The Canadian Encyclopedia) and by the end of the 1920s, this number had increased to 1.62 million cars (of all types). Urban roads and highways increased the connectivity between urban centers and helped in the development of remote areas. By 1946, there were almost 40,000 km of paved roads, and 20 years later, almost 150,000 km of roads had been paved, with approximately a third of them urban (The Canadian Encyclopedia).

Decentralization of population and activities continued to accelerate during the 1960s, but to a lesser extent in Canada compared to US cities (Mercer and England, 2000; Jackson, 1985; Ghorra-Gobin, 1994). After the exodus of wealthy populations and factories, middle classes, commercial activities and standardized personal services followed the movement to the periphery. Cities grew and expanded, facilitated by the democratization of the automobile and the construction of highways.

The financing of the highway system gave the option to locate further from the center, resulting in urban sprawl.

Canadian cities have increased their footprint at a much higher rate than demographic growth and economic nuclei developed at the periphery. This has been the result of public authorities financing the highway system although in a lower proportion than their neighbour(s), thereby encouraging higher centrality. "Conventionally, Canadian cities are characterized as having more vital central cities and being more compact and less dispersed than their US counterparts" (Mercer and England, 2000: 57). Although transport accounts for urban structure evolution (Chapain and Polèse, 2000; Ingram, 1998; Knox, 1998 among others), other factors are necessary to explain the constant tendency for urban areas to sprawl as well as the general internal structure of cities. Income increase, local conditions and "the population factor" must also be considered (Polèse, 1998; Ingram, 1998 among others).

Rising incomes increase housing options for households. But this does not explain the discrepancies concerning housing options between cities and between regions with regard to where the different socioeconomic classes of the population choose to live. Population needs, expectation and aspirations can explain, in part, these differences. For some, suburbs are socially valued and represented as ideal living environments whereas others consider proximity to work and cultural facilities, in an environment with architectural and urban qualities, as more desired than suburban neighbourhoods. *Gentrification* – often linked to young professionals working in the new economy and associated to a "return to the city" movement – was an important process of the regeneration of inner-city neighbourhoods in large cities throughout Canada (as in the United States and other regions around the globe). The fact remains that, despite similar demographic and *life cycle* characteristics – the significant stages in life such as getting married, having children, getting divorced, which guide the demand for space and the features of the living environment – the two neighbouring countries represent substantial differences. Canadian cities not only show greater centrality but also lower urban *segregation* – a greater concentration of a population group (compared to the general distribution within the city) based on poverty, ethnic profiles, religion, etc. In Canada, important waves of immigrants helped maintain demographic dynamics in urban centers (Fillion and Bunting, 2000). In the United States, racial tensions and crime rates in city centers during the 1960s and 1970s led to significant residential movement from the white, middle-class towards the periphery (a population movement known as "white flight") leaving major urban centers to fall into states of impoverishment and degradation (see Walks and Bourne, 2006; Jackson, 1985).

Figure 6. Different Canadian urban forms

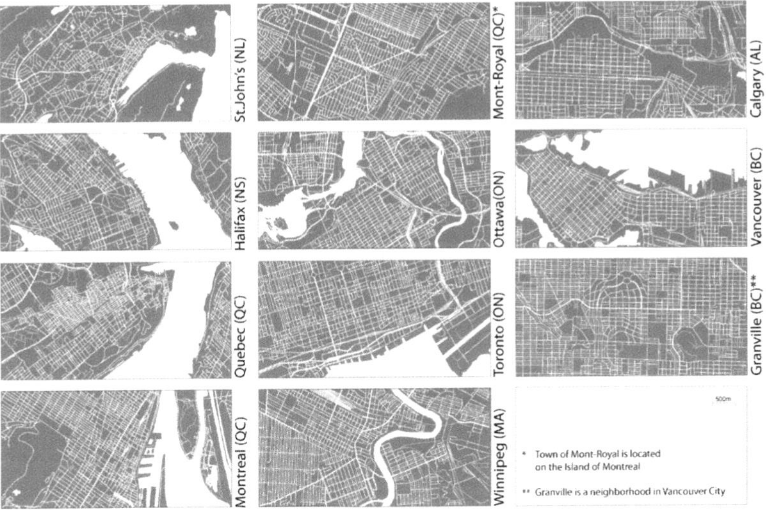

Sources: Based on Google maps data 2012.

Building Places

As previously noted in this chapter, the rapidity and scale in which Canada's urban cores have been evolving in the last century have had a dramatic impact on the population's way of life. Qadeer (1997) uses the example of trees as a source of neighbourhood battles: Canadian cities are reflections of the country's multi-cultural history, and planning issues such as how public trees should be maintained might seem trivial, but they reveal the deep complexity of building and using urban spaces by a diverse population. Citizens' needs and wants express cultural differences, of course, but they reflect socioeconomic differences as well. Aiming at equitably accommodating the needs of divergent groups is complex; public versus private interests often reveals power struggles, even clashes. Built in a set historical context, the city changes both physically and in terms of its representations; historically this dynamic relationship between the built environment, the services to the population and inhabitants is sometimes inconsistent. There is, as British geographer Doreen Massey (1994) explains it, a construction of real-world geographies and cultural specificity of definitions such as gender, class, etc.; this has brought, in various cities throughout the country, different responses. Some are city specific; others are linked with the Canadian planning system. Therefore, although there is a common "story of concentrated growth in the nation's big cities 'Canada's urban reality' conceals important, subtle variations" (Wyly, 2011: 2). This section proposes to observe how the Canadian municipal system has shaped cities from coast to coast, and offers an overview of urban changes and conflicts as well as city representations and appropriations.

The Shaping of the Canadian Municipal System

The Canadian planning system relies on municipal and regional governments. The desire for a municipal organisation was mentioned in the 1839 Durham report: Lord Durham considered the need for people to settle their local problems, and to be involved in issues regarding their social and economic environment (Tolley and Young, 2001). In the 1840s and 1850s, provincial acts established frameworks for local governments, allowing them to raise taxes and enact by-laws. However, the Constitution Act of 1867 did not clearly state the legal standing of municipalities, but established the subordination of municipalities to provincial government – each province, for instance, was responsible for making laws related to the municipalities that provinces can form, transform and disband. With the development and the urbanisation of Canada, the number of rural and urban municipalities increased. Financing, administrative and sometimes political difficulties prompted provinces to amalgamate municipalities in order to form new (bigger) ones. Three waves of such fusions occurred in the early 1900s, the 1960s and recently in the 2000s (e.g., amalgamations of the Ottawa-Carleton region into the new city of Ottawa, and the Greater Toronto Area becoming the mega-city of Toronto). In 2001, the Canadian census established the number of amalgamations at 27, containing nearly two-thirds of the country's population. Of these, from the West coast to the East coast, Vancouver, Edmonton, Calgary, Winnipeg, Toronto, Ottawa, Montréal, Québec and Halifax are currently playing a pivotal role in the country's economic and political life (Collin and Thomàs, 2004).

Thomas Adams and Shaping Canadian Planning

Before World War I (1914-1918), planning was essentially the work of concerned groups of citizens who organized themselves into "city planning commissions" and "civic improvement leagues" (Simmins, 2012). Concerned mainly with health and poverty issues, they were sympathetic to reforms inspired by international examples and, in particular, the British Garden City and the American City Beautiful movements that would, it was believed, improve the life of inhabitants. The implementation of the Commission of Conservation by the Minister of the Interior, Cliffort Sifton, in 1909, marked the beginning of Canada's modern planning history. The work and words of one of the Commission's town-planning advisors, Thomas Adams, largely influenced the first half of the 20th century.

Thomas Adams was born near Edinburgh in 1871. A prominent town planner of the time, he adhered to Ebenezer Howards' ideal of the Garden City movement, as well as the works of Patrick Geddes in seeking to better understand human needs and problems in order to propose city efficiency and functionality. One of the founders (and the first president) of the British Town Planning Institute, Adams had international recognition. Approached by then advisor to the Commission of Conservation, Dr Charles Hodgetts, during an international city planning conference in Toronto in 1914, Adams agreed to become town-planning adviser for the Commission that same year. Between 1914 and 1921 (the year that marked the demise of the Commission of Conservation), Adams actively participated in the development of a country-unified town planning philosophy. Amongst his ideals, he regarded planning as a combination of art (in terms of urban design) and scientific procedure, notably through the creation of comprehensive legislation. He made his opinion widely known in several publications of *Town Planning and Conservation of Life journal*, for example, and wrote the first Canadian planning text *Rural Planning and Development* in 1917. Adams also had a significant influence on the creation of institutional and professional structures such as the Civic Improvement League (1915) and the Town Planning Institution of Canada (1919). He also ran a professional practice during his stay in Canada, completing the planning of the town of Kipawa (now Témiscaming, in the province of Québec), the reconstruction scheme following the 1917 explosion of the Richmond district in Halifax (Nova Scotia), and perfecting his in-depth knowledge of Canadian towns. Adams remained in Canada after the demise of the Commission of Conservation in 1921, working for two years for the National Parks Division on different projects. He then moved to the United States where he became director of the regional plan for New York City and surrounding areas. Thomas Adams died in England in 1940. The planning principles he developed, promoting a more efficient city, can still be found in today's Canadian planning, in particular with respect to zoning laws and regulations.

Local Government

Provincial legislations define the status, role and management of municipalities. Moreover, provincial legislations define financial resources and have to approve most municipal borrowing. As a consequence, despite being decentralized, democratically elected authorities, municipalities have relatively limited legal, fiscal, and political power: they cannot intervene in any area of provincial jurisdiction, and have few interactions with the federal government (Collin and Thomàs, 2004; Seller, 2008).

Although the powers allocated to municipalities vary from province to province, and sometimes from one type of urban municipality to another – the recent creation of new, big cities through major municipal mergers has, for instance, increased the power of metropolitan areas –, the powers of local governments revolve around three main axes. The first axis is the implementation and monitoring of local by-laws. Municipalities have to deal with issues such as land occupancy and development; as we have seen previously, they have to prepare zoning plans that will allocate areas for specific uses and describe the characteristics specific to these areas. Municipalities will therefore implement by-laws that would, for instance, permit (or preclude) any activities or characteristics they decide are suitable for those areas. They deal with issues such as nuisances due to light and noise, local traffic, environment, built heritage, etc.

The second set of powers in the provision of local public services deal with such issues as the management of local police, planning and development services, education management, or public utilities. Through a certain number of committees (varying by municipality), the local government can tackle the preservation of green spaces, built heritage, economic development, etc. However some issues aren't specific to one city, and some problems in cities are not solely municipal or local problems (e.g., issues such as poverty or health). It has proven necessary to establish functional relationships between the federal and/or the provincial governments and municipalities and interest groups. One example of such collaboration is the appointment under Prime Minister Trudeau, in March 1971, of a Minister of State for Urban Affairs whose mandate it was to "plan, coordinate and develop new urban policies; integrate federal urban priorities with other federal policies and programs; and develop coordinating intergovernmental relationships" (Tolley and Young, 2001: n.p.). The Ministry was later dismantled (in 1980), and a Minister of State for Infrastructure and Communities was put in place in 2004 (later renamed the Minister of Transport, Infrastructure). Federal involvement in municipal affairs is usually through joint federal-provincial programs, financing services delivered by local governments.

The third category of power refers to the financing and collecting of municipal taxes, albeit with no real autonomy (Tolley and Young, 2001). Municipalities' revenues come from three different sources: real estate and property-related taxation, transfers from higher levels of government for specific purposes, and user-fees for municipal services. According to Collin and Thomàs (2004), almost 85% of municipal revenues are derived from local sources, and government transfers have significantly declined over the decades. Despite this, municipalities' fiscal powers are limited. They cannot, for instance, run a deficit budget, and they must have provincial approval before borrowing for long-term budgeting.

Municipalities have implemented many tools to help them in their daily activities and decision-making process. Schuster and de Monchaux (1997: 5-6) have summarised these tools of government action into five categories:

1. Implementing policy through direct provisions such as ownership and operation (i.e., "The local government will do X");
2. Regulating actions of other actors, such as private individuals, through a legal framework (i.e., "You must (or mustn't) do X");
3. Providing incentives or disincentives in order to help certain actions to happen or to be prevented (i.e. "If you do X, the local government will do Y");
4. Establishing, allocating or enforcing property rights (i.e., "You have a right to do X, and the local government will enforce that right");
5. Informing the public and actors in order to influence their actions ("You should do X" or "You need to know Y in order to do X").

Local government practices have colonial (mainly British, but with some French influences in the province of Québec) and pre-colonial legacies (namely indigenous traditions). We can infer that, because of the increase in urban population numbers and the rising influence of some urban cores, various actors play an important part in urban government and governance. The creation of decentralized structures has given more input to local neighbourhoods and boroughs; Complaints about restrictions on the municipality's decision-making autonomy have led to a search for alternatives to traditional urban planning and municipal systems. For example, Alberta (in 1994), British-Columbia (in 1999), and Ontario and Québec (since 2000), have given broader powers to localities (Sellers, 2008). Human capital is more and more recognised as an important part of the urban planning system as it influences the evolution of the city.

Zoning Philosophy

Thomas Adams believed that, in order to be efficient, a city should be divided into different parts, each designed with its specific function in mind. For example, residential areas should be home to a healthy community life, hence amenities that allow for such a lifestyle should be available in these areas. In the case of industrial and commercial areas, transportation and communication must be considered. For Adams, the key to modern society was better land-use development, and this idea laid the foundation for zoning.

Zoning is the division of land into zones of similar (or permitted) use. In an urban area, zones were defined as industrial, commercial, residential, institutional, recreational, etc. Early zoning practices were used along with early public health initiatives to prevent, for example, the use of noxious and/or polluting industries in residential areas, and were later intrinsic to planning practices. In the Canadian legislature, provincial bodies oversee the control of land use. This provincial responsibility has been part of Canadian legislation since the end of the 19th century (Hodge, 1984). Provinces can determine the uses and characteristics of areas throughout their entire territory. Most provinces had implemented early planning acts by the first decades of the 20th century, where zoning played an important part (e.g. New Brunswick's Act Relating to Town Planning of 1912, Ontario's Cities and Suburbs Plans Act of 1912, Manitoba's Town Planning Act of 1916). Each province has empowered municipalities and regions to control the use of land within their boundaries though the enactment of bylaws. Zoning decisions are based on development plans and it is considered at the heart of development control because it will influence both immediate and long-term land use goals (Thomas, 2012). In Canada, zoning affects not only the land itself but also all that is constructed on the land thus zoning laws have two main sets of objectives. The first set of objectives of zoning laws concerns the use of the land itself: by either separating incompatible uses or bringing similar uses closer together, zoning aims for better use of the land, protecting it from uncontrolled development and bettering the quality of life of its inhabitants. The second set of objectives has the goal of better controlling the intensity of development itself. Each zone is, of course, designated for certain uses, but in addition to this general designation a series of characteristics regarding any buildings to be built is also specified, such as height, density, etc.

Although zoning is rather straightforward, its practice has raised many issues over the year. Historically, zoning had been used for undeveloped areas in order to foresee the area's urban future. Moreover, zoning is most effective in relatively homogeneous areas. However, it is difficult to strictly apply it to existing built areas, especially in traditional urban cores where multifunctionality prevails. In many cities, for example, central areas previously dedicated to one function, such as industries, are being gentrified. Other zoning options therefore have to be implemented, such as *spot zoning* – classifying use for one single site differently from the surrounding uses – or nonconforming use (i.e., the right to maintain said use as long as the use remains unchanged).

There is a reciprocal relationship between zoning and transport. By separating or putting uses closer together, or by controlling the intensity of development, zoning determines the distance to cover between activities and possible options for transport modes.

Cities and People

There are three levels of urban and planning actors: local government personnel, private actors and interest groups. In Canada there are 881,926 local government personnel, accounting for 35% of all three-levels of government personnel (Sellers, 2008). These employees' roles and actions depend primarily on their position within a hierarchal system. Some are elected, others are hired by individuals; some have decision-making powers while others have none. The second level of actors, private actors – such as property, business and commercial owners and individuals – also have an influence on the city's future by using their electoral or economic powers, although the degree to which they can impact their urban environment varies considerably.

Finally, there are interest groups lobbying for specific issues. Their range of actions can be community or privately oriented, impacting a specific geographical area or a particular urban issue. Their financing can originate exclusively from private sources, or can be partly or totally subsidised by public funding. Moreover, some exist in a very specific time- and space-framed context, highlighting issues some cities have had to deal with; conflicts community groups have been willing to take on in order to increase the quality of life of its residents is one such issue. Although not specific to Canada, these battles are characteristic of a shift in city planning and town organisation. One example is the Stop Spadina campaign in Toronto in the early 1970s. During the post-World War II era, many highway projects, often crossing downtown areas, were proposed to facilitate rapid transit in an increasingly automobile oriented society. These projects were also considered critical for the development of new areas. In 1962, the construction of an expressway to downtown Toronto was officially approved but it required the partial or total demolition of neighbourhoods such as Rosedale and Forrest Hill Village. Opposition to the project grew rapidly, with locals and public figures such as Jane Jacobs actively opposing the construction of the highway. The battle lasted for over 30 years, with some set backs as portions of the expressway were built. In 1996, the all controversy ended with the sale of land that had been expropriated for the construction of the highway but never used for it. Another example of a community fight is the Milton-Park case in Montréal. Located in the downtown area, Milton-Park was, in the late 1960s, a vibrant albeit relatively poor neighbourhood. Its prime location, just a few blocks east of the McGill University campus, made it particularly interesting for redevelopment. The construction of La Cité, a mixed-use complex consisting of several high-rises, was authorised. The project required the displacement of the local population and the demolition of several dozens of buildings, however. Again, the local community, an "island of blue-collar workers, roomers, students and artists, political and social activists, and immigrant families in the midst of a sea of gentrification" (Backer, 1993: 277), rallied around a few ad hoc leaders, halting the reconstruction process. Today, the six-block area has been transformed into co-operatives and non-profit housing associations.

These outspoken critics of the traditional top-down city planning of Canadian cities, combined with increased pressure to better manage the environment (AUMA, 2009), have recently been encouraging a shift from traditional urban government rule to local governance – in other words, a collective approach to the future of local urban environments. Historically, successful municipalities were economically viable. Today, they strive to be sustainable, financially, environmentally and socially. Thus municipal community relations have become essential to better collaborative governance, and urban planning is not the only factor influencing the transformation of the physical components of a city (Qadeer, 1997). Community actors, with the support of pluralistic planning, strive to provide better responses to the needs and wants of residents and others. Two examples help illustrate these new tendencies in Canada's planning system: the Vancouver Greenways Program, launched in 1995, and the creation of the Office de consultation public de Montréal (Montréal Public Consultation Office) in 2002. In 1991, the city council appointed an Urban Landscape Task Force to produce a report on sustainable use and management of the city's urban landscape. The report recommended the development of a citywide system of greenways, and the Vancouver Greenways Program was then implemented. Among the many objectives of the program, there is a clear desire to involve the local population by creating opportunities to use and transform public green spaces (e.g., by encouraging the public to propose solutions to problematic intersections) thereby helping to sustain the environment (e.g., safer paths for pedestrians and bicyclers). Montréal's Public Consultation Office is another example of the increased role of public participation in city planning. It was created according to the City of Montréal's Charter, with the mandate of conducting public consultations concerning any project modifying the urban planning master plan. Moreover, if asked by the public via a petition, it can also hold public consultations on any urban issue (e.g., urban agriculture).

City Representation and Appropriation: the Case of Built Heritage

Finally, the city is a cultural construction. It is as much a construction on paper, of ideas and values, as a construction of buildings and functions (Morisset, 1998). The way people imagine their environment, and the way they perceive and use it are all part of a city's global image. In the 1960s and 1970s, major city projects (for example, Toronto's Spadina Speedway) have illustrated how people are attached to their environment, or specific aspects of it, and they are willing to fight with the established authorities in order to maintain what they consider to be a better quality of life. This has lead researchers like the American urban planner Kevin Lynch (1960) to explore how individuals' perception of the physical urban form is influencing their experience of the city and their interpretation of the urban landscape: "Every citizen has had long associations with some part of the city, and his [*sic*] image is soaked in memories and meanings" (Lynch, 1960: 1). Through the construction of mental maps, Lynch analysed people's legibility of urban places (based on their recognition of paths, edges, districts, nodes and landmarks), and this method is still used today. And because people are emotionally rooted in their surroundings, constantly creating social and personal connections, the urban environment must be considered as more than simply an object (Alexander, 1977). The sense of place, which allows for a sense of belonging, participates in a multi-layered cultural geography (Julien and Jébrak, 2008).

The example of built heritage will help demonstrate this notion of a sense of place. As a city evolves over time, and as its socioeconomic characteristics change due to numerous factors, so does its cityscape: buildings are being built, others are slightly or significantly modified or destroyed. Some, though, gain significant value and become part of the city's cultural landscape. Since the late 1970s, the history of built heritage in Canada has followed a similar path as other Occidental countries, with an increased number of measures to preserve and monitor historical structures, and to educate and inform the public and stakeholders (Cullingworth, 1987).

Again, Canada's built heritage legacies originate from two approaches: the British private trust approach and the French governmental approach. Canada's built heritage protection system is divided between federal and provincial competences. The federal government's first attempts at preservation took place when British troops left Canada in 1871 and the management of military installations (and later their preservation, monitoring and opening to the public) was required. A first campaign was conducted for the preservation of Québec City fortification walls and the reconstruction of the city's gates. Targeting both natural and cultural heritage, the federal government later implemented the National Monuments and Historic Sites Commission (active since 1913) and enacted the Canada National Parks Act in 1930. Parks Canada manages national parks and national historic sites and administers the Canadian Register of Historic Places. However, federal designation has no real legal effect: the Canadian constitution gives exclusive responsibility of property rights to provinces.

Each province operates under its own legislation, delegating powers related to planning to, for instance, local governments (e.g., the approval or disapproval for the destruction of a designated building, the designation of built heritage at the municipal level, etc.). Definitions of what is representative of one's provincial legacy therefore vary from one province to another. Moreover, local preservation structural and legal frameworks have been implemented at different times, and have followed different paths (in terms of what is representing one's past for instance. In the province of Québec, for instance, the French legacy of the 17th and 18th centuries was the main component of what was considered built heritage up until the 1970s. This was based on a 1922 law stating that designations had to be approved by Québec's executive committee on recommendation from the Historic Monuments Commission. Churches, commemorative monuments and engravings, and mills were given heritage designation. In the 1950s and early 1960s, tougher amendments were made to the Historic Monuments, Sites and Objects Act (later renamed the Historic Monuments Act; it was replaced in 1972 by the Cultural Act). A major change potentially affecting Québec's cities came in the creation of historical districts in the 1950s (which lead to the constitution of the Old Montréal Historic Site).

However, the application of built heritage designations revealed weaknesses in the system. Savage destruction of privately owned buildings, sometimes quite newly built by heritage standards, prompted citizens, interest groups, and municipalities alike to participate in the debate concerning the definition of built heritage. The famous 1973 failed campaign to preserve Montréal's Van Horne Mansion (built in 1870 by an unknown architect, and later remodelled by Edward Colonna) is an example of the ever-changing nature of heritage. The mansion once housed the American-born head of the Canadian Pacific railroad company, William Cornelius Van Horne (1843-1915), who held one of the biggest fortunes in Montréal, and was an avid art collector. The house's location on Sherbrooke Street West in the downtown core made it extremely desirable for investment purposes. The promoter who bought the land and house wanted to demolish the building with the intention of building a more profitable high-rise in its place. At the time, Montréal's built heritage was overseen by the Jacques-Viger Commission. The organisation was, and continues to be, responsible for monitoring the development in the historical districts of Old Montréal and other important sites across the city. The Commission didn't support the demolition of the building. However, the City of Montréal's permit agency granted authorisation to the promoter. Thus ensued a legal and media battle involving Québec's Cultural Assets Board which requested the provincial Ministry of Cultural Affairs to consider the designation of the mansion as an official heritage site, and citizens – who actively campaigned for the recognition of the mansion's important architectural qualities. In the end, the campaign to protect the mansion failed, and the provincial government decided not to preserve the building as it did not represent Québec's architecture. A demolition permit was issued (Drouin, 2005). For Drouin, this event opened a public and province-wide debate on as the definition of heritage, and how and why certain sites should be preserved. Such debates occurred throughout Canada, revealing the many approaches to, and opinions about, heritage. "The heritage of the built environment is not a result of hap-hazard survival, but rather the outcome of individual and group consciousness relating to a sense of place. The built environment as it has survived is a cultural construction, its appearance and meanings dependant on a complex process of selection, protection and intervention" explain Phelps *et al.* (in Negussie, 2004: 204).

More recently, the notion of cultural landscape has emerged as a way to comprehend built heritage as a whole, linked to both the physical reality of the environment and the reflection of the values of whatever social group is dominant at the time (Negussie, 2004): Heritage perception, representation and appropriation vary according to people's interests, political outlook and socio-economic characteristics of the city. Choices made in identifying what is heritage and how we preserve or transform it will, in the long-term, impact the collective memory and how people want to present the city to others: "Within the context of planning in historic environments, a dichotomy exists between preserving the past for its intrinsic value and the need for development in response to changing societal values" (Nasser, 2003: 467-468).

Built heritage is a palimpsest because its numerous layers of discourses, decisions and actions participate in the creation of a collective memory and of an urban identity. Built heritage helps understand how city representation and appropriation play an important part in Canadian urban cores and their future. The desire to maintain a certain cultural landscape associated with the city's past has had an impact on how planning has been conducted in some cities. Buildings, whether designated as such or not, can be landmarks. A neighbourhood, facing significant redevelopment schemes along with the disappearance of some of its historical structure (regardless of their aesthetic or historical value), could experience a rupture in the overall sense of place. For instance, some cities have encouraged the preservation of facades (called facadism) while allowing for the destruction or redevelopment of the structure behind the facades (e.g., Red River College in Winnipeg, Alcan Hotel in Montréal, etc.). Although this practice has its critics arguing that it stagnates city architectural innovations or progress, for instance, some see in it an interesting way of maintaining some traces of the past, which can support city marketing initiatives. Historical areas and World Heritage Sites also participate in the economic life of the city, having an impact, for example, on tourism and tourism related activities. The World Heritage Sites in Québec City attract thousands of tourists from within and outside Canada every year. Visitors come for the history and the culture; the city thus has had to intervene in the creation and maintenance of public spaces in order to better accommodate the tourist gaze (information panels, guides and maps, information centre, etc.). These tourist activities attract certain types of commercial activities (restaurants, souvenir shops, artisans' markets, etc.) that will in turn have an impact on the urban fabric (e.g., pedestrian and car traffic patterns, nocturnal activities, etc.), and on the overall characteristics of the neighbourhood and the city (e.g., property values, levels of air, sound and visual pollution, etc.).

Conclusion

Canadian cities are different from one another in their development, their socio-economic characteristics, and their urban forms. But they are also very similar: the city, and its geographies, both physical and imaginary, contributes to the construction of the identity of its inhabitants and their social relations. In return, each individual participates in the construction of the city both directly and indirectly by building it and by creating narratives around it. These are the key characteristics of urban places.

Review Questions

- Since when Canada is considered an urban nation?
- What are the three largest cities in Canada?
- What influence the location choice for the first permanent settlements in Canada?
- Can you explain how different transport modes shaped the Canadian urban system and urban forms?
- What are the forces that influence the urban forms?

- Can you explain in a few words the contribution of Thomas Adams un Canada's urban planning
- What are the three main axes around which revolve local government's power?
- Can you give a definition of zoning?
- There are many critics for the top-down approach in planning. Can you give an example illustrating these critics?
- Cities are palimpsests. How built heritage itself is also a palimpsest?

Further Readings

Bunting, Trudi and Pierre, Filion (2000), *Canadian cities in transition: the twenty-first century*, Don Mills, Ont.; New York: Oxford University Press.

Davies, Jonathan S. and David, L. Imboscio (eds.) (2009), *Theories of Urban Politics*, 2nd ed., Los Angeles: Sage.

Denhez, Mark (1997), *Heritage Strategy Planning Handbook: An International Primer*, Toronto: Dundurn Press.

Fischler, R. (2004), "The Problem, or Not, of Urban Sprawl", *Policy Options*, February 2004.

Fowler, Edmund P., and David, Siegel (eds.) (2002), *Urban Policy Issues: Canadian Perspectives*, 2nd ed., Toronto: Oxford.

Garcea, Joseph, and Edward C. LeSage Jr. (eds.) (2005), *Municipal Reform in Canada: Reconfiguration, Re-Empowerment, and Rebalancing*, Toronto: Oxford.

Hall, Tim and Heather Barrett (2012), *Urban Geography*, Routledge Contemporary Human Geography, London: Routledge.

Hiller, Harry (2010), *Urban Canada*, Don Mills: Oxford University Press.

Lennard, S.H., S. von Ungern-Sternberg, and H.L. Lennard (eds.) (1997), *Making Cities Livable*, International Making Cities Livable Conferences, California: Gondolier Press.

Polèse, Mario, and Richard Stren (eds.) (2000), *The Social Sustainability of Cities: Diversity and the Management of Change*, Toronto: University of Toronto Press Inc.

Tindal, Richard C., and Susan Nobes Tindal (2009), *Local Government in Canada* (7th ed.), Scarborough, Ont.: Thomson Nelson Learning.

Websites

Canada e-Book. Statistics Canada: http://www43.statcan.ca/r000_e.htm

City of Ottawa: http://www.ottawa.ca

The Atlas of Canada. Gouvernment of Canada. Natural resources Canada: http://atlas.nrcan.gc.ca

The Canadian Encyclopedia: http://www.thecanadianencyclopedia.com

The Companion Website for The Cultural Landscape: An Introduction to Human Geography 8th ed.: http://wps.prenhall.com/esm_rubenstein_humangeo_8/

The virtual library of the Canadian Parliament: http://www.parl.gc.ca/common/Library.asp?Language=E

Urban Poverty Project 2007. Canadian council on social development: http://www.ccsd.ca/pubs/2007/upp/

References

Alberta Urban Municipalities Association (2009), "Local Governance: A Short Review of Changes in Various Jurisdictions", *AUMA Future of Local Governance Research Paper*, [http://www.auma.ca/live/digitalAssets/33/33581_Future_of_Local_Governance_Research_Paper_041509.pdf], page retrieved April 20, 2012.

Alexander, Christopher (1977), *A Pattern Language: Towns, Buildings, Construction*, Oxford, UK: Oxford University Press.

Allain, Rémy (2004), *Morphologie urbaine: géographie, aménagement et architecture de la ville*, Paris: Colin.

Bacher, John C. (1993), *Keeping to the Marketplace. The Evolution of the Canadian Housing Policy*, Montréal, Qc.; Kingston, Ont.: McGill-Queen's University Press.

Beauregard, Ludger (1984), "Géographie historique des côtes de l'île de Montréal", *Cahiers de géographie du Québec*, 28(73-74): 47-62.

Bourget, Charles (1995), "Le Vieux Québec: un plan d'inspiration médiévale", *Cap-aux-Diamants: la revue d'histoire du Québec*, 42(1995): 10-13.

Bourne, Larry S. (2000), "Urban Canada in Transition to the Twenty-First Century: Trends, Issues, and Visions", p. 26-51 in Bunting, T.E. and P. Filion (2000), *Canadian cities in transition: the twenty-first century*, Don Mills, Ont.; New York: Oxford University Press.

Bunting, Trudi and Pierre Filion (2000), *Canadian cities in transition: the twenty-first century*, Don Mills, Ont.; New York: Oxford University Press.

Burgess, Ernest (1925), "The Growth of the City: An Introduction to a Research Project", in *The trend of population*, Publications of the American Sociological Society, XVIII: 85-97.

Cadwallader, Martin T. (1996), *Urban Geography: An Analytical Approach*, Upper Saddle River, N.J.: Prentice Hall.

Chapain, Caroline and Mario Polèse (2000), "Le déclin des centres-villes: mythe ou réalité? Analyse comparative des régions métropolitaines nord-américaines", *Cahiers de Géographie du Québec*, (44)123.

City of Vancouver (2006), Downtown Eastside Community Monitoring Report 2005/06, 10th ed., City of Vancouver: City Clerk's Office.

Collin, Jean-Pierre, and Mariona Thomàs (2004), "Metropolitan Governance in Canada or the Persistence of Institutional Reforms", *Urban Public Economics Review*, No. 2, pp. 13-39.

Cullingworth, J. Barry (1987), *Urban and Regional Planning in Canada*, New Brunswick, N.J., Center for urban policy research.

Curley, Alexandra M. (2005), "Theories of Urban Poverty and Implications for Public Housing Policy", *Journal of Sociology and Social Welfare*, Vol. 32, No. 2, pp. 97-119.

Cuthbert, Alexander R. (2006), *The form of cities: political economy and urban design*, Malden, MA; Oxford: Blackwell Pub.

Drouin, Martin (2005), *Le combat du patrimoine à Montréal (1973-2003)*, Montréal: Presses de l'université du Québec.

Filion, Pierre, Bunting, Trudi and Len, Gertler (2000), "Cities and Transition: Changing Patterns of Urban Growth and Form in Canada", pp. 125 in Bunting, T.E. and P. Filion (2000), *Canadian cities in transition: the twenty-first century*, Don Mills, Ont.; New York: Oxford University Press.

Filion, Pierre (2010), "Growth and decline in the Canadian urban system: the impact of emerging economic, policy and demographic", *GeoJournal*, 2010(75): 517-538.

Hall, Tim and Heather Barrett (2012), *Urban Geography*, Routledge Contemporary Human Geography, London: Routledge.

Hiller, Harry (2010), *Urban Canada*, Don Mills: Oxford University Press.

Hodge, Gerald (1998), *Planning Canadiant communities: an introduction to the principles, practices and participants*, 3rd ed., Toronto: ITP Nelson.

Ingram, Gregory (1998), "Pattern of metropolitan Development: What Have we Learned?", *Urban Studies*, 35(7): 1019-1035.

Jackson, Kenneth T. (1985), *Crabgrass frontier: the suburbanization of the United States*, New York; Toronto: Oxford University Press.

Julien, Barbara, and Yona Jébrak (eds.) (2008), *Les temps de l'espace public urbain: Construction, transformation et utilisation*, Montréal: Multimondes, coll. Cahiers de l'Institut du patrimoine.

Kellas, Hugh (ed.) (2010), *Inclusion, collaborative and urban governance*, Vancouver: The University of British Columbia, City Limits collection.

Knox, Paul and Steven, Pinch (2000), *Urban social geography: an introduction*, Harlow, England; New York: Prentice Hall.

Knox, Paul L. (1994), *Urbanization: an Introduction to Urban Geography*, Englewood Cliffs, NJ: Prentice Hall.

Lavoie, Yolande (1981), *L'émigration des Québécois aux États-Unis de 1840 à 1930*, Québec: Documentation du conseil de la langue française.

Lees, Loreeta (2004), "Urban geography: discourse analysis and urban research", *Progress in human geography*, 28(1): 101-107.

LeGates, Richard T. and Frederic, Stout (1996), *The city reader*, London; New York: Routledge.

Ley, David (1996), "The New Middle Class in Canadian Central Cities", in Caufield, Jon and Linda, Peake (ed.), *City Lives and City Forms: Critical Research and Canadian Urbanism*, Toronto, Buffalo, London: University of Toronto Press, p. 15-32.

Lynch, Kevin (1960), *The Image of the City*, Boston: the MIT Press.

Macionis, John J. and Vincent, N. Parrillo (1998), *Cities and Urban Life*, Upper Saddle River: Prentice Hall.

Massey, Doreen (1994), *Space, Place and Gender*, Minneapolis: University of Minnesota Press.

McCann, Larry and Jim, Simmons (2000), "The Core-Periphery Structure of Canada's Urban System", p. 76-96 in Bunting, T.E. and P. Filion (2000), *Canadian cities in transition: the twenty-first century*, Don Mills, Ont.; New York: Oxford University Press.

Morisset, Lucie K. (1998), *Arvida, cite industrielle. Une épopée urbaine en Amérique*, Sillery, Qc.; Brest, France: Septentrion et Institut de géoarchitecture.

Nasser, Noha (2003), "Planning for Urban Heritage Places: Reconciling Conservation, Tourism, and Sustainable Development", *Journal of Planning Literature*, 17(4): 467-479.

Negussie, Elene (2004), "What is worth conserving in the urban environment? Temporal shifts in cultural attitudes towards the built heritage in Ireland", *Irish Geography*, 37(2): 202-222.

Olson, Sherry (2000), "Form and Energy in the Urban Built Environment", p. 224-243 in Bunting, T.E. and P. Filion (2000), *Canadian cities in transition: the twenty-first century*, Don Mills, Ont.; New York: Oxford University Press.

Pacione, Michael (2001), *Urban Geography. A global perspective*, London; New York: Routledge.

Parks Canada (2007), *Lachine Canal National Historic Site of Canada*, [Online: http://www.pc.gc.ca/lhn-nhs/qc/canallachine/natcul/index_E.asp].

Polèse, Mario (2008), "Les nouvelles dynamiques regionals de l'économie québécoises: cinq tendances", *Inédit/Working paper*, No. 2008-7, Montréal: INRS Urbanisation Culture et Société.

Qadeer, Mohammed A. (1997), "Plurastic Planning for Multicultural Cities", *Journal of the American Planning Association*, 63(4): 481-494.

Rose, Damaris (2010), "Local state policy and 'new-build gentrification' in Montréal: the role of the 'population factor' in a fragmented governance context", *Population, Space and Place*, 16(5): 413-428.

Ryan, Mary P. (1990), "Everyday Space: Gender and the Geography of the Public", essay in *Women in Public*, Baltimore: Johns Hopkins University Press.

Sellers, Jeffrey (2008), "North America (Canada and the United States)" in *Decentralization and Local Democracy in the World*, pp. 231-51, Barcelona: United Cities and Local Governments.

Statistics Canada (2008), Québec 1608-2008: 400 ans de recensements, Available online http://www.statcan.gc.ca/pub/11-008-x/2008001/article/10574-fra.htm#a6.

Stelter, Gilbert (ed.) (1990), *Cities and Urbanization: Canadian Historical Perspective*, Toronto: Copp, Clark, Pitman.

Schuster, J. Mark, John de Monchaux, and Charles A. Riley II (eds.) (1997), *Preserving the Built Heritage: Tools for Implementation*, London: University Press of New England.

Timmer, Vanessa, and Nola-Kate, Seymoar (2005), *The Livable City*, The World Urban Form Forum 2006, Vancouver Working Group Discussion Paper, Vancouver: International Centre for Sustainable Cities.

Tolley and Young (2001), Municipalities, the Constitution, and the Canadian Federal System, Reference number BP-276E.

United Nation, Department of Economics and Social Affairs (2012), The World Urbanization Prospects, the 2011 Revision, [online: http://esa.un.org/unpd/wup/index.htm].

Walks, R.A. and L.S., Bourne (2006), "Ghettos in Canada's cities? Racial segregation, ethnic enclaves and poverty concentration in Canadian urban areas", *The Canadian Geographer/Le Géographe canadien*, 50(3): 273-297.

Wyly, Elvin (2011), "Recent Changes in the Canadian Urban System", *Project background paper*, [http://www.geog.ubc.ca/~ewyly/g350/cansystem.pdf], page retrieved, May 26, 2012.

Websites

Canadian Geographic, The Canadian Atlas Online: http://www.canadiangeographic.ca/atlas/.

Geogratis, Atlas of Canada: http://geogratis.cgdi.gc.ca/.

Parks Canada, Lachine Canal National Historic Site of Canada: http://www.pc.gc.ca/lhn-nhs/qc/canallachine/natcul/natcul2/c.aspx.

The Canadian Encyclopedia: http://www.thecanadianencyclopedia.com.

PART II

REGIONAL PERSPECTIVES

CHAPTER 7

Atlantic Canada

Norm CATTO, Sheridan THOMPSON-GRAHAM, and Kelly VODDEN

Memorial University of Newfoundland

Figure 1. Communities in the Atlantic Provinces

a)

b)

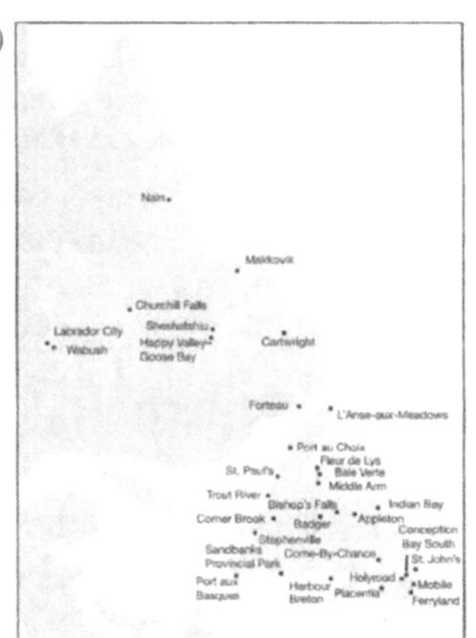

FIGURE 1: Communities in the Atlantic provinces: a) New Brunswick, Nova Scotia, and Prince Edward Island; and b) Newfoundland and Labrador

Left: New Brunswick, Nova Scotia, and Prince Edward Island. Right: Newfoundland and Labrador. (Dept. of Geography MUN, Vasseur and Catto 2008)

Introduction

Atlantic Canada includes the provinces of New Brunswick, Newfoundland and Labrador, Nova Scotia, and Prince Edward Island (Fig. 1). Maritime Canada (or the Maritimes) includes New Brunswick,

Nova Scotia, and Prince Edward Island, but not Newfoundland and Labrador.

Newfoundland and Labrador has the largest area of the four provinces (405,720 km^2), more than three times the land area of the three Maritime Provinces combined, and extends from 60°23' N (Cape Chidley) to 46°37' N (Cape Pine). Labrador encompasses 294,330 km^2. With an area of 111,390 km^2, the island of Newfoundland is the fourth largest island in Canada.

Of the three Maritime Provinces, New Brunswick is the largest at 73,440 km^2, Nova Scotia has an area of 55,490 km^2, and Prince Edward Island is the smallest, at 5590 km^2. The southernmost point in Atlantic Canada is Cape Sable, NS (43°28' N). With the exceptions of Churchill Falls, Labrador City and Wabush, no community in Atlantic Canada is more than 200 km from the nearest marine shoreline. The physical geography of Atlantic Canada has had a major impact on land use and economic activity throughout the region.

Physiography

The physiography of Atlantic Canada can be divided into three major regions: the Atlantic Coastal Lowland, the Atlantic Highlands, and the Canadian Shield. The Atlantic Coastal Lowland encompasses eastern New Brunswick, Prince Edward Island, the Northumberland Strait coastline of Nova Scotia, parts of the Gulf of St. Lawrence coastline of Newfoundland, and southernmost Labrador along the Strait of Belle Isle (Fig. 2). Relief is very low, with all the terrain lying below 200 m ASL. Coastlines are dominated by sandy beaches, barrier islands, and gently sloping shelves. Most areas are marked by low-energy beaches and longshore drift transport parallel to the shoreline.

Prince Edward Island lies entirely within the Atlantic Coastal Lowland, and is characterized by planar to gently rolling topography, with many flat plains. Much of the topography is the result of post-glacial downcutting by the numerous small streams, however, many are now diminished in volume or diverted due to agricultural activity.

A gently eastward-sloping plain is present along most of the Gulf of St. Lawrence coast of New Brunswick. Tidal ranges along the Gulf of St. Lawrence coast of New Brunswick vary from 1.5 to 3.5 m. Salt marshes are developed in most tidally-influenced lagoons and estuaries (Fig. 3). A distinctive feature of the coastline is the barrier islands developed by north to south sediment transport.

Along the west coast of Newfoundland, a narrow coastal plain discontinuously borders the western margin of the Long Range Mountains. The boundary between the coastal plain and the geologically older Long

Range Mountains coincides roughly with the limit of post-glacial marine inundation following deglaciation. Narrow, gently southeastward sloping plains occur in southern Labrador along the Strait of Belle Isle from L'Anse-au-Clair to Pinware, and adjacent to Battle Harbour.

The Atlantic Highlands, representing the northeastern extension of the Appalachian Mountains, form the physiography of western and northern New Brunswick, the margins of the Bay of Fundy, most of Nova Scotia, and most of the island of Newfoundland. The Atlantic Highlands are rolling to rugged uplands, with coastlines that are deeply indented, dominated by cliffs and high-energy gravel beaches, and steep offshore bathymetry (Fig. 4).

In New Brunswick, the Atlantic Highlands extend from southwest Moncton to Charlotte County, and along the western shore of the Saint John River north to Gaspé. Fundy National Park is encompassed in the westernmost portion of the highlands, the Caledonian Hills. Nova Scotia's Bay of Fundy is bordered by North Mountain providing spectacular views of the eastern part of the Annapolis Valley and the Avon River.

While the central axis of the Annapolis Valley is a flat plain, the southern margins of the Annapolis Valley as well as the Avon River lowland are bordered by South Mountain. The physiography has been modified by glaciation, and numerous drumlins, to 50 m in length and 15 m in height, resemble eggs and cigars scattered across the landscape.

Cape Breton Island, also within the Atlantic Highlands, is an area marked by irregular hills, rolling plateau (e.g. Cape Breton Highlands National Park), and steep escarpments. The segments of the Cabot Trail leading from sea level to plateau summit are a challenge to vehicles, and represent the highest relief evident from any highway in the Maritimes.

In Newfoundland, the Atlantic Highlands encompass all of the terrain with the exception of the coastal plain along the Gulf of St. Lawrence (Fig. 5). The terrain is marked by short, fault-created valleys separated by low, rolling uplands veneered and blanketed with glacial sediment. The alignments of embayments and valleys reflect the underlying rock structure and jointing pattern. Much of the coastline is marked by numerous small islands and skerries are present. Beaches are restricted to sheltered coves, and generally are marked by high energy, dominated by cobbles and pebbles.

Figure 2. Atlantic Coastal Lowland

Figure 2a: Tracadie, PEI. The terrain generally has low relief. Barrier islands are developed along the coastline of the Gulf of St. Lawrence. Photograph by Gail Catto.

Figure 2b: The combination of limited relief, good podzolic soils, and favourable climate encourage agriculture, particularly potato farming. Farm northwest of Charlottetown, PEI. Photograph by Norm Catto.

Figure 3. New Brunswick salt marshes

Figure 3a: St. Louis lagoon, Kouchibouguac National Park, New Brunswick; Salt marshes are common along the Gulf of St. Lawrence coastline of New Brunswick. Photograph by Norm Catto.

Figure 3b: Shediac Bay, New Brunswick; Many salt marshes have been partially or totally drained, for use as agricultural areas or for ocean-side homes or cottages. Photograph by Norm Catto.

Figure 4. Atlantic Uplands of Nova Scotia

4a. Gravel beach flanked by bedrock cliffs, north of Presqu'ile, Cape Breton Highlands National Park. Photograph by Norm Catto.

4b. View southward from North Mountain at "The Lookoff", Annapolis Valley. Photograph by Norm Catto.

Also within the Atlantic Highlands is the Northern Peninsula. In Gros Morne National Park, the rugged topography of the Long Range

Mountains grades into rolling slopes, flat-topped to gently undulating plateaux, and deep fjords. The Atlantic Highlands extend southwestward to include the Annieopsquotch and Anguille Mountains.

Physiographically and geologically, Labrador is part of the Canadian Shield. Most of Labrador is marked by rolling topography, developed as the Precambrian bedrock of the Canadian Shield was gradually eroded (Fig. 6). Virtually all of the erosion occurred prior to the most recent glaciation. Deeper erosion is confined to a few isolated valleys and fjords.

Western Labrador is marked by relatively low relief, with generally less than 200 m elevation difference between the deepest valleys and the adjacent summits. Linear lakes and river valleys have developed along faults, folds, and weak bedrock units. The construction of the dam at Churchill Falls created the tenth largest lake in Canada, the Smallwood Reservoir (6,527 km^2). Most of southeastern Labrador lies within the physiographic region of the Canadian Shield. In contrast to western Labrador, there are no large, elongate lakes in this region outside of the Churchill River valley.

The Mealy Mountains are rugged, marked by irregularly – aligned short, steep streams and steep cliffs. The maximum elevations of the summits reach 1,190 m. More than 20 cirques, carved by former glaciers, are present in the headwaters of the Etagulet and English Rivers, with floor elevations at 600 m or greater. No active glaciers are present.

Northern Labrador is separated by two subdivisions are: the George Plateau (similar to western Labrador region), and the Torngat Mountains. The Torngat Mountains are a series of rugged northwest-southeast ranges, cut by deep east-west fjords (Fig. 6). Fjords extend up to 15 kilometres inland from the open coast, and main valleys dissect the ranges, extending across the drainage divide and Labrador/Québec boundary. Relief along the coastline is approximately 1,200 m, and individual peaks rise above 1,500 m.

Bedrock Geology

In Atlantic Canada, the oldest Precambrian bedrock units are present in the Canadian Shield of Labrador, and the core of the Northern Peninsula of Newfoundland. Younger units, ranging from latest Precambrian to Palaeozoic, are exposed throughout Newfoundland and the Maritimes.

Figure 5. Atlantic Uplands of Newfoundland

5a. Caplin Cove, Avalon Peninsula. Photograph by Norm Catto.

5b. Lark Harbour, Humber Arm. Photograph by Norm Catto.

Figure 6. Labrador's terrain is the result of glaciation acting on the pre-existing bedrock landscape

6a. Molson Lake, western Labrador, showing Precambrian bedrock overlain by glacial deposits, including an esker. Photograph by Norm Catto.

Labrador

The Precambrian history of Labrador is marked by numerous episodes of mountain formation, each of which is termed an orogeny. As well, igneous intrusions (such as granites) associated with younger orogenies frequently penetrate and metamorphose older rocks, producing a complex pattern of igneous and metamorphic units of differing ages. After much research, Labrador has been divided into several Structural (geological) Provinces (Fig. 7), each of which is defined on the basis of the ages of rock assemblages and on the nature of folding and faulting.

The Nain Structural Province extends along the coastline from Nachvak Bay southward to the Kanairktok River (Fig. 8). Rock units include gneisses, chert, anorthosite, norite, and other igneous and metamorphic rocks, ranging from 3,800 to 1,305 million years old. Ramah chert has been used for the manufacture of artefacts throughout the human occupation of Labrador. Norite, an igneous rock, contains nickel sulphides. The Voisey's Bay nickel mine is currently exploiting a large norite intrusion.

Figure 6. Labrador's terrain is the result of glaciation (cont'd)

6b. The glaciated Torngat Mountains flank Iron Strand (Hutton) Beach in northern Labrador. Photograph by Norm Catto.

The rocks in westernmost Labrador are assigned to the Superior Province of the Canadian Shield. These granites and gneisses are 2,700-2,650 million years old. Most were further metamorphosed or modified during later orogenic events. Interspersed with the wide belts of alternating resistant gneiss and granite are weaker volcanic and sedimentary rocks. The weaker rocks weather to produce valleys, locally producing a characteristic corrugated topography.

The Southeastern Churchill Province extends across northern and northwestern Labrador. Several rock units are present. Deformed volcanic, metamorphic, and igneous rocks form part of a broad arc that extends northwards through Nouveau Québec, formed between 2,100 and 1,840 million years ago and subsequently deformed during the tectonic collision between the Superior and Nain provinces.

The Makkovik Province is located to the south of the Nain Province. Granites and gneisses, up to 1,860 million years old, are common rock types.

The Grenville structural province is the largest and youngest structural province of the Canadian Shield represented in Labrador (Fig. 8). Most of the province is dominated by granites and related rock types, and acidic, highly metamorphosed gneiss and schist derived from the metamorphism of pre-existing sedimentary and igneous rocks during several Precambrian orogenic events. Orogenic activity began in the Grenville Province approximately 1,600 million years ago, and culminated between 1,050 and 960 million years ago. At this time, a now-destroyed continent to the southeast (where the Appalachians of Gaspé are now) was thrust against the Grenville Province rocks, causing metamorphism and mountain-building. The style of deformation was in some respects similar to that occurring in the modern Himalayan Mountains of India and Nepal, but 960 million years of weathering and erosion has ground away the peaks, leaving only the roots of the mountain system exposed as the Grenville Province.

Newfoundland and the Maritimes

The bedrock geology of Newfoundland and the Maritime Provinces reflects several orogenic mountain-building events from the later parts of the Precambrian (beginning 1,000 million years ago) to the end of the Palaeozoic (245 million years ago). Younger Precambrian sedimentary, igneous, and metamorphic rocks are present in several locations in Atlantic Canada, including the northern shore of the Bay of Fundy in New Brunswick, northern Cape Breton Island, central Newfoundland, and the Avalon Peninsula (Fig. 9).

Folding and faulting of latest Precambrian rock units is responsible for the topography of much of eastern Newfoundland. Plate tectonic activity during the late Precambrian resulted in the formation of a mid-ocean rift valley system to the west of the present location of the Avalon and Burin Peninsulas. This ocean, a predecessor of the modern Atlantic, is referred to as the 'Iapetus Ocean'. About 565 million years ago the Avalon and Burin Peninsulas formed part of a continent that also included northwesternmost Africa, the majority of Europe, England, southern Scotland, and Ireland. The climate was temperate to subtropical, allowing some of the earliest multi-celled metazoan organisms to flourish in the warm seas and to subsequently be preserved as fossils in rock at Mistaken Point Ecological Reserve, southeastern Newfoundland.

Figure 7. Geological Structural Provinces, Labrador

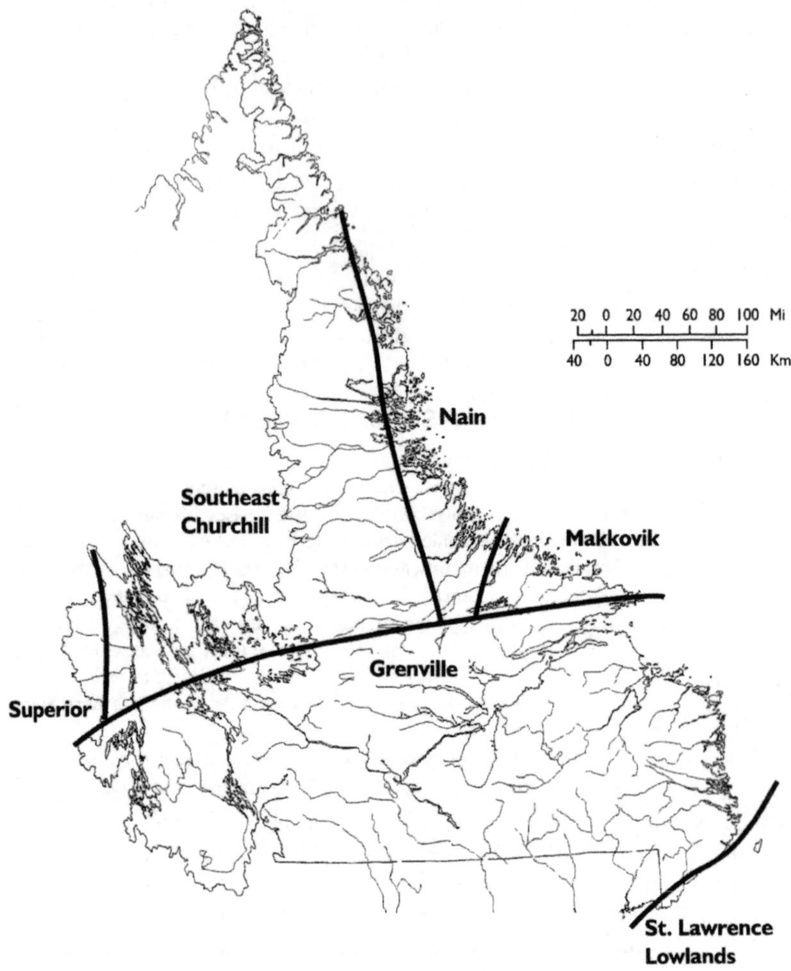

Source: Norm Catto, DELT, Memorial University.

Figure 8. Geological units, Labrador

8a. Gneiss bedrock of the Southeastern Churchill Province, in the northern Torngat Mountains, is subject to ongoing weathering and erosion. Photograph by Norm Catto.

8b. Eroded granites and metamorphic rocks exposed at Muskrat Falls, Churchill River, lie within the Grenville Province. Photograph by Norm Catto.

Orogenic events during the subsequent Palaeozoic Era (565-245 million years ago), including the Taconic (Ordovician) and the Acadian (Silurian-Devonian), shaped the terrain of Newfoundland and the Maritimes. Sedimentary sandstones and shales were deposited, buckled, faulted, and metamorphosed by the mountain-building events, and intruded by granite and other igneous rocks. In Nova Scotia, gold is found in metamorphosed sedimentary rocks near Lunenburg and in the Rawdon Hills.

At Gros Morne, belts of early Palaeozoic sedimentary rock are present along the coast. At Cow Head, extensive exposures represent the collapsed margin of a coral reef that grew when Gros Morne lay in the Southern Tropics, ca. 500 million years ago. Other sedimentary rocks, of Cambrian and Ordovician ages, contain brachiopods, graptolites, and trilobites.

During the latter part of the Devonian period and throughout the Mississippian period (ca. 370-320 million years ago), sandstones, shales, salt, anhydrite, and gypsum were deposited overlying the older rocks in southwestern Newfoundland, parts of Cape Breton, and central Nova Scotia. Gypsum and anhydrite (dehydrated gypsum), used to manufacture wallboard, are mined at several large quarries near Brooklyn, NS. Younger sandstones, siltstones, and conglomerates of Pennsylvanian age (320-290 million years old) extend westward from the Gulf of St. Lawrence shoreline to cover approximately half of New Brunswick. These iron-rich rocks, formed along the flanks of the Appalachian Mountains, are soft and easily eroded.

In comparison to most other areas of Atlantic Canada, the geology of Prince Edward Island is straightforward. The island is underlain by sedimentary rocks which developed at the feet of the Appalachian Mountains during the Pennsylvanian and Permian periods, 300-270 million years ago. During this time, Prince Edward Island lay in the heart of the supercontinent Pangaea, and was subjected to a continental climate similar to that which prevails today in western Texas.

The youngest rocks exposed on the surface in the Maritimes comprise North Mountain, Nova Scotia. These basalts, of Triassic and Jurassic age, formed in a mid-ocean rift (similar to the modern Mid-Atlantic Rift) where the Bay of Fundy is today. Basalts poured out from submarine volcanoes, covering the ocean floor. The resulting volcanic rocks subsequently hardened into the North Mountain Basalt and uplifted to their present elevation.

Figure 9. Examples of Bedrock types

9a. Metamorphosed sedimentary rocks, Brigus South, NL. Photograph by Norm Catto.

9b. Volcanic Rhyolite, near St. Andrews, NB. Photograph by Norm Catto.

9c. Devonian granite, Black Brook, NS. Photograph by Norm Catto.

9d. Permian sandstone, Cap Egmont, PEI. Photograph by Norm Catto.

Glaciation

Atlantic Canada was glaciated several times during the Quaternary. The oldest Quaternary sediment is the Bridgewater till, exposed near Bridgewater NS. The age of this deposit is in excess of 300,000 years. Younger glacial sediments (150,000 years) are preserved in several locations.

The most recent glaciation occurred between 20,000 and 10,000 years ago. Glacial ice covered all of Atlantic Canada, with the exception of the highest mountain peaks in NL. The large Laurentide Glacier expanded from its source area in western Labrador to cover much of northeastern North America. Other glaciers developed in several centres on the island of Newfoundland, New Brunswick, mainland Nova Scotia, Cape Breton Island, and the area of the Gulf of St. Lawrence between Prince Edward Island and the Îles-de-la-Madeleine. The floor of the Gulf of St. Lawrence was exposed above the lowered sea level, as much of Earth's water was incorporated in the large glaciers. During the height of glaciation, sea level in the Gulf of St. Lawrence and the Grand Banks was as much as 100 m below the present position.

A succession of glacial advances covered the terrain. In south-central New Brunswick, glacial ice from the Gaspereau Centre expanded radially, including southward across the Bay of Fundy to cover the Annapolis Valley of Nova Scotia, transporting basaltic pebbles and boulders from North Mountain. The glaciers also expanded eastward, reaching the modern Gulf of St. Lawrence coastline at Kouchibouguac National Park. Cape Breton Highlands National Park was also covered by glacial ice.

Two ice sheets expanded to cover all of Prince Edward Island. Western King's County and Queen's County, including Stanhope and Rustico, was covered by glacial ice which advanced northwards from Nova Scotia, crossing Northumberland Strait. This ice advance can be recognized through landforms and striations, and also through the presence of boulders of older sedimentary and igneous rock derived from the Antigonish-New Glasgow area of Nova Scotia.

In Newfoundland, several independent glacial centres developed over the Avalon Peninsula, in central Newfoundland, the Northern Peninsula, and the Annieopsquotch Mountains. Ice expanded to cover the entire island, extended into the Gulf of St. Lawrence to merge with glaciers from the Maritimes and the Québec North Shore, and reached 50 km offshore in the North Atlantic Ocean.

Figure 10. Glacial Features

10a. Rounded solidified pieces of the Bridgewater glacial till are found on beaches in southwestern NS. Photograph by Norm Catto.

10b. Glacial till, ca. 20,000 years old, along Malpeque Bay, PEI, is subject to erosion. Photograph by Norm Catto.

10c. Alpine glaciation carved deep troughs and fjords in the Torngat Mountains of Labrador. Photograph by Norm Catto.

10d. Erosional roches moutonées, such as this one near Brigus, are common throughout Newfoundland. Photograph by Norm Catto.

The Laurentide glacier in Labrador expanded to the east and southeast, covering all of Labrador with the exception of the highest peaks of the Mealy and Torngat Mountains. Local glaciers developed in both mountain areas. Laurentide ice interacted with the glaciers from the Maritimes and Newfoundland along the southern margin.

Deglaciation began approximately 15,000 years ago, and most coastal areas were free of ice by 12,000 years ago. Although remnant glaciers persisted through the Younger Dryas cold event, they did not readvance, and final ablation of all the glaciers in Maritime Canada and on the island of Newfoundland occurred prior to 10,000 years ago. In Labrador, the large Laurentide glacier gradually retreated westward, and the final deglaciation of westernmost Labrador occurred about 7,000 years ago.

Sea Level Change

As the ice began to retreat, isostatically depressed areas were flooded by marine waters. Along the Gulf of St. Lawrence coast of New Brunswick, marine waters reached elevations up to 45 m above modern sea level. In western Prince Edward Island, the rise in sea level was sufficient to flood the land between Malpeque and Bedeque Bays, covering Summerside, and also flooded the low terrain between Egmont and Cascumpec Bays to the west (Carleton area). Prince County consisted of two islands at this time, one extending from North Cape to O'Leary, and a second surrounding Mount Pleasant and Wellington, in addition to small areas of land surrounding Kensington and Kinkora. Sea level reached 30 m above present in Terra Nova National Park, 75-100 m above present along the coastline at Gros Morne, approximately 150 m above present on the tip of the Northern Peninsula, and 135 m above present west of modern Lake Melville in Labrador (Fig. 11).

As the land recovered from removal of the weight of glacial ice, it began to rebound, and the marine waters receded. In most of coastal Labrador, this process is still occurring. Elsewhere in Atlantic Canada, however, sea level has been rising for the past 6,000-3,000 years. Currently, rates of sea level rise are approximately 1 mm/a at St. Anthony, 2 mm/a in northern New Brunswick and Gros Morne, and 3 mm/a in Nova Scotia, PEI, and southern Newfoundland.

Climate

Almost every community in Atlantic Canada has a variety of sayings to express the concept of changeable weather and climates that occasionally puzzle outsiders. The influence of the ocean on climate is a governing fact of Atlantic Canadian life.

Prevailing wind directions vary locally, and combined with ocean currents and local topographical effects, the result is to produce somewhat different climates in adjacent areas. Regionally, mid-boreal (Köppen-Geiger Dfb) climates prevail throughout the Maritimes and southern Newfoundland. Northern Boreal or taiga (Dfc) climates mark the Northern Peninsula and southern and central Labrador. Tundra (Et) climates are found in the Torngat Mountains, and on the highest summits of the Mealy Mountains.

The Cape Shore of the Avalon Peninsula, along the east coast of Placentia Bay, is an example of a mid-boreal climate in an exposed area. The Cape Shore is marked by relatively short, cool, and wet summers, with moderately mild, wet winters (January to March), long springs, and relatively short autumns. Daily mean temperatures in Placentia are approximately -3 °C in January and 17 °C in July. Mean annual precipitation varies with location between approximately 1,000 mm and 1,400 mm. In coastal areas, snow represents approximately 15-25% of the precipitation. Freezing drizzle is common, with an average of 35 hours per year. Easterly and southwesterly winds alternate during the summer, and southwesterlies dominate during the winter. Fog days average 206/year with at least 1 hour of fog, with a total annual sunshine of approximately 1,500 hours, one of the lowest totals in Canada.

The climate of the Annapolis Valley of Nova Scotia also can be classified as mid-boreal, although the sheltering influence of topography is evident. In valley areas, daily mean temperatures in January and July are -5 °C and 19 °C, respectively. The mean annual precipitation in the Annapolis Valleys is approximately 1,300 mm, of which 20% (250 cm) falls as snow with precipitation totals in the highland areas similar. Snow cover persists for approximately 95 days in wooded areas of the Annapolis Valley.

Although Prince Edward Island is generally considered to have a mild summer climate, the seasonal temperature values differ little from those of other southern boreal regions. At Malpeque, January and July mean temperatures are -7 °C and 17 °C, respectively. The area receives approximately 1,200 mm of precipitation in a typical year, of which 15-20% falls as snow. Snow cover persists for up to 120 days per year in wooded areas, but snow depths seldom exceed 60 cm. Temperature and precipitation in this region produces a humid moisture regime, with adequate moisture in the soil for plant growth throughout most of the growing season.

Figure 11. Evidence of changes in sea level is present throughout Atlantic Canada

11a. A raised marine beach at Trout River, NL, indicates that sea level was about 50 m higher than present 12,000 years ago. Photograph by Norm Catto.

11b. Marine clay at Springdale, NL, indicates sea level was 75 m higher than present 12,000 years ago. Photograph by Norm Catto.

11c. This granite pebble was dropped from an iceberg into marine clay approximately 10,000 years ago, near Gull Island Rapids, Churchill River, Labrador. Photograph by Norm Catto.

11d. Currently, sea level is rising in most of Atlantic Canada. This Black Spruce stump, 14C dated at 380 AD, was killed as sea level rose, contaminating its root system with salt water. Photograph by Norm Catto.

Western Labrador's climate has been described as having only three seasons. Winter is noteworthy, with a January daily mean temperature of -22 °C, and temperatures fall below -40 °C on several occasions during a typical winter. The prevailing west-northwesterly winds in western Labrador blow from the interior of the continent, creating dry, cold winters marked by calm days and minimal humidity. 40% of the precipitation occurs as snow. Summers are cool, with a July daily mean temperature of 13 °C in Labrador City.

In northern Labrador, the Torngat Mountains lie along a climate boundary. Coastal areas such as Nachvak Fjord have a rigorous northern boreal climate (Köppen-Geiger Dfc), modified only marginally in summer by the influence of the cold Labrador Current. Inland regions, higher summits above ± 900 m, and the terrain north of Seven Islands Bay to Cape Chidley are under the influence of a true arctic tundra (Et) climate.

North Atlantic Ocean

Geography and life in Atlantic Canada are strongly influenced by the North Atlantic Ocean. In common with the other ocean basins, the water of the northern North Atlantic circulates vigorously. The processes which drive this circulation include wind stress, generated by the prevailing winds; pressure gradients on the sea surface. In the northern North Atlantic, all of these are of importance.

Surface Currents

Surface currents in the northern North Atlantic are driven by the combination of prevailing winds (the westerlies), and pressure gradients on the surface. The net result is to establish a gyre system, involving clockwise rotation (deflection to the right) in the latitudes south of 60ºN. The Gulf Stream is the major current associated with this gyre system, and transports approximately 55 million m^3/s of water northeastward at 10-15 km/h off the continental shelf of North America to the Grand Banks, and across the Atlantic to southern Ireland.

The Gulf Stream does not flow in a uniformly straight line. As all currents do, it meanders, producing loops along its margins. At Cape Hatteras, North Carolina, the Gulf Stream is deflected seaward, and follows the continental shelf/continental slope break to the Grand Banks.

Along the coastline of North America, a cold counter-current runs to the south from the Gulf of St. Lawrence to Cape Hatteras. This cold current acts to cool the shorelines of Nova Scotia, and allows upwelling to periodically occur in offshore shallow areas such as the Scotian Shelf and Georges Bank. Nutrients brought to the surface in these areas promote phytoplankton growth, which in turn provides suitable environments for fish and shellfish.

As the Gulf Stream proceeds across the North Atlantic, it branches into several currents. Although most of the water flows to the southeast, as the Canaries Current and the currents encircling the Sargasso Sea, approximately 10 million m^3/s of water flows northeast along the western coasts of Ireland, Scotland, and Norway. This current, referred to as the North Atlantic Drift off Ireland and Scotland, and as the Norwegian Current further north, forms one component of the northern loop of the gyre system, marked by counter-clockwise flow (deflection to the left). The waters of the Norwegian Current gradually are diverted westward along the margin of the Arctic Ocean pack ice limit, flowing along the Iceland and Kalaallit Nunaat (Greenland) coasts. Augmented by cold, fresh water from the glaciers, the flow returns southward along the Labrador coast as the Labrador Current.

The Labrador Current transports approximately 6 million m^3/s of water southwards at a velocity of 1-2 km/h (measured offshore of Belle Isle), making it among the slowest currents in the ocean. The peak velocity is usually reached in late summer.

Tides

In Atlantic Canada, the semi-diurnal tide progresses southwestward, requiring approximately 4 hours to travel from Cape St. Francis to Yarmouth (Fig. 13, 14). Once the tide enters a restricted embayment, such as the Bay of Fundy, its progression is slowed as the water begins to interact frictionally with the bay floor and the coastline. In the funnel-like Bay of Fundy, the effect is to increase the range of the tide (from approximately 6 m at the bay mouth to 12-14 m in the Minas Basin), while simultaneously slowing its velocity. In macrotidal areas, such as the Bay of Fundy, the high tidal range produces a characteristic suite of geomorphic features.

Waves

Along the Northern North Atlantic, wave activity is responsible for shaping the majority of the coastal landforms. Although bedrock features are largely the products of pre-existing geology and climate-induced frost weathering, wave action accounts for the majority of sedimentary landforms and contributes substantially to coastal erosion of unconsolidated cliffs.

A wind blowing at 5.1 m/s (10 nautical miles per hour, or 10 knots) can theoretically produce waves 2.2 m high with a velocity of 8.6 m/s, but the wind velocity must be maintained for at least 11 hours and the fetch must be at least 129 km. For gale-force winds at 40 knots to produce waves of 25.8 m height and 28 m/s velocity, they must operate constantly for 65 hours over a minimum fetch of 2,590 km. As waves of this height have been recorded by ship's captains off the coast of Newfoundland, these theoretical conditions can be met, but they are relatively rare. In most cases, the heights and wave velocities actually produced are far less than the theoretical maxima, although still impressive to coastal observers.

Historical

Understanding of the basic events behind the settlement of Atlantic Canada is essential to a study of the modern geography of the region. Although European visitation began when Norse Vikings first settled at L'Anse-aux-Meadows at the tip of the Northern Peninsula in approximately 1000 AD, and Portuguese and Basque fishermen and whalers utilized the coastal areas of Newfoundland, southern Labrador, and Cape Breton, the shaping of settlement patterns throughout the region largely occurred between the time of the first arrival of European settlers in the early 1600s and the mid-1800s.

Maritime Canada is the traditional territory of the Mi'kmaq peoples. Mi'kmaq lands, or Mi'kma'ki, stretch across eastern and coastal areas of New Brunswick, Nova Scotia, and Prince Edward Island. Early peoples made their home along the coast and riverbanks and included the Mi'kmaq but also the Wolastoqiyik or Maliseet people of western New Brunswick and the Saint John River valley; and the Passamaquoddy in southwest New Brunswick.

The Aboriginal population of what is now Labrador included the Inuit and Innu (and later the Metis). The Beothuk people occupied the island portion of the province until they were killed, lost to disease or starvation, integrated with Mi'kmaw[1] communities and/or displaced from their territories during the first two centuries of permanent European settlement (Pastore, 1997; Baker, 2003), finally disappearing in the early 1820s. Mi'kmaq peoples also occupied the southern and central areas of the island, serving as middlemen in the fur trade and allies to the French in the 17th and 18th centuries.

1 According to Native Council of Nova Scotia (2007) "Mi'kmaw" is the singular of Mi'kmaq and an adjective when it precedes a noun (e.g. Mi'kmaw people or Mi'kmaw treaties).

Figure 12. Semi-diurnal tides in Atlantic Canada

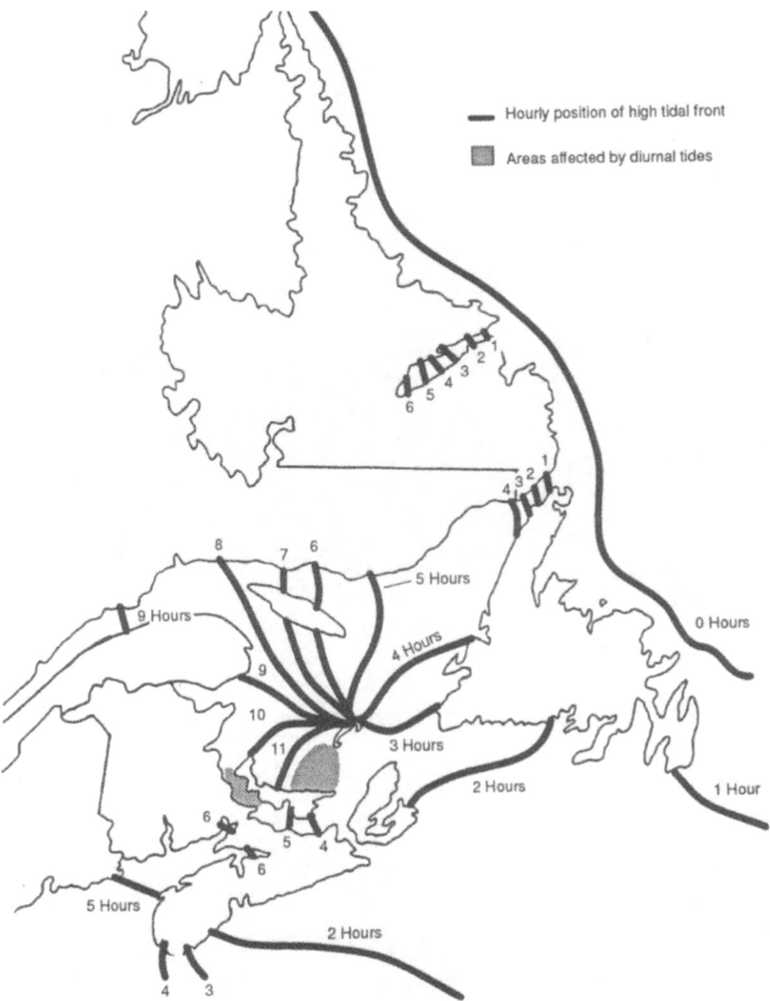

Map redrawn by Norm Catto and DELT, Memorial University.

Figure 13. Tidal range in the Bay of Fundy varies from 6 m to more than 14 m

13a. Tidal Channel exposed at low tide, Port Williams NS. Photograph by Norm Catto.

13b. Incoming high tide at Annapolis Royal, NS. Photograph by Norm Catto.

13c. Conglomerate eroded by tidal action, Hopewell Rocks Provincial Park, NB. Photograph by Norm Catto.

13d. Grounded fishing vessels awaiting high tide, Digby Neck, NS. Photograph by Norm Catto.

The settlement history of the Maritimes and Newfoundland is linked both to the power struggle between Britain and France, and socio-economic conditions in the European countries (including Ireland). After the initial voyages of John Cabot (1497-1498), Gaspar Corte-Real (1500), and Jacques Cartier (1534), the first attempt at settlement was that of Samuel-de-Champlain, in 1604. Champlain established his first colony, with some military but no geographical logic, on Ile-St-Croix, now in Maine, on the St-Croix River along the New Brunswick border. The establishment of a settlement on a small island with limited resources proved to be a mistake, and what was left of the group under the leadership of the Sieur-de-Monts moved to the Nova Scotia mainland at Port-Royal (near Annapolis Royal) in the autumn of 1604. Although de Monts tried to keep the colony viable, social gestures such as the "Ordre de bon temps" (Order of Good Cheer) could not disguise the increasing death toll among the hapless, unprepared colonists. The winter of 1604-1605 was not particularly brutal, but it was more severe than the settlers from temperate coastal France were used to or could cope with. As soon as possible, the ragged colonists retreated to France. When Champlain returned to establish a new colony, he chose the site of Québec City, and the Maritimes were temporarily bypassed.

The first settlements in Newfoundland also encountered difficulties, as colonists unfamiliar with the climate attempted to cope at the first settlement at Cupids on Conception Bay (1609), along the open Atlantic shoreline south of St. John's. Colonization of the southern part of the Avalon Peninsula began with Sir William Vaughan, a Welsh courtier and poet who decided that Newfoundland represented the solution to economic problems and overcrowding in Wales. Vaughan acquired the Ferryland-Renews area in 1616, and named it "New Cambriol" (from Cambria, the Roman name for Wales). The choice of the Ferryland-Renews area was based on knowledge gleaned from Basque and French fishermen, resulting in place names with Basque roots such as Aquaforte (Ria de Aguea, fresh water) and Fermeuse (Ria Formoso, beautiful river).

In 1617, Vaughan sent a group of colonists to Aquaforte, where they spent the winter in summer huts built by seasonal fishermen. The survivors were rescued the following summer and moved south to Renews. After a further winter at Renews and an attack by English pirates, the colony was abandoned. Vaughan retained ownership of the Renews-Cappaheden area, but sold Fermeuse and Ferryland. George Calvert (Lord Baltimore) bought the Ferryland area from Vaughan in 1620, and sent out a group of colonists in 1621.

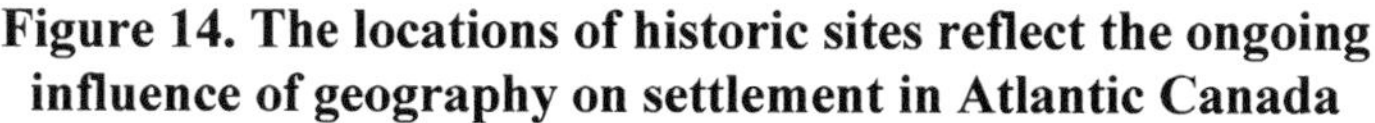

Figure 14. The locations of historic sites reflect the ongoing influence of geography on settlement in Atlantic Canada

14a. L'Anse-aux-Meadows, NL, was the original site colonized by Europeans in Atlantic Canada. Photograph by Gina Noordhof.

14b. The reconstruction of the French settlement at Port-Royal, NS, is located north of Annapolis Royal. Photograph by Norm Catto.

The settings of Cupids (above, 14c) and Ferryland (below, 14d) indicate the importance of the fishery in the establishment of communities in Newfoundland. Photographs by Norm Catto.

The conflict between the French community of Plaisance (now Placentia; above, 14e) and the English community of St. John's (below, 14f) was only resolved by the Treaty of Utrecht in 1713. Photographs by Norm Catto.

Although initially Protestant, Calvert converted to Catholicism, which made him more sympathetic to the idea of settling English and Irish Catholics. Unlike Vaughan, Calvert visited his colony, arriving in 1627 to find a viable group of 100 settlers. However, after the winter of 1628-1629, Calvert and his family decided that they did not wish to stay, and left to found a new colony in Maryland. About 30 of the original colonists remained.

After Calvert's abandonment of the colony, governmental and commercial monopoly rights were granted to David Kirke, who had defeated Champlain in 1629 and deported all of the residents of Québec to France. When Québec was handed back in the peace treaty of 1632, Kirke petitioned for compensation and was awarded the Colony of Avalon. Kirke established his headquarters at Ferryland and remained there through 1651. He established the first fish merchant business located in Newfoundland, setting the pattern for merchant-harvester relations that would persist for almost 300 years. At that time, Ferryland was the largest centre in Newfoundland, with established trading links to ports in England, Ireland, and Massachusetts. Kirke was deposed as governor in 1651 but continued his business from Ferryland. Gradually, other English settlements were established throughout eastern Newfoundland, including St. John's, Harbour Grace, Trinity, and Bonavista.

Placentia (Plaisance), the largest French settlement in Newfoundland, was initially used by the Basques ca. 1540, and was founded as a French community ca. 1660. Unlike the English colonies, Plaisance was always intended primarily as a military base, rather than a fishing port or colonial venture. The inhabitants, however, persisted in conducting themselves in non-military ways, including fishing and commercial occupations, and even traded with the 'enemy' in St. John's and New England. Although French vessels licensed by the king of France were supposed to have a monopoly on trade, Plaisance merchants traded primarily with Québec and Boston. New England merchants visited several times annually, bringing less expensive and higher quality goods than the French monopolists, and receiving fish in return. Warfare, triggered by events in Europe, periodically disrupted matters: in the winter of 1696-7, the Sieur d'Iberville led a series of overland attacks from Placentia, destroying all the English settlements with the exception of Bonavista.

Settlement from France gradually filtered into the Maritimes through the 1600s as Nouveau France became more established. The major centres of settlement developed around Port-Royal in the Annapolis Valley; Grand Pré, and the Avon River area in central Nova Scotia; Beausejour along the Nova Scotia – New Brunswick border, in the swampy Tantamar Marshes; and on southwestern Prince Edward Island (then known as Ile-St-Jean). French merchant Nicolas Denys established Cape Breton's first permanent European settlement in the mid-1600s. Resource utilization proceeded rapidly, as the colonists consulted, studied, and inter-married with the Mi'kmaq. Despite inter-marriage, population growth was slow, and by 1713 there were probably fewer than 2,000 agricultural colonists in the area. Around this time, the name "Acadie" or "Acadia" came into use to describe the region.

The 1713 Treaty of Utrecht confirmed English possession of the Avalon Peninsula. France was required to abandon Placentia and relocate to the newly established Louisbourg in Cape Breton. Although residents were given the option of remaining and declaring allegiance to the British, only four men did so, all of whom were deeply involved in the trade with New England. The site was garrisoned by the English, but the majority of settlers were Irish Catholics involved in the fishery. English merchants preferred Irish Catholics because they were less likely to leave once landed, had few political options or abilities to cause difficulties. France had lost the fishing infrastructure necessary to pursue the Grand Banks cod stocks, but retained rights to use coastal facilities on the Gulf of St. Lawrence coast of Newfoundland. This "French Shore" arrangement lasted until 1904.

Acadia was never regarded as a significant colonial possession by the French government, and was largely left outside of the political endeavours of Québec. As a result, Acadia had poorly developed military infrastructure, and developed as a series of agricultural communes with little communication with or interest in the outside world. In contrast to Québec, no concerted effort was made by the French government to establish colonies or promote emigration to Acadia. As a result, France was willing to cede formal possession of southern Acadia (mainland Nova Scotia and Prince Edward Island) to Britain in 1713 but retained Cape Breton and fortress Louisbourg, primarily as security for ship access to the St. Lawrence River. French-speaking residents were permitted to remain in southern Acadia, although some migrated to New Brunswick. An undefined boundary developed across the Tantamar Marshes separating the British area (Fort Lawrence) from the French-speaking Acadian area to the north (Fort Beausejour).

When French-British hostilities seemed imminent once again, in 1755, British officers ordered the deportations of Acadians from Grand Pré, Annapolis Royal, and other localities throughout the Annapolis and Avon Valleys. Many Acadians left previously settled areas to establish new homes in more isolated, hopefully less militarily-interesting regions, such as the Gulf of St. Lawrence coastline of New Brunswick, the St. Mary's Bay area of Nova Scotia, Iles-de-la-Madeleine (Magdalen Islands), the isolated Cheticamp shore of western Cape Breton Island, the northwestern shores of Prince Edward Island, and the Port-au-Port Peninsula in westernmost Newfoundland. The deportations succeeded in eliminating the Acadian population of the Annapolis-Avon heartland, but also in spreading it into other regions of Atlantic Canada.

Although war did ensue, in the immediate aftermath little changed. Halifax and Annapolis Royal remained the only significant British settlements in the former Acadia, apart from each other psychologically and economically. French-speaking Acadian communities continued to develop through their own resources and natural increase throughout the Gulf of St. Lawrence coastal areas, where the nearshore fishery was adequate and the farming conditions good. The rockbound Atlantic coast of Nova Scotia remained virtually unsettled. Louisbourg was destroyed as a fortress and garrison, but fish harvesters remained throughout southern Cape Breton and at Cheticamp, which remains largely inhabited by Acadian descendants today.

For several years after its possession was confirmed by the 1763 Treaty of Paris, Britain did little towards settling its newly acquired domain of Acadia. Interest was finally sparked by the necessity for re-locating the Scottish population displaced by the agricultural clearances in the Highlands, difficulties resulting from local problems in Ireland, the problems posed by over-population in the Hanoverian and Hessian areas of Germany, and the requirement to resettle and compensate those Americans who had remained loyal to George III throughout the American Revolution. All of these populations had to be accommodated in British territory, and Nova Scotia, New Brunswick, and PEI lay largely vacant and waiting.

Most immigrants came from impoverished regions – Highland Scots, Irish, West County English, Cumberlandshiremen. Scottish settlement was concentrated in Cape Breton Island's central Bras d'Or valley and along the 'Gaelic shore' north of Sydney, and in the Pictou and Antigonish areas of Nova Scotia. St. John's Island (as Ile-St-Jean had been renamed after 1763) was populated predominantly by Irish immigrants, with west county English settlements in the western parts. Governor Lawrence, in addition to encouraging Irish settlement, also introduced potatoes for the first time to Atlantic Canada. Settlement proceeded much more slowly on St. John's Island than in mainland Nova Scotia, due in part to the expectations and speculative activities of absentee landlords, and in part due to the absence of economic opportunities other than farming. In 1796, St. John's Island was re-named "Prince Edward Island", and was detached politically from Nova Scotia to form a separate colony.

Figure 15a. The Clock Tower, constructed 1803, symbolized the British presence in Halifax

Photograph by Norm Catto.

Figure 15b. Acadian settlers displaced from Nova Scotia migrated to form new communities in Atlantic Canada, such as Mont-Carmel in western PEI

Photograph by Norm Catto.

After the American Revolution "United Empire Loyalists" settled throughout the coastal regions of Atlantic Canada (except along the Acadian Coast of New Brunswick), but were concentrated in Nova Scotia, in Charlotte County and along the Saint John River Valley in New Brunswick. The result was to establish a pattern in New Brunswick with distinctly Acadian communities along the Gulf of St. Lawrence coastline from Cap-Pele north to Baie-des-Chaleurs, and inland to the Madawaska area north of Grand Falls-Grand Sault. English-speaking communities developed along the Saint John River valley from Saint John north to Grand Falls-Grand Sault, west to Charlotte County, and northeast along the Kennebacasis River to Sussex and Moncton. Even today, a NW-SE line drawn through Grand Falls-Grand Sault (the only Canadian community with an official bilingual name) and Moncton largely separates predominantly English-speaking communities to the south and west from French-speaking communities to the north and east.

After 1800, relatively few new settlers arrived in Maritime Canada. The ethnic makeup of the region was essentially complete. Small groups of Irish continued to arrive throughout the nineteenth century, but most victims of the potato famine chose Ontario or the United States. Scots continued to arrive until ca. 1840, as clearances proceeded. Many Cape Breton residents are descended from Highlands and Islands Scots that arrived in the 1800s, but again most chose to head further to the west or south. No campaigns were undertaken after 1800 to entice immigrants to the Maritimes, and no organized 'companies' or associations sprung up comparable to those that operated in Ontario and the United States, and later in Western Canada. Virtually all of those who came in this phase were mercantilist or 'service sector' professionals, rather than farmers or fish harvesters. Immigration of African-Americans, both directly from Africa and from the United States, occurred prior to the US Civil War, but the population was largely concentrated in the Halifax area. This history of ethnic development left its mark on Maritime Canada. Residents developed a strong sense of local identity, often not extending beyond the local community. They also developed a sense of xenophobia in many instances, coupled with and fuelled by feelings of isolation and a sense of being 'bypassed'. Traditions were maintained strongly, and attitudes hardened without the arrival of new people with different customs and ideas.

Regarded as a source of fish by merchants based in England and as a suitable area for training sailors by the British Royal Navy, permanent settlement was initially discouraged (e.g., by banning women from accompanying the seasonal fishing expeditions). However, populations gradually became established in fishing communities along the Avalon and Bonavista Peninsulas, subsequently spreading to the Burin Peninsula, the southern coast, and northeastern Newfoundland. Immigration in a formal sense was never encouraged by Britain, and the limited agricultural and manufacturing potential together with the absence of large land grants to absentee landlords as seen in the Maritimes, acted as substantial deterrents throughout the 1800s.

Most arrivals to Newfoundland between 1600 and 1800 came from three regions: southern Ireland; west-country England; and the Channel Islands of Jersey and Guernsey. Generally, communities developed along separate ethic-religious lines: most of the Irish migrants were Catholic, whereas the English and "Jerseymen" tended to be Protestant. Irish-dominated communities developed in the southern Avalon Peninsula, along Placentia and St. Mary's Bay, and the Atlantic shoreline. West-country English communities dominated northern Conception Bay, Trinity Bay, and northeast Newfoundland. "Jerseymen" formed communities in central Conception Bay, and through fishing operations also established communities along the Strait of Belle Isle and southern Labrador. In communities with both religious groups represented, including St. John's, distinctive Irish and English settlements developed, with considerable antagonism a recurring feature. Throughout the 1800s, the birthrate was higher among the Irish descendants, but the English groups continued to hold a dominance of political and economic power, assisted by government officials sent from Britain. Religious divisions remained significant well into the 20th century: it was only in 1995 that the various religious school boards throughout Newfoundland and Labrador were amalgamated into a single publically-funded and administered school system.

In addition to the ethic-religious division, a socio-economic division developed between the larger communities, particularly St. John's, and the smaller fishing outports after 1800. Merchants based in St. John's exerted substantial economic control over fishermen in the outports, as the merchants controlled both the prices they paid for cod and the prices they charged for fishing gear, food, and other necessary supplies. The manifestations of this "truck system" resulted in fishermen becoming progressively indebted to the merchants over time, creating a cycle of economic control. As the socio-economic situations of St. John's ("town") and the outports ("bay") diverged, resentment and misunderstandings between the "townees" and "baymen" grew. Echoes of this distinction have persisted: a substantial number of St. John's residents voted against joining the Canadian Confederation in 1949, while sentiment was more in favour in rural communities. The ongoing economic impacts of petroleum exploitation have also benefited St. John's and the surrounding suburban communities, while the outport communities have seen fewer direct benefits.

Figure 16. Ramea, in western Newfoundland, is an active "outport" fishing community

Photograph by Norm Catto.

Contrasting Urban Landscapes: Halifax and St. John's

The extent at which urban development has taken and continues to take place throughout each of the Atlantic Provinces varies, depending on both physical and social characteristics such as topography, available resources, historical settlement, economics, and current migration patterns. Patterns of urban development vary both in place and time throughout the urban landscape of Atlantic Canada. One way to exemplify the distinct and colourful landscape within the Atlantic region is to focus on two of Canada's oldest capital cities: Halifax, NS and St. John's, NL.

Figure 17. St. John's (top), and Halifax (bottom)

17a. Photograph by Norm Catto.

17b. Source: Natural Resources Canada.

The cities were founded with different purposes and motives. St. John's was developed as a base for the cod fishery, beginning in the early 1600s. Located on the easternmost harbour of Newfoundland, St. John's is an (almost) ice-free port, sheltered from the North Atlantic but with good access to the fishing grounds. It also served as a convenient trans-shipment port for repackaging fish sent from the outports, and goods sent from Europe and the eastern USA. Although there was, and still is, a military presence in the city, St. John's was never developed as a major base. The layout of the downtown streets, with the main streets paralleling the harbour apron and following topographic benches, developed in response to the practical needs of the fisheries merchants and associated businesses. Cross-streets are short and steep: horses and people could not negotiate long, steep inclines.

Halifax was developed primarily as a Royal Navy base in its formative years (ca. 1740), a counter to the French Fortress Louisbourg. Its position was seen as part of the overall eastern security net for British North America, protecting the colonies and fishery trade of New England and serving as a staging area for amphibious assaults on Louisbourg and Québec. The grid pattern evident in downtown Halifax, below the military fortification of Citadel Hill, initially was a product of British military surveyors. Pre-existing communities, such as Boston and St. John's, are not built on grids. The contrast is especially evident between Halifax and Boston, as both communities are built in almost identical topographical settings, on the flanks of steeply sloping drumlin hills of glacial origin. Towns in Atlantic Canada which either had a significant British military presence at their founding or were laid out by retired British military engineers are marked by grid systems, including Halifax, Saint John (NB), Charlottetown, and Fredericton. Towns without an initial dominant military presence, such as St. John's, Moncton, and Sydney, are marked by street patterns that developed in accordance with the dominant economic activities (fishing, railroad development, and mining, respectively).

Although civil government has come to be a major component of the economies of all Atlantic capitals, unlike Fredericton (established as a governmental counterweight to the mercantile-commercial community of Saint John) or Charlottetown, neither Halifax nor St. John's was initially established as a capital city. The original governmental buildings (NS Legislature and Colonial Building, respectively) were positioned at the fringes of the downtown areas as they existed, rather than in the centre of the urban area (as at Charlottetown, Fredericton, and Ottawa). When the Colonial Building was replaced after Confederation, the new Confederation Building was constructed in an analogous position, on the fringe of the urban area ca. 1955.

The history of urban re-development shows marked differences between the cities. St. John's suffered from a sequence of fires throughout the 1800s, culminating in the Great Fire of July 8th, 1892. Unfortunately at the time the city did not have an organized fire department, and the fire spread throughout the city, killing three people and leaving 11,000 people homeless. The estimated property damage was $350 million (2009 equivalent), with less than 40% covered by insurance. After much deliberation, the Newfoundland government and the St. John's Municipal Council agreed on a rebuilding scheme for the city. The proposed changes included widening and straightening of many downtown streets. However, government agencies decided against restructuring the city on a grid pattern, as such an undertaking would require a significant amount of time and money.

With much of its residential and commercial structures demolished, the reconstruction era became the foundation for many of the present day registered heritage structures that were either built or re-built after 1892. The distinctive attached row houses of downtown St. John's are a legacy of the Great Fire. One of the most notable reconstruction architects was John Thomas Southcott. Well known for his Second Empire-styled buildings that had distinctive mansard roofs and bonnet-topped dormers protruding from the concave-curved roof surface, Southcott's works received recognition as being the "Southcott style" contributing to the annual Newfoundland Historic Trust presentation of the Southcott Award for excellence in the restoration of heritage structures.

St. John's did not develop as a significant outlet for North American trade, and the population remained relatively small and dominated by the traditional merchant families until Confederation in 1949. Consequently, no substantial redevelopment of the downtown and harbour area occurred, and the pattern of relatively low buildings, attached structures, and interspersed residential and commercial uses in the downtown (with business premises at street level and residential accommodation on upper floors) persisted. Beginning in the 1960s, pressure grew for downtown redevelopment, with the construction of individual office towers and large hotels. However, these developments were opposed by many residents, and the downtown core did not develop into the typical high-rise dominated pattern characteristic of most large North American cities.

Figure 18. Downtown St. John's

Distinctive mansard-style roofs, bonnet-topped dormers, and bay windows characterize attached row houses on Prescott St. (18a, above) and Gower St. (18b, below). Photographs by Norm Catto.

Figure 18. Downtown St. John's (cont'd)

Much of the downtown shopping area along Water St. is characterized by attached 3-storey flat-roofed buildings, with the ground floors used for commercial premises and the upper stories for residential accommodation (18c, above). Below (18d) the Anglican Cathedral, like all buildings in the downtown, was reconstructed after the Great Fire of 1892. Photographs by Norm Catto.

No substantial urban renewal efforts have been undertaken since 1949. The spread of single-family detached houses to suburban communities has occurred, as elsewhere in Atlantic Canada, but very few high-density complexes, apartment blocks, and condominiums have been constructed.

Halifax, in contrast, developed a commercial centre along its harbour, based on Trans-Atlantic trade, initially in timber for ship construction. The requirements of the Royal Navy initially limited expansion of the commercial dockyards, but additional facilities gradually were developed to the north and west of the original port, and across the harbour in Dartmouth to the east. Throughout the 1840s, it became evident to many residents that future prosperity depended upon modernization, on transportation, and on communication. In Nova Scotia, Halifax, as the best harbour with the best infrastructure, was the focus of the railway system. Lines were to be built radiating outwards and inland from Halifax; there was no point in building a line from Halifax to Lunenburg, from wharf to wharf, because goods traveling that way would go by sea. Although the relative costs of rail and water shipping varied from time to time and place to place, water transport always dominated rail transport in cost terms. Shipping goods by rail was between 5 and 20 times as expensive as shipping them by water. These economics still hold; grain is shipped from Manitoba to Toronto most economically by loading it onto lake freighters in Thunder Bay. The further inland a port is, the better are the economics. If Montréal was ice-free throughout the year, Halifax would suffer a substantial drop in business. There were no substantial topographical obstacles in southern Nova Scotia, and so the lines could go wherever one chose. Downtown Halifax thus began to develop as a Trans-Atlantic port city, and the harbour area and adjacent Central Business District was under constantly ongoing redevelopment.

Although fires periodically occurred in Halifax, the dominant event triggering redevelopment was the Halifax Explosion of December 6, 1917. The explosion resulted when the Norwegian vessel SS Imo collided with the French freighter SS Mont Blanc, which was carrying 2,500 tonnes of explosives intended for the battlefields. The resulting explosion, the largest human-created blast prior to the development of the atomic bomb, killed almost 2,000 people, injured more than 9,000, and destroyed 2 km^2 of the city directly north of the downtown. An initial financial contribution of $350 million (2009 equivalent) made by Canadian, British and other agencies initiated re-development of the downtown and adjacent area.

As a result of the effects of the explosion and ongoing commercial growth, downtown Halifax developed a Central Business District with high-rise structures, more closely resembling other Canadian urban centres than the other cities in Atlantic Canada, such as St. John's. Immigration throughout the 1800s and early 1900s contributed to the establishment of distinctive ethnically-diverse neighbourhoods. Differences in socio-economic status and problems of marginalization caused difficulties on occasions, notably in the mid-1960s when the predominantly African-Canadian neighbourhood of Africville was demolished to allow commercial waterfront development and bridge construction, with the former residents displaced elsewhere in Halifax. The periodic efforts at development and urban renewal have thus produced a greater mixture of architectural styles in the urban core of Halifax than is the case in St. John's.

The two cities thus differ substantially from each other, but are also constructionally and culturally distinct from other Canadian urban areas. Both cities have continued to grow in population and economic importance relative to their provinces. More than 40% of the population of Nova Scotia lives within the newly expanded.

Halifax Regional Municipality (HRM), and about one-third of the population of Newfoundland and Labrador lives on the northeast Avalon Peninsula (including St. John's and directly contiguous communities). Each metropolitan area contributes approximately half of the provincial GDP.

Throughout rural Atlantic Canada, out-migration is the dominant form of population movement. However, those that migrate from outside Atlantic Canada to HRM and the northeast Avalon appear to be primarily composed of those in 'creative' (knowledge based professionals) and service sector jobs. Both cities exceed national averages in terms of percent employed in the 'creative' sectors of the economy and in measures of 'bohemia' (e.g. arts, music). Substantial migration from rural areas of Atlantic Canada to both cities is also occurring, with the largest growth concentrated in suburban communities.

Population

In 2009, the four Atlantic provinces had a combined population of 2,337,600 residents or 6.9% of all Canadians (up from 2,257,555 in 2006 census but growing at a slower pace than the remainder of the country). The region includes both the most sparsely populated (NL) and most densely populated (PEI) provinces in Canada.

Figure 19. Halifax

19a (above) Downtown Halifax includes a revitalized port area, with refurbished limestone block warehouses flanking the harbour and high-rise buildings along Barrington Street. 19b (below) Industrial and port services are concentrated in Dartmouth (shown here) and the Port of Halifax, south of the downtown. Photographs by Norm Catto.

Table 1. Demographic parameters including migration and age distribution statistics for the four Atlantic provinces

	NB	NL	NS	PEI	Canada
Pop. 2009-07-01	749,500	508,900	938,200	141,000	33,7 M
% change July 2008-July 2009	+0.3%	+0.5%	+0.2%	+1.1%	+1.2%
Pop. density (km^2)	10.3	1.3*	17.0	24.4	3.2
% Urban (2001)	50%	58%	56%	45%	80%
% in largest CMA	17.6% Moncton	36.5% St. John's	42.1% Halifax	41.1% Charlotte-town	15.1% Toronto
% Ages 0-14	15.3%	14.9%	15.0%	16.3%	16.6%
% Ages 15-64	69.2%	70.4%	69.2%	68.1%	69.5%
% Ages 65+	15.5%	14.7%	15.8%	15.6%	13.9%

*Newfoundland 4.4 persons/km^2; Labrador 0.1 persons/km^2.

Atlantic Canada is a relatively homogeneous society, with only 3% of the population of Aboriginal identity and 3.8% of immigrant origin (vs. 3.8% and 19.6% respectively in Canada). NL has the smallest proportion of its population made up of immigrants (1.7%) and NS the highest (5%). As in other provinces, the immigrant population is concentrated in urban centres, although rural communities have begun to look to immigration as a possible solution to shrinking populations.

The population in NB is dominantly concentrated in two belts. One major concentration of population extends along the Saint John River, including the capital of Fredericton (86,000 in the metropolitan area); the largest port in the province, Saint John (126,000 in the metropolitan area); and the northern forestry-based communities of Grand Falls-Grand Sault (5,600) and Edmundston (16,600). The other major locus of population is located adjacent to the Gulf of St. Lawrence, including the cities of Moncton (132,000 in the census metropolitan area), Miramichi (18,000; formerly the separate communities of Chatham and Newcastle), Bathurst (13,000), and Campbellton. New Brunswick is the only province in Canada where bilingual services are routinely available in all large communities.

The population in NL is dominantly concentrated in the northeast Avalon Peninsula, within the St. John's CMA (186,000 in 2008). Other major centres on the island include Corner Brook (26,600), Grand Falls-Windsor (13,600), Gander (10,000), and Stephenville (6,600). Closure of pulp-and-paper mills in Grand Falls-Windsor and Stephenville had adverse effects. Growth in the northeast Avalon (St. John's CMA), both in population and economic activity, stands in contrast to the remainder of the province. Only 5 of the 20 regional economic zones are projected to increase in population by 2019.

In 2007, Labrador had a total population of 26,890 people. Overall, the population has declined by approximately 5% since 2000. Spread over a land area of more than 290,000 km^2, the population density of 0.1/km^2 is among the lowest in Canada. The largest centres of population are Happy Valley-Goose Bay (7,600), Labrador City (7,230), Wabush (1,755), and Nain (1,020) according to the December 2009 estimates of the NL Government. North West River and Sheshatshiu had a combined population of approximately 1,530. No other centre has a population in excess of 800 people.

In Nova Scotia, the Halifax Census Metropolitan area's population of 395,000 represents more than 42% of the provincial total. The Cape Breton Census Agglomeration, which includes Sydney (106,000), Truro (45,000), Kentville (27,000), Yarmouth (26,000), Amherst (9,500), Digby (8,000), and Antigonish (4,200) are other major centres of population. As in Newfoundland, most of the population is concentrated along the coastlines. The rapid growth rates in metropolitan Halifax, as well as communities within commuting distance (including Kentville and Truro), contrasts with declines in Cape Breton and elsewhere in rural NS.

PEI has the highest population density of the Atlantic Provinces, and the lowest proportion in urban areas. Charlottetown and environs account for ~58,000 of the island's residents. Summerside (16,200) is the province's only other city; no other community has more than 2,000 people.

The Economy

Foundations

As in other regions of Canada, early settlers were attracted to Atlantic Canada by the lands, waters and resources that had long supported the area's original peoples. Newfoundland's main economic role in its early Trans-Atlantic history was the production of fish. Crews from Portugal, Spain, France and Britain travelled to the island to fish along its coasts during the summer months, returning to Europe in the fall. Increased competition and the growing international trade in salt cod encouraged the transition to a more sedentary fishery by the early 1800s, with early settlement patterns shaped by access to fisheries resources and shore space for fishing facilities. Lack of good soil and a relatively short growing season worked against agricultural development, but early settlers grew root vegetables, kept livestock, hunted and trapped, harvested wild berries and cut wood to supplement their livelihoods and goods supplied on credit by fish merchants. Today, self-provisioning activities such as hunting, fishing, wood-cutting, barter, and unrecorded cash exchanges comprise a significant component of incomes in some rural areas.

As the Maritime provinces developed in the early 1800s, they appeared to be headed for relative prosperity, comparable to that evident in New England. The timber industry was booming as a result of the increased demand for shipbuilding materials during the Napoleonic Wars. Agricultural settlement spread up the Saint John River, throughout coastal Nova Scotia, and along Prince Edward Island. Less than thirty years later, the prospects looked different. The agricultural industries were in decline. Many farmers had left for Ontario and Ohio. Most of the large white pines had been logged. Prince Edward Island's ship building industry was collapsing, New Brunswick's economy weak, and Nova Scotia's tied to the garrison at Halifax. To this day, much of the region has failed to regain the advantage it once held over Ontario and Québec.

This early period reflects a number of deep-seated challenges that continue to exist in what is chronically Canada's lowest performing regional economy. These include: a legacy of resource dependency; heartland-hinterland relationships, first with Europe and later with central Canada, the eastern USA, and the provinces' own urban centres; significant demands for infrastructure investment; and a relatively sparsely populated settlement pattern with lower levels of education. Capital city specialization has also played a role. For instance, Halifax was located on the best harbour in the Maritimes, and should have developed rapidly as a focus for commercial business. Its role as the major British fleet base on the western North Atlantic seaboard, however, meant that waterfront land was required for military purposes, and led the town's merchants to think in terms of military supply and defense rather than exports and industrial development.

Numerous regional development programs and policies have been put in place in an attempt to address resulting regional inequities, including the introduction of the equalization program in 1957 to ensure access to reasonably similar levels of provincial government services and levels of taxation across the country. These measures have created dependency not only on natural resources but also on government programs and transfers.

Today

Natural resource industries remain important for many Atlantic Canadians and their communities, particularly in rural areas, and reliance on these sectors exceeds the Canadian average. Primary sectors accounted for 7% of Canada's GDP in 2008 but 13% in Atlantic Canada (see Table 2). In terms of employment, primary resource extraction also ranks higher in all Atlantic provinces than in Canada as a whole.

The economy of Atlantic Canada is dominated by service industries. Despite high rates of primary sector employment, the service sector also dominates to a greater extent than elsewhere in Canada, with the exception of Newfoundland and Labrador. Important health, education, social assistance, and public administration service sectors are supported by government transfers and natural resource revenues.

Both rural and urban economies are sensitive to changes in global markets. Resource-based rural and smaller communities are facing international competition, particularly in fish processing and paper manufacturing, as well as domestic competition with agricultural products, while urban-based service sector enterprises struggle to capture their share of growing global markets. Reduced economic opportunities in traditional sectors have led to urbanization and out-migration to more prosperous parts of the country, although the latter trend has slowed. Losses in the age 15 to 34 segment of the population have left employers unable to secure new recruits other than older workers with limited formal education and restricted options for alternative employment or retraining. Local tax bases have also been undermined.

Table 2. Labour force characteristics, November, 2009

	NB	PEI	N. Scotia	NL	Canada
Participation rate	64.8	69.1	64.9	59.5	67.2
Unemployment rate	8.8	11.7	9.5	15.9	8.5
Employment rate	59.1	61	58.8	50	61.4
GDP per capita (2008,$)	36,579	33,034	36,454	61,608	48,011
Median income (2006, $)	22,900	23,800	23,900	20,500	26,500

Source: Statistics Canada, CANSIM, table 282-0087 and 384-0002, 2006 Census.

In all four Atlantic provinces, employment has grown in recent years, although at a pace below the Canadian average. Unemployment rates continued to be among the highest in the country, as did rates of seasonal work and reliance on employment insurance. In Newfoundland and Labrador, the rate was nearly triple the national average. By 2009, unemployment had fallen while Canada's unemployment rate rose, narrowing but leaving a still substantial gap (Table 2). GDP more than doubled from 2000, debt levels fell and in 2009-2010, for the first time since Canada's equalization payment system began, Newfoundland and Labrador became a "have" province, no longer eligible for equalization benefits. While the province continues to lag behind in employment rate, labour force participation, median incomes and reliance on social transfers to individuals, new wealth and employment gains have contributed to significant improvements in economic well-being.

New Brunswick's economic fortunes fluctuated from the Maritime province with the highest unemployment rate during the 1980s to the lowest (although still high by national standards) for several years in the 1990s and late 2000s, primarily due to improvements in the export positions of wood products, aquaculture and shellfish industries. Per capita GDP was highest among the Maritime provinces but lower than oil-rich NL in 2008. With the largest population and metropolitan area, Nova Scotia had consistently outperformed New Brunswick economically until quite recently. Nova Scotia has a higher proportion of its work force employed in the service sector and a lower proportion in primary resource industries.

Table 3. Gross domestic product (GDP) at basic prices, by North American Industry Classification System (NAICS) and province, 2008 (dollars x 1,000,000)

	NB		PEI		NS		NL		Atlantic Prov.		Canada	
	millions of chained (2002) dollars/% of total (all industries)											
Goods producing industries	5,398	25.6	936	24.9	6,465	24.0	8,365	45.9	21,164	30.2	363,625	29.5
Manufacturing	2,353	11.2	359	9.6	2,577	9.6	851	4.7	6140	8.8	171,906	13.9
Agriculture, forestry, fishing and hunting	670	3.2	310	8.3	829	3.1	371	2.0	2180	3.1	27,410	2.2
Mining, oil and gas extraction	219	1.0	0	0	880	3.3	5,768	31.7	6867	9.8	56,230	4.6
Service producing industries	15,704	74.4	2,821	75.1	20,469	76.0	9,848	54.1	48,842	69.8	869,154	70.5
Total (all industries)	21,1		3,7		26,9		18,2		70,0		1,232,7	

Source: http://www.statcan.gc.ca/pub/13-016-x/13-016-x2009002-eng.pdf.

Prince Edward Islanders generated the lowest per capita GDP in Canada in 2008. The province's unemployment rate was the highest in Maritime Canada and in 2008 PEI was the largest recipient of the transfer on a per capita basis. Average hourly wage rates over a 10-year period have been the lowest in the country.

Regional differences are evident within all provinces. In Nova Scotia, HRM and the Avon-Annapolis areas are generally economically healthy. The south shore, southwest shore, and Antigonish-New Glasgow are weaker, and Cape Breton Island lags behind. Individual variations in short-term economic positions can be accounted for by considering a variety of local factors.

Table 4. Employment by major industry groups, seasonally adjusted, by province (Nov. 2009)

	NB		PEI		NS		NL		Atlantic Canada		Canada	
	Thousands of employees/% of total (all industries)											
Goods producing industries	84.4	22.9	18.1	25.5	89.5	19.7	43.3	20.1	235.3	21.2	3,714.5	22.0
Manufacturing	33.8	9.2	5.5	7.7	34.5	7.6	11.6	5.4	85.4	7.7	1,769.4	10.5
Agriculture	7.4	2.0	4.1	5.8	7.6	1.7	0.6	0.3	19.7	1.8	317.8	1.9
Forestry, fishing, mining, oil and gas	11.2	3.0	2.2	3.1	13.4	2.9	14.7	6.8	41.5	3.7	307.5	1.8
Service producing industries	283.4	77.1	52.9	74.5	365.3	80.3	172.0	79.9	873.6	78.8	13,159.4	78.0
Total (all industries)	367.8		71		454.8		215.3		1108.9		16,873.9	

Source: Statistics Canada, CANSIM table 282-0088.

Cape Breton Island has some natural disadvantages compared to mainland Nova Scotia: its soil is poorer for agriculture, and its climate more rigorous. It was bypassed during the initial phase of the development of Nova Scotia, following the fall of Louisbourg. Its topography made the construction of railways and other infrastructure difficult; railway construction also suffered from political interference. Exploitation of its primary raw material, coal, and its primary industrial product, steel, have suffered from competition and changing industrial practices: today, there are no active collieries or steel mills in "industrial Cape Breton".

Rural residents are more likely to be seasonally employed or unemployed, to rely on government transfers, and have lower levels of education and lower median wages than urban residents. The gap between rural and urban regions continues to grow as labour markets transform from a rural, resource-based economy to one primarily relying on skills, knowledge and technology to compete in the global marketplace. Segments of the population such as the elderly, single parents, women, immigrants and Aboriginal communities also have distinctive sets of socioeconomic conditions. Rural men in PEI are over-represented in primary industries and women in low-paying processing jobs. When communities secure funding for small-scale ventures, men are more likely to receive related paid employment and women unpaid volunteer responsibilities.

Fish, Food and Forests

Fisheries, agriculture, and logging and forestry combined represent a greater proportion of GDP in Atlantic Canada than in the rest of the country. Fishing constitutes the largest component of this category in NS and NL, whereas agriculture is more significant in NB and PEI. Seafood exports in NS exceed NL in value (primarily due to shellfish production), but volume is higher in NL. In total, nearly 80% of Canada's fish harvesters reside in the Atlantic region. Unlike Canada's Pacific Region fishery, which has experienced a sharp and continuing decline in employment, the Atlantic fishery has proven remarkably resilient despite a difficult period of transition from cod to shellfish.

Fifty years ago, the Newfoundland and Labrador economy was dependent on the fishery. Emphasis changed from salt cod to frozen fish product and a more centralized, technology-intensive fishery and catches, fishing capacity and employment reached historic levels. By 1991, Northern cod spawning stock biomass had fallen to 10% of its 1962 abundance level. In 1992, the federal government imposed a moratorium on the east coast groundfish fishery, disrupting the lives of over 30,000 Atlantic Canadians that depended on the fishery for their livelihood in what has been referred to as "the largest single layoff in Canadian history". Overfishing, water temperature, seal predation, reduction in food sources and the destruction of habitat are among the factors blamed. Over $3 billion were spent between 1990 and 1998 on programs to minimize the socio-economic disruption. Research indicates that these programs did not adequately prepare people for work in other fields or significantly reduce the numbers of people dependent upon the fishery. Cutbacks in both the federal Employment Insurance program and government services during the same time period have posed difficulties for rural economies.

Since the early 1990s, the fishing industry has moved from groundfish (such as cod) to shellfish (particularly crab and shrimp). The sector is worth more today than prior to 1992, but employs fewer individuals. Significant problems remain in the industry, which is widely considered to be in need of restructuring. In the Maritimes and Labrador, increased Aboriginal involvement in the fishery has been an important step in recognizing Aboriginal rights and title, but has come with resistance and tensions as scarce fisheries resource rights change hands. Despite changing circumstances, most Atlantic regions continue to have some degree of reliance on fishery-based activities, both historical/traditional and based on adaption and current investment as communities and regions strive to be part of the fishery of the future – one more focused on conservation and on a wider range of marine resources than in the past.

One such resource is farm-raised fish products. Aquaculture is of growing significance in the region, including farming of Atlantic salmon, rainbow trout, mussels, and oysters, as well as species such as char, striped bass, haddock, halibut, and winter flounder. New Brunswick is the region's aquaculture leader, having grown from approximately $1 million annually in salmon product sales to a gross aquaculture output of over $100 million annually since 1990.

Agriculture in the region is a diverse, highly integrated sector representing 3.8% of the farms in Canada and 1.6% of Canada's agricultural land area. In 2008 the value of crop and animal production (excluding processing) to the Atlantic economy ranged from 0.3% (NL) to 5.2% (PEI) of provincial GDP. Nearly half of the region's farms are located in NS, where the main agricultural products include fruit, dairy cattle and milk production, beef, berries and nuts, and nursery and tree production.

Agriculture and tourism are the major industries in PEI. Upwards of 90% of the province's land base has agriculture potential. Major farm types in PEI include beef, potatoes, and dairy cattle operations. PEI is Canada's leading potato producer and third largest carrot producer. Psychologically, the potato in PEI is comparable to the cod elsewhere in the northern North Atlantic. The sandy texture of the island's podzolic soils is highly conducive to potato farming. With the removal of the beech-oak-sugar maple Acadian Deciduous Forest, the supply of humic matter to the soil has gradually been declining so farmers have compensated by adding humus and nutrients artificially, adding to farming costs. Rural populations and agricultural employment have declined in the province since the 1930s. Between 1991 and 2006, the number of farms in PEI dropped by 28%, with declines particularly in beef, dairy, and hog farms. Contributing factors include a strong Canadian dollar, high prices for feed, fertilizer, and fuel, international competition, and centralized buying patterns. Changes in climate such as warmer, wetter summers and increasing demand for organically produced products are among the factors that could reshape the industry.

Foreign-owned and export-oriented forestry dominates the Atlantic forest industry, driven in large part by the pulp and paper industry. Forestry is a major employer in NB, and has been an economic mainstay throughout the province's history. Forestry represents a small portion of employment in PEI and NL, and a much smaller proportion of the NS economy than agriculture, although employment levels are similar in fishing and forestry sectors. Mechanization has meant declining employment benefits per cubic metre of timber harvested.

Mines and Energy

Access to a large variety of mineral resources has made mining, oil and gas extraction important contributors to the economy, particularly in NL and to a lesser extent in NS and NB. There are no mineral resources in commercial quantities in PEI.

In NL mining, oil and gas extraction were worth $5.8 billion in 2008, representing 31.5% of GDP (vs. 4.6% in Canada). Metal ore production (notably iron in western Labrador) is significant. The Iron Ore Company of Canada (IOC), now Canada's largest iron ore pellet producer, began producing iron ore from its Carol Lake project in 1962. Labrador West now contains the two largest mines in NL: IOC (Labrador City) and Wabush Mines (Wabush) account for ~60% Canada's iron ore exports.

Mining in Labrador experienced a further boost with the discovery of a large nickel-copper-cobalt deposit at Voisey's Bay. The deposit is mined by INCO, and development involves the construction and operation of a mine and concentrator in Voisey's Bay, Labrador. Processing facilities are under construction at Long Harbour, Newfoundland. Other mining activity includes production of labradorite building stone (Nain), gold (Baie Verte Peninsula), slate (Random Island), and antimony (Beaver Brook, SW of Gander).

NL has been reaping a windfall from its three offshore oil developments on the Grand Banks, particularly since the price of oil increased after 2006. The 1979 discovery of a vast offshore oil and natural gas field alongside the important Grand Banks fishery led to provincial-federal negotiations and the 1985 Atlantic Accord, which ensured employment and spinoff benefits. In 2007, after ten years of operation, oil production was valued at $10.3 billion and the province collected $1.5 billion in royalties. With a fourth, major offshore oil project, metals processing and a major hydroelectric project planned, continued natural resource-based growth in oil, minerals, and hydroelectric industries is expected into the future. These developments are helping to curtail population losses and bring workers back from the oil fields of Alberta. Others work offshore or in camp environments in Alberta and elsewhere around the world and send earnings home, creating an additional remittance economy. Despite growing fiscal dependence on mining, oil, and gas, the sector constitutes just 2% of employment, with oil industry benefits concentrated in the capital region and minerals extraction in Labrador.

Extensive hydrocarbon exploration and development offshore of NS resulted in the launch of the Sable Offshore Energy Project and Deep Panuke natural gas project, which is expected to begin producing in

2010. NS mining operations include gypsum extraction in Cape Breton, the Annapolis and Avon River valleys, gold from Waverley and the Rawdon Hills and rare and valuable marble from the Denys Basin of Cape Breton.

In the 19th century it was Cape Breton coal that drove Nova Scotia's mining economy. After a century of coal mining and steel making, which made it an economic centre in Atlantic Canada, more than 20,000 jobs (24% of the workforce) were lost when coal mines began closing in the 1960s due to the high costs of extraction. When Sydney's Dominion Steel Corporation announced that its steel plant would close in 1967, the province opted to take it over, beginning two decades of major industrial operations operated as crown corporations. Despite provincial ownership and subsidies the Sydney coke ovens stopped firing in 1998 and Cape Breton's last coal mine, once employing 4,000 residents, closed in 2001 and left behind the hazardous waste site known as the Sydney Tar Ponds. Cape Breton Island also lost the Marble Mountain limestone quarry in 1991, which had operated since 1869.

Mining is also important in northern New Brunswick, where lead and zinc production occurs, and in the Sussex area, where one of the world's largest potash deposits is located. Saint John, New Brunswick has become a major energy hub for the region. It is the home of Canada's largest oil refinery, an LNG terminal and regasification plant, and the Point Lepreau Nuclear generating station, the only one in the Atlantic Provinces.

There is tremendous variability between the Atlantic Provinces in terms of current energy mix for electricity production. Whereas electricity in NB is produced from coal, oil, nuclear, and hydro, in NL 97% of electricity production is from hydro. Nova Scotia relies on coal and oil. PEI does not have hydroelectric power generation capacity and purchases most of its electricity from NB as well as generating small amounts from wind turbines and oil. Wind generation has become the fastest growing renewable energy supply in Atlantic Canada, with wind farms constructed in all provinces.

Manufacturing

Historically a resource periphery, manufacturing is less prevalent in the region than elsewhere in the country. The region's manufacturing sector is strongly connected to natural resources, particularly to food production. Food (including seafood), wood and paper products make up over two-fifths of the region's manufacturing output. Enterprises are smaller and more capital intensive on average than at the national level but invest less in research and development. In the manufacturing sector, four factors contribute to the productivity gap between Atlantic

Canada and Canada as a whole: less innovative effort, particularly in high-tech industries; fewer economies of scale; lower educational attainment of the workforce; and greater seasonality of production.

In NB, wood products are the province's most important manufacturing sector in terms of GDP contribution. Pulp and paper markets weakened through the 1990s and into 2010 while costs rose, the dollar remained high, competition increased from low-cost foreign producers, and wood supply shrunk. Reduced profits led to numerous mill closures throughout Atlantic Canada. Several lumber mills also shut down or operated at reduced capacity as the number of firms in the Atlantic forest industry dropped in the mid-2000s.

All four provinces employ significant numbers in seafood production. The remaining firms have survived through adaptation, innovation and diversification but concerns exist about over-reliance on snow crab, crab quota cuts, price wars, technological change, short seasons and reliance on Employment Insurance.

Processing of agri-food products is a significant and vibrant part of the Maritime economy. Processing plants benefit from a strategic location and year-round transportation links to major international markets, and low relative costs. The industry is characterized by a broad range of crops and livestock, and significant emphasis on adding value. In NS, food products rank first in GDP, followed by aerospace products and parts, seafood and wood products. The region is the "wild blueberry capital of the world" and a major producer of frozen vegetables and French fries, due to the success of companies such as McCain Foods, Cavendish Farms and Oxford Frozen Foods.

Services: Tourism and Technology

Public sector and retail services make up the bulk of service sector production in Atlantic Canada. Two segments within the service-producing industries of growing importance to the region are tourism and technology. Cape Breton's economy has bounced back, at least in part, from the collapse of the coal economy by diversifying and creating small businesses. With significant increases in services, tourism and information technology, the Island's unemployment rate fell through the early 2000s.

The information and communications technology (ICT) sector has grown continuously in Atlantic Canada since 1997. A large share of these revenues was generated by telecommunications services. During this period the region experienced the highest annual growth for ICT revenues in Canada, growing in all three main sub-sectors (manufacturing, telecommunications services, and software and computer services).

Almost half of all ICT workers were employed in the telecommunications services industries (including call centres), while another 43% worked in the software and computer services.

Tourism is a major component in the economies of all four Atlantic Provinces and the largest industry in PEI. Following the loss of the groundfishery, rural Atlantic coastal communities in particular turned to tourism as an option for economic revitalization in the 1990s. The accommodation industry dominates tourism employment in predominantly rural areas while the food and beverage sector is the leading tourism employer in urban regions. Iceberg viewing has become a popular form of tourism along the Newfoundland coastline.

NS has traditionally targeted the broadest audience, with the greatest number of programmes. NB and NL have increased efforts, with positive results. Most of the gains are attributable to increased tourism from central Canada and the northeastern USA. NB has performed very well in the regional market, attracting visitors from elsewhere in Atlantic Canada and from New England, but has done less well in other parts of North America and overseas. In contrast, NL draws its visitors from a broader area, with greater increases in European visitor numbers. PEI has previously orchestrate more focused tourist campaigns, targeted at particular regional or cultural groups. More than 1 million people visit the Prince Edward Island National Park – Cavendish area annually. Tourists outnumber islanders 8:1 during the summer months. There is a danger that tourist influxes of this magnitude can result in erosion of the local culture and distortion of socio-economic patterns – that the island will fall into a "tourist trap". This danger is already apparent at Cavendish, marked by ever-increasing numbers of tourist attractions without connection to Prince Edward Island's history, culture, or geography.

Marine transportation includes fishing harbours and ferry services help support the tourism sector. Recreational harbours are currently growing in significance. Ferry services are an important component of the provincial economic system, used by tourists, residents and industry. International cruise ship traffic has been increasing in eastern Canada and traffic at the region's major cruise ports. By 2025, the number of passengers is expected to increase to 9.1 million by air and 882,000 by cruise ship.

The Atlantic Provinces are faced with difficult circumstances and future socio-economic decisions. Newfound prosperity in NL, for example, masks underlying issues such as resource depletion, continued staple dependence, inequitable wealth distribution, and labour shortages in the midst of high unemployment. Residents of rural and small-town NL continue to fall behind their urban counterparts in economic well-being, educational attainment, access to health care, and housing. Significant and growing inequalities also exist within rural and small towns, often exacerbated by fisheries and transition policies. Increasingly, labour shortages are being reported, linked to low wages, inability to compete in national labour markets and lack of remaining workers with appropriate skills.

As transfer payments from the federal government systematically decrease, the Atlantic provinces will have to generate more of their own revenue and continue to diversify their economies. The tendency to specialize within regions increasingly integrated into a regional economy seems at first glance to be contradictory to the need for diversification, but specialization in different branches of economic activity will lead to diversification in the region as a whole. Although political union, particularly among the three Maritime Provinces, shows no signs of revitalization, economic union is progressing rapidly on a de facto basis, regardless of political will or intervention. The initiative to develop "the Atlantic Gateway" as a major trade corridor illustrates this trend.

Conclusion

Atlantic Canada's physical environment is very much a product of the underlying geological landscape that has undergone immense geomorphholological changes for hundreds-of-thousands of years. Physical processes, such as plate tectonics, climate and the scouring, gouging effects of glaciations, have swept over Canada's eastern region resulting in a distinctive landscape. The physical features of this landscape vary in type and scale and include; mountain ranges, valleys, plains, and coastlines whose composition ranges from cobble stones, sandy beaches to rocky cliff lines.

Today, these eastern coastline provinces maintain their economical existence independently through industries such as agriculture, transportation, tourism and energy technologies, but initially their economic viability was primarily a product of their location next to the world's second largest ocean. Although Atlantic Canada shares its physical borders with Québec and the United States, historically its greatest cultural influence has come from the European communities from across seas.

In response to Atlantic Canada's uninhabited space, and the diverse nature of the physical landscape, settlers of various origins: French, English, Scottish, Irish, and Spanish migrated to this region contributing to the cultural diversity and sense of community found throughout various parts of Atlantic Canada today. Each province has lived out its own history in terms of community evolution and economical challenges, which in turn not only contributes to their distinct and individual character but invariably will add to the success of their future as provinces within Canada.

Review Questions about the Region and Websites for Answers

- Name three locations with medium to high frequency of storm surges in Atlantic Canada.

  http://atlas.nrcan.gc.ca/site/english/maps/environment/naturalhazards/storm_surge/storm_surge.

- How does the risk of forest fire in Atlantic Canada compare to other parts of Canada? Where did the largest known fire in eastern North America occur?

  http://atlas.nrcan.gc.ca/site/english/maps/environment/naturalhazards/forest_fires.

- Discuss how coastal sensitivity to sea level rise varies in different areas of Nova Scotia.

  http://atlas.nrcan.gc.ca/site/english/maps/climatechange/potentialimpacts/coastalsensitivitysealevelrise.

- Where is the largest Mi'kmaq (or Micmac) community in Newfoundland and Labrador?

  Newfoundland and Labrador Heritage Web Site Project – http://www.heritage.nf.ca.

- Name five National Historic Sites in Atlantic Canada and briefly describe reasons why these sites have received this designation.

  http://www.pc.gc.ca/progs/lhn-nhs/index_E.asp.

- Which Atlantic province experienced the greatest growth in population from 1996 to 2006?

  http://www.cid-bdc.ca/english/index.html.

- What is the origin of the name given to Grand Manan Island, New Brunswick?

  http://new-brunswick.net/new-brunswick/names/names2.html.

- Outline some of the reasons and events leading to the resettlement Africville Nova Scotia.

  http://www.africville.ca/resettlement/.

- Why is July 2, 1992 a significant date for residents of Newfoundland and Labrador?

  http://archives.cbc.ca/on_this_day/07/02/ or http://www.heritage.nf.ca/society/moratorium_impacts.html.

– What is "the Atlantic Gateway" and why is it significant to the region's economy?

http://www.atlanticgateway.gc.ca/index2.html.

References

ACZISC Secretariat and Canmac Economics (2006), *The Ocean Technology Sector in Atlantic Canada*, Prepared for Atlantic Canada Opportunities Agency.

Beale, E. (2008), As Labour Markets Tighten, Will Outmigration Trends Reverse in Atlantic Canada?, Atlantic Provinces Economic Council, May 2008.

Colman-Sadd, S.P., Hayes, J.P., Knight, I. (1990), *Geology of the island of Newfoundland*, Map 90-01, Geological Survey Branch, Department of Mines and Energy, Government of Newfoundland and Labrador.

de Peuter, J., and Sorensen, M. (2005), *Rural Prince Edward Island – Profile: A Ten Year Census Analysis 1991-2001*, Government of Canada, Rural Secretariat.

DFO: Fisheries Licences (2002), Retrieved December 20, 2009, from http://www.dfo-mpo.gc.ca/stats/commercial/licences-permis-eng.htm.

Environment Canada (2005), *Canadian Climate Normals, 1971-2000*, Ottawa: Atmospheric Environment Service, Environment Canada.

Environment Canada (2005), *National Climate Data and Weather Archive*, Retrieved December 15, 2009, from www.climate.weatheroffice.ec.gc.ca.

Flanagan, K. (2009), *Poverty Reduction Policies and Programs Prince Edward Island Social Development*, Report Series, Retrieved on December 10, 2009, from http://www.ccsd.ca/SDR2009/Reports/PEI_Report_FINAL.pdf.

Grant, D.R. (1989), "Quaternary geology of the Atlantic Appalachian region of Canada", in Fulton, R.J. (ed.), *Quaternary Geology of Canada and Greenland*, Geological Survey of Canada, Geology of Canada, 1, 393-440.

Hamilton, L., and Butler, M. (2001), "Outport adaptations: Social indicators through Newfoundland's cod crisis", *Human Ecology Review*, 8(2), 1-11.

Handcock, G. (1989), *Soe longe as there comes noe women: Origins of English Settlement in Newfoundland*, Newfoundland History Series, #6, Breakwater Press, St. John's, Newfoundland, Canada.

Hertzmann, O. (1997), "Oceans and the Coastal Zone", in W.G. Bailey, T.R. Oke and W.R. Rouse, eds., *The Surface Climates of Canada*, McGill-Queens University Press, 101-123.

Holden, M. (2008), *Canada's New Equalization Formula*, Retr. Dec. 27, 2009: http://www2.parl.gc.ca/content/LOP/ResearchPublications/prb0820-e.htm#fn2.

House, J. (2006), *Oil, fish and social change in Newfoundland and Labrador: Lessons for British Columbia*, Lecture to the Special Centre for Coastal Studies, Simon Fraser University, Burnaby, BC.

Industry Canada (2008), *ICT Sector Regional Analysis – Atlantic Canada*, Retrieved January 3, 2010, from http://www.ic.gc.ca/eic/site/ict-tic.nsf/eng/it07185.html.

Lepawsky, J., Phan, C., and Greenwood R. (2008), *Metropolis on the margins: Talent attraction and retention to the St. John's city-region, Newfoundland*, Retr. Oct. 22, 2009: http://www.utoronto.ca/isrn/publications/NatMeeting/NatSlides/Nat08/Lepawsky_et_al_Metropolis_on_Margins.pdf.

MacIver, D.C., Meyer, R.E., *Decoding Canada's Environmental Past: Climate Variations and Biodiversity Change during the last Millennium*, Downsview: Atmospheric Environmental Service, Environment Canada.

McAleer, C. (2007), *Rural Poverty in Prince Edward Island: Some considerations for Island Women*, Presentation to the Standing Senate Committee on Agriculture and Forestry on Rural Poverty, PEI Advisory Council on the Status of Women.

Natural Resources Canada (2010), *Atlantic Maritime Ecozone*, Retrieved January 31, 2010, from http://ecosys.cfl.scf.rncan.gc.ca.

NB Profile (2009), Retrieved December 20, 2009, from http://www.new-brunswick.net/new-brunswick/overview.html.

Newfoundland and Labrador Heritage Web Site Project (2000), *Forest Industries*, Retrieved October 22, 2009, from Newfoundland and Labrador Heritage: http://www.heritage.nf.ca/society/forestry.html.

Nova Scotia, Canada (2009), *Economic Benefits from Offshore Petroleum Activity*, Retrieved December 23, 2009, from http://www.gov.ns.ca/energy/oil-gas/offshore/economic-benefits.asp.

Nova Scotia Agricultural College (NSAC), Processing Carrot Research Program, Retrieved November 10, 2009, from http://nsac.ca/pas/instind/pcrp/Research/PCRP%20handout.pdf.

*Nova Scotia Museum of Natural History (2009), *The Natural History of Nova Scotia*, Retrieved on December 3, 2009, from http://museum.gov.ns.ca/mnh/nature/nhns/index.htm.

*Sharpe, A., (2003), *The Canada-Atlantic Canada Manufacturing Productivity Gap: A Detailed Analysis*, Retrieved on December 4, 2009, from http://www.csls.ca/reports/acoa2003.pdf.

Shaw, J., Piper, D.J.W., Fader, G.B.J., King, E.L., Todd, B.J., Bell, T., Batterson, M.J. and Liverman, D.J.E. (2006), "A conceptual model of the deglaciation of Atlantic Canada", *Quaternary Science Reviews*, 25: 2059-2081.

Shaw, J., Taylor, R.B., Forbes, D.L., Solomon, S. (2001), *Sea level rise in Canada*, in Brooks, G.R. (ed.), "A synthesis of geological hazards in Canada", *Geological Survey of Canada Bulletin*, 548, 225-226.

Statistics Canada, Retrieved November 10, 2009, from http://www40.statcan.gc.ca/l01/cst01/agrc25b-eng.htm.

Statistics Canada (2009), *Canadian Business Patterns Database*, Retrieved December 13, 2008, from http://www.statcan.gc.ca/ads-annonces/61f0040x/index-eng.htm.

Statistics Canada (2006), *Canada's Changing Labour Force, 2006 Census: The provinces and territories*, Retrieved November 10, 2009, from http://www41.statcan.ca/2007/4005/ceb4005_000-eng.htm.

Stea, R. (2004), "The Appalachian Glacier Complex in Maritime Canada", in Ehlers, J., and Gibbard, P.L. (eds.), *Quaternary Glaciations – Extent and Chrnology, Part II*, Elsevier, Amsterdam, 213-232.

Vasseur, L., and Catto, N.R. (2008), *Atlantic Canada; in From Impacts to Adaptation: Canada in a Changing Climate 2007*, edited by D.S. Lemmen, F.J. Warren, J. Lacroix and E. Bush, Government of Canada, Ottawa, ON, p. 119-170.

Wardle, R.J., Gower, C.F., Ryan, B., Nunn, G.A.G., James, D.T., and Kerr, A. (1997), *Geological Map of Labrador; 1:1 million scale*, Government of Newfoundland and Labrador, Department of Mines and Energy, Geological Survey, Map 97-07.

Wright, M. (2001), *A Fishery for Modern Times: The State and the Industrialization of the Newfoundland Fishery, 1934-1968*, Oxford University Press, Don Mills, ON.

CHAPTER 8

The Territories of Québec

Geography of a Territory Undergoing Transformation

Isabelle THOMAS-MARET, Paul LEWIS and Sandra BREUX

The Institute of Urban Planning, Faculty of Environmental Design, Université de Montréal

Introduction

Québec is, from most points of view, a case apart in the Canadian context, even if it shares a lot with the rest of Canada. What image can we construct of the geography of Québec apart from the history that characterizes it? At first sight, two characteristics in particular stand out. First, Québec extends over an immense territory, for the most part unoccupied or very sparsely occupied, but nonetheless marked by human presence. And, secondly, an urban structure marked by the primacy of Montréal, which brings together half, or close to half, of the population of Québec. In addition to Montréal, a network of small and medium cities polarizes the regions that often experience difficulties, as shown with the demographic decline. The Québec territory is in a period of profound transformation in the wake of globalization, as well as the over exploitation of natural resources.

The objective of this chapter is to sketch out a geographical portrait of Québec. This is an exercise in regional geography, a region defined by the limits of the province of Québec. Québec constitutes a geographically coherent whole that can be broken down into several regions; this was what the government did at the end of the 1960s by outlining ten administrative regions that correspond in fact to functional regions, more than to historical or natural regions, in accordance with Manzagol's (1996) typology. At the same time, however, it must be recognized that if Québec constitutes a coherent whole, it shares characteristics with other Canadian regions, especially Ontario and the Maritime provinces.

More specifically, the thrust of our approach is to present the geography of Québec while emphasizing as much its assets as its limits. The chapter is composed of 4 parts. In the first part, we give an overall view of the Québec territory starting with a physical description, in order to understand its specificities and characteristics. We also show that despite the resources available, the exploitation of this richness has not always been satisfactory as the occupation of the territory plainly shows. This is the reason for which, in the second part, we outline the main components of the territorial occupation. Our viewpoint applies essentially to Québec's urban population. A geographical and historical breakdown portrait of the urban population comes first, and then the demographic evolution of Québec cities is set out in detail. Particular attention is paid to the political and administrative organization. The different sections draw out the dynamics at play between metro regions, second-tier cities and the resource regions. Following this, in the third section, we turn to the phenomena of metropolization and urban decentralization, with respect mainly to the distribution of employment. Two sections are dedicated to the cases of Montréal and Québec City. The dynamics between metropolitan regions and medium-sized cities are explored more in depth. Finally, the last part deals with the peripheral regions, emphasizing as much their assets as their difficulties; their demographic evolution is presented and we pay particular attention to their perspectives of development along with the measures implemented to counter their decline. At the completion of this analysis, the conclusion explores the challenges facing Québec at the start the 21st century.

1. The Québec Territory: Immenseness and Nordicity

The objective of this part is to sketch out the portrait of the province, focusing on its physical geography. Far from being exhaustive, this section aims to give a quick outline of the major traits of the territory under study, while underlying its main specificities.

Québec is located in the Eastern part of Canada. It is bordered to the South by the United States, to the West by Ontario, to the North by the Arctic Sea (and Hudson's Bay) and to the East by the Atlantic Ocean (and Labrador Sea), which via the St. Lawrence River gives access deeply into the continent, and the Maritime provinces, essentially New Brunswick, Newfoundland and Labrador. From South to North, Québec extends over a little more than 17 degrees of latitude and, from East to West, over 22 degrees longitudinally. In terms of area, Québec is the largest province in Canada, taking up 1,667 million km^2, that is, close to one fifth of the total area of Canada.

Overall, it can be said that the Québec territory displays four specific characters. First, the Québec territory is strongly marked by the presence of water which covers approximately 12% of the total geographic area; in Québec, 1 million lakes and 130,000 watercourses have contributed to the myth of the abundance of water that has marked debates in recent years. The most important water feature is the St. Lawrence, whose source is in the Great Lakes, at the heart of the continent. The river covers more than 1,200 km – 1,000 in Québec alone. Utilized for commercial navigation, the St. Lawrence Seaway was developed 50 years ago jointly by Canada and the United States.

Secondly, with a population of approximately 7 million inhabitants, the density of the Québec territory is very low, in the order of 4.7 persons per km^2. At the same time however, it must be understood that this statistic is not very significant since the northern part of the territory is very sparsely populated – barely 35,000 persons, while close to three quarters of the Québecois live more or less within a maximum of 100 kilometres within either banks of the St. Lawrence River.

Thirdly, the Québecois climate is marked by nordicity. Because of its geographic size, as well as its location in the northern hemisphere, Québec experiences substantial climactic variations from one region to another; temperature varies as a function of latitude, topography as well as proximity to the ocean. In the south, the Québec climate is temperate, with four distinct seasons. The more one goes toward the North, the colder the temperature. There are essentially four climactic regions in Québec:

- Humid continental, South of the 50th degree of latitude;
- Subarctic, between the 50th and 58th degrees of latitude;
- Arctic, in the extreme North;
- Maritime, for the Gaspé peninsula and the Magdalen Islands.

The Laurentian plateau, which is in fact part of the Canadian shield, covers close to 95% of Québec; in the South, it ends with the Laurentians, a chain of mountains presenting rolling hills, beautiful forests and numerous lakes, essential features of recreational and touristic development. Three other significant geological divisions can be identified, from the South to the North:

The Appalachians, a chain of mountains that extends into the United States of America.

The St. Lawrence lowlands, a plain that borders the St. Lawrence River where the best farmland in Québec is to be found.

The Hudson Bay lowlands with sparse vegetation and permafrost.

Figure 1. The geological divisions of Québec

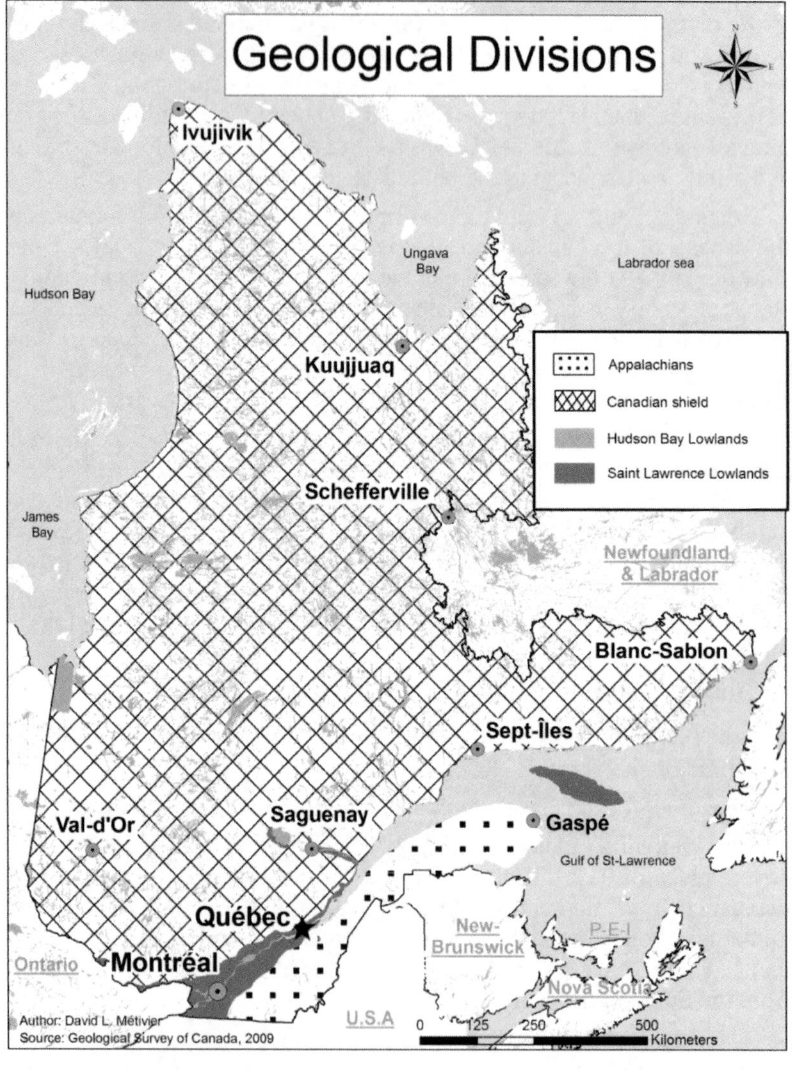

Source: David L. Métivier (Geological Survey of Canada 2009).

Finally, the Québec territory is immense, richly endowed with resources that have contributed to and shaped the economy to the point where most of the land corresponds to what we sometimes call resource regions, as if there were nothing there to be found but resources. In fact, this is almost true, in the sense that these regions are sparsely populated;

and the main activities of these populations are in the exploitation of resources. Nonetheless, the exploitation of resources has not really been planned, which explains at least in part why many regions of Québec experience problems of access to resources. Hence abundant resources have not implied sustainable development.

Summary

The characteristics of the Québec territory feature the richness and the potentiality of this geographic area. Nonetheless, the exploitation of these specificities has not always brought forth adequate development. The conditions of occupation of the territory may explain in part the development of natural assets; this question we now address.

2. Territorial Occupation: the Urban Perspective

The population is unevenly distributed in the Québecois territory: a good part of the territory is unpopulated. Actually, the urban population is very strongly concentrated in Montréal which accounts for half of the Québecois population, and in the 5 other metropolitan areas that make up Québec. Outside of the metropolitan areas, the territory is sparsely populated, with barely 2.4 million inhabitants (compared to 5.2 million inhabitants for the 6 major urban areas of Québec) (ISQ[1], 2010). In the remainder of the territory there are very few or even no cities of importance, which is not without consequences for the availability of services. Regions are threatened by the propensity to consolidate the services in large centres, especially in the context of cutbacks in public services intended to ensure the sustainability of such services. This is particularly noticeable in the health domain; the reorganization of the system tends to concentrate more and more functions in large centres, Québec City and Montréal, and, in a lesser measure, Sherbrooke and Trois-Rivières. The present section is dedicated to an analysis of the occupation of the territory: geographical and historical distribution of the urban population on one hand, demographic portrait of contemporary Québecois cities on the other hand, as well as the current political and administrative organization of the territory.

2.1. An Uneven Distribution of the Urban Population

The population of Québec is estimated to be 7,788,800 people as of 1st January 2009, while it was 7,718,400 at the beginning of 2008. This amounts to an increase of 70,400 inhabitants in a year, which corresponds to an annual rate of growth of 9.1 per thousand" (Institut de la

---

1 ISQ: Institut de la Statistique du Québec. Compiled from data available on the ISQ site (viewed 14 March 2010). http://www.stat.gouv.qc.ca/.

Statistique du Québec – ISQ, 2009: 11). This population is however unevenly distributed over the territory (Figure 2).

Figure 2. Relief map of Québec

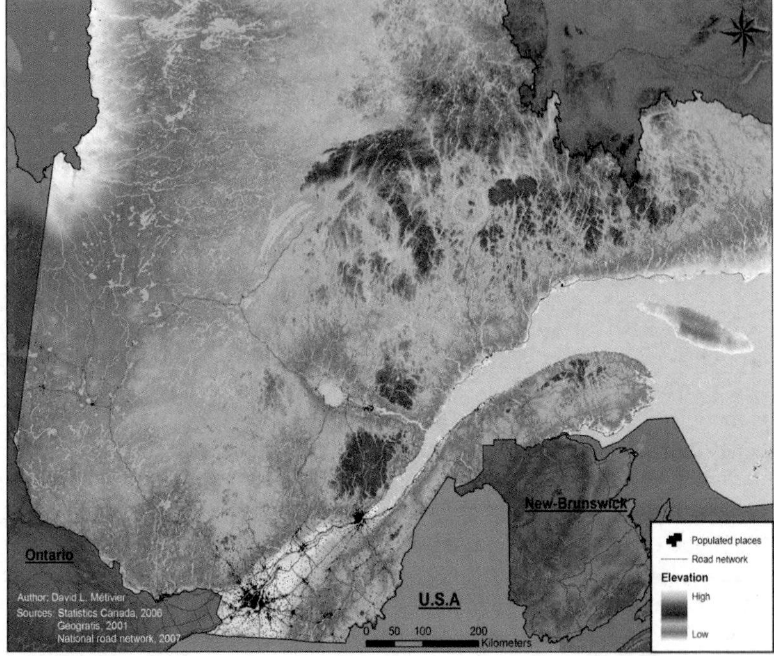

Source: David L. Métivier (Statistics Canada 2006, Geogratis 2001, and National road network 2007).

Taking a first look at the Québec territory, we can see how the geography of Québec was a decisive factor in the location of cities across the centuries. Indeed the first cities of "la Nouvelle-France" were naturally established on the banks of the St. Lawrence. For a long time, the river has been the backbone of the transportation system as well as the most efficient means to travel between settlements in the colonial period. Soil fertility in the St. Lawrence lowlands made it an ideal location to establish human settlements that, at this period, were largely dependent on agriculture. It is in these conditions – agriculture and port activities – that Québec City was built, the first authentic town of New France. Perched on the summit of a steep cliff – where the river narrows, the location and the site afford a readily defensible position from threats and control the accessibility to the St. Lawrence river and to a large number of trading posts in New France. Thus the city became the colony's main port.

The river has been the structuring axis of the population settlement in the Québec territory. Several seigneuries were granted and a number of more important establishments on either bank of the St. Lawrence river developed between Tadoussac and Deux-Montagnes. Trois-Rivières and Montréal were the next sizeable towns built in the colony, respectively in 1634 and 1642. Québec City remained the hub and the most populated city of the colony up until the early 19th century, at which time it was outshined by Montréal in the urban ranking. Montréal's predominance was further emphasized after the ditching of the St. Lawrence waterway in 1844 that allowed boats to navigate to the port of Montréal, which progressively replaced Québec City as the main port in Québec.

It was also during the 19th century that the first wave of occupation of the Québec hinterland began. Industrialization and the building of railways fostered new regions to develop. On the North bank of the river, forest product industries gave rise to new settlements close to the major rivers in Mauricie, Saguenay-Lac-Saint-Jean, the Outaouais and in the Laurentians. Product transformation facilities were mostly located on the South bank of the river, which fostered the development of new regions such as Beauce, Bois-Francs and the Eastern Townships. The 20th century bought with it a second wave of expansion of the ecumene – permanently inhabited areas, in the back country. The economic crisis of the 1930s heightened this new wave of land settlement in far regions such as the Lower St. Lawrence, the Gaspé area, Abitibi-Temiscaming and the North Shore where the economy thrived on the harvesting, development and transformation of natural resources (Bouchard, 2006). "Up until the beginning of the 20th century, most of the territories within today's administrative regions of Québec witnessed significant increases in population, especially in the agricultural ecumene. Following this, however, the pace slowed down except in the strongly urbanized regions of the St. Lawrence valley where the population continued to increase, at least until the 1930s' Crisis" (Courville, 2000: 325).

The organization of cities in the organization of the Québec territory did not waver, even during the 1939-1945 pre-war period when the increase in the population sagged. The after-war period changed the situation and specific regions experienced genuine demographic expansion thanks to the exploitation of their natural resources; the North Shore is a case in point. To a large extent, this explains the contrast in the distribution of the population among regions: differences between regions are actually quite significant. Among the least populated regions today, the North of Québec accounts for some 41,479 inhabitants while Abitibi-Temiscaming comprises 145,886 inhabitants (Table 1, Figure 3). A similar observation can be made in the most populated regions. Where the regions of Montréal, Québec City, Montérégie and the Laurentians are among the most populated, each region has its own personality: for example, the region of Montréal is three times more populated than the Laurentians.

Table 1. Demographic portrait of regional Québec, 2009

Administrative region	Population	Area (km^2)	Density	Net migration 2008-2009	Demographic outlook 2031-2006 (±%)
-01-Lower St. Lawrence	200,756	22,184	9.0	-456	-1.3
-02-Saguenay-Lac-Saint-Jean	273,264	9589,8	2.8	-659	-7.0
-03-Capitale-Nationale	687,810	18,638	36.9	2,761	11.6
-04-Mauricie	262,399	35,451	7.4	121	5.5
-05-Estrie (Eastern Townships)	307,389	10,194	30.2	40	11.4
-06-Montréal	1,9 M	498	3,827	-19,463	12.1
-07-Outaouais	358,872	30,503	11.8	1,437	23.7
-08-Abitibi-Témiscamingue	145,886	57,339	2.5	-392	-2.7
-09-North Shore	95,704	236, 699	0.4	-452	-11.6
-10-Nord-du-Québec	41,479	718,228	0.1	-384	6.8
-11-Gaspésie-îles-de-la-Madeleine	94,067	20,272	4.6	-61	-1.3
-12-Chaudières-Appalaches	403,011	15,070	26.7	463	8.9
-13-Laval	391,893	246	1,593	2,882	28.6
-14-Lanaudière	457,962	12,313	37.2	4,390	37.9
-15-Laurentides	542,416	20,559	26.4	4,335	34.00
-16-Montérégie	1,4 M	11,110	126.6	4,861	21.5
-17-Centre-du-Québec	230,685	6,920	33.3	577	12.3

To this day, the populated areas of Québec are mainly concentrated in the Southern part of the province and in the St. Lawrence lowlands. The larger urban clusters are also concentrated there. Moreover, it is estimated that the metropolitan region of Montréal accounts for close to 58% of the Québecois population (Bouchard, 2006). The weight of cities in the total population of Québec has also enormously changed. The population, that was almost 80% rural towards 1867, was more than 80% urban in 2001 (Polèse and Shearmur, 2003: 31). The 6 largest urban agglomerations of the province, that is, Québec, Gatineau, Sherbrooke, Saguenay and Trois-Rivières, account today for close to 70% of the total population of Québec (Bouchard, 2006).

Figure 3: Distribution of the Québec population in 2001

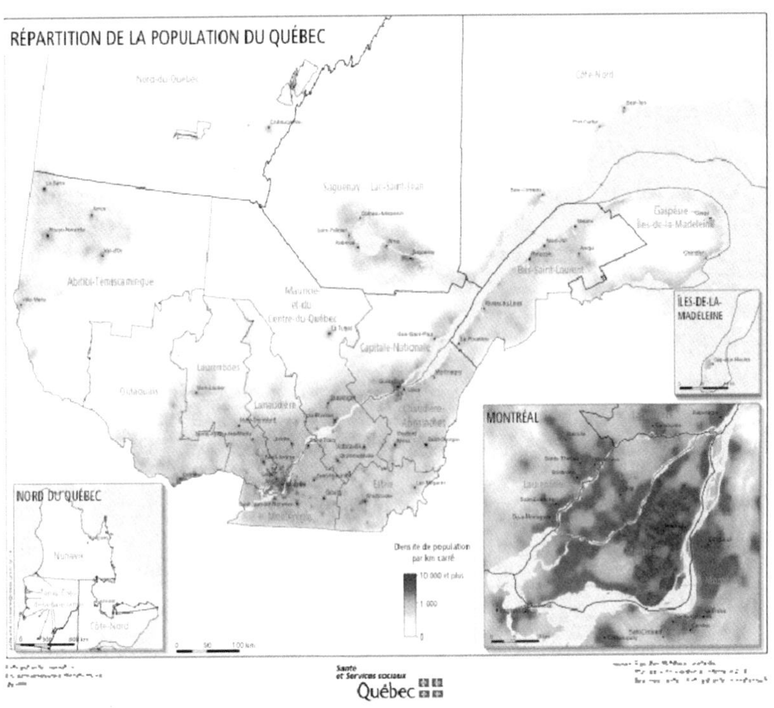

Source: Government of Québec.

The current portrait and the demographic outlook emphasize the weight of two regions: Montréal and Québec City. These regions can be defined as metropolitan areas, which according to Statistics Canada are entities "made up of a large urban zone (called a city core) as well as socially and economically integrated neighbouring urban and rural sectors. The population of a metropolitan area has at least 100,000 inhabitants, according to the results of the previous census". The metropolitan areas of Montréal and Québec City are uniquely made up of supralocal bodies called metropolitan communities, whose objective is to plan and to coordinate the actions within the specific territories, especially land-use planning. The metropolitan community of Montréal (CMM) brings together as much as 3.5 million inhabitants and 82 municipalities, while that of Québec brings together 71,000 inhabitants and 28 municipalities.

The demographic forecast for Québecois regions noticeably emphasizes the growth of the areas bordering these two metropolitan entities (Estrie, Chaudières-Appalaches, Lanaudière, Laurentides, Centre-du-

Québec, Montérégie, Laval) (Table 1). *A contrario*, the regions most distant from these two centres show negative or relatively weak net migration figures. By and large, the demographic sketch of the Québecois cities outlines these phenomena.

2.2. *The Cities of Québec: Demographic Portrait of the Large and Medium Sized Cities*

At the top of the urban ranking are two large cities differentiated from one another by a difference of close to 1,160,000 people, Montréal and Québec City with 1,667,700 and 508,349 inhabitants respectively. While the position of Montréal gives it the status of a metropolis, Québec City, the provincial capital, is an important economic and administrative hub.

Then come the medium sized cities that, in Québec, are first defined by the population criterion, that is, between 20,000 and 200,000 inhabitants approximately; to this measure is added a function of centrality and regional outreach (Simard, 2003). Overall, "upper level" (Bruneau, 1989) medium sized cities retained their status over the course of recent decades; Gatineau, Saguenay, Sherbrooke and Trois-Rivières remain at the top of the Québecois demographic ranking in their category, in which are now included the cities of Laval, Longueuil and Lévis (Table 2). However, in the case of the latter cities, the centralism criterion may be questioned. Medium-sized cities with less than 100,000 inhabitants are, for their part, more numerous and demographic disparities between each of these cities is less significant.

Medium-sized cities of Québec can be classified into two spatial systems: on one hand, a denser central region bringing together many urban entities in close proximity to each other, and, on the other hand, a vast peripheral area where centres are more distant and less numerous (Figure 3). As a function of their location, Pierre Bruneau (1989) distinguishes three categories of medium-sized cities: perimetropolitan medium-sized cities (Sorel, St. Hyacinthe, St-Jean-sur-Richelieu), intermetropolitan (Granby, Drummondville, Victoriaville) and peripheral (Val d'Or, Rimouski, Rivière-du-Loup).

The configuration of the transportation network contributes to reinforce this tendency. Historically, the city-forming factor was largely influenced by the St. Lawrence River and its tributaries, all of which offered themselves as a perfect transportation network. The road network and highway system were subsequently derived from the waterways as well as in the form of axes spreading out from the two main metropolitan areas, that is, Québec City and especially Montréal.

Table 2. Demographic portrait of the ten major cities, 2009

Cities	Population
Montréal	1,667,700
Québec	508,349
Laval	391,893
Gatineau	256,240
Longueuil	234,003
Sherbrooke	153,384
Saguenay	143,564
Lévis	136,066
Trois-Rivières	129,519
Terrebonne	102,827

Source: ISQ, 2010: 1.

The two largest cities also benefit from the leading airport facilities in the province. As such, the Québecois transportation network contributes to a centripetal or concentration movement of both the population and its activities. It is noted as well that the Québecois population is still mainly concentrated along the axis of the St. Lawrence Seaway, but for the rest, the extent of territorial unoccupancy underlines a form of originality which that is reinforced by a discernibly singular political and administrative organization.

2.3. Political and Administrative Organization

Territorial organization in Québec is probably unique: indeed most of the Québec territory is unoccupied and unorganized (Figure 3). Two reasons explain such specific character: the physical geography of the province on one hand, as mentioned above, and on the other hand, history. Public lands cover almost 92% of the total area of the province, a real fountain of natural reserves. Conversely, agricultural and urban zones represent only 2% of the total area[2].

Historical circumstances also explain the current organization of the Québecois territory. Despite the absence of municipal directions under the French administration[3], a flourishing economy, population growth, and the extension of land settlement, followed by the rebellions of 1837-1838, highlighted the need for a political and administrative structure in Lower Canada. The Durham Report, the orders of Sydenham's special

---

2 Ref.: Government of Québec website, 2009, http://www.gouv.qc.ca.

3 One expert writes: "where central structures are believed to have beeen well established under French administration, this was not the case for local structures. Only clergy-ran parishes and the 'régime seigneurial' would partly offset any inconsistent, unfrequent or embryonic attempt to establish *communal* [municipal] management in Nouvelle-France" (Baccigalupo, 1990: 6).

council, as well as the Lower Canada Municipal and Road Act of 1855 laid the foundations for that structure; thus, even before the birth of the Canadian Confederation, there were more than 600 municipalities in Québec. The construction of the Confederation, which granted the provinces the possibility to avail themselves of municipal structures, emphasizes how necessary those were and at the turn of the 20th century Québec had more than 1,000 municipalities, spread out over roughly one third of its territory (Collin *et al.*, 2003; Proulx *et al.*, 2005: 28-32).

The number of municipalities did not cease to increase in the 20th century and during the 1970s the birth of urban centres implied the creation of administrative regions (Proulx *et al.*, 2005: 40).

Based on the ancient dioceses, the work of geographers and historians, as well as a study of the outreach zones of the regional development hubs, administrative regions were marked out in 1966 around the main urban centres; the 10 first formal administrative regions were officially implemented in 1968. A reorganisation of these regions has since taken place, particularly in 1988 in an effort to harmonize administrative regions with regional county municipalities (MRCs), which brought the number of administrative regions to 16. Since the 1977 splitting of Bois-Francs and Mauricie, there are now 17 administrative regions in Québec.

Attempts to rationalize the organization of the Québecois territory did not prevent the establishment of new municipalities and in the early 2000s, Québec was the province with the largest number of municipalities in Canada. All previous attempts to reduce their number failed. At the beginning of the 21st century however, the provincial government, of PQ allegiance, decided to force amalgamation in order to improve – among other things – efficiency in the municipal system. For lack of public support, this top-down attempt was partially deconstructed by the new Liberal government elected in 2003. The latter authorized municipalities to dissociate themselves under certain conditions. Among the consequences of this policy was an increase in the complexity of the existing political system and the creation of new entities at the local level[4].

There are three administrative or governance levels in Québec: a local level, the municipalities; a supra-local level, regional county municipalities, and metropolitan communities; and finally a regional level designated under 'administrative regions' (Table 3). Despite actions to counter municipal fragmentation, Québec remained in 2006 the province of Canada which still has the largest number of municipalities, i.e.,

4 For more information of the subject, read Trépanier (2008); one can also consult the MAMROT (Ministère des Affaires Municipales, des Régions et de l'Occupation du territoire du Québec) website at www.mamrot.gouv.qc.ca.

1,115[5], over a rather small portion of its territory, while this figure varies between 8 and 829 in the other Canadian Provinces or Territories (Rivard and Collin, 2006).

Table 3: Three levels of governance (Québec)

Level	*Designation*[6]	*Number*
Local	Municipalities[7]	1,115
Supra-local	Regional County Municipalities (MRC) Metropolitan Communities (CMM)	86 MRCs[8] 2 CMMs
Regional	Regions	17

Summary

Three findings arise from the spatial and demographic analysis of Québec. First, the activities and the population of Québec are scattered over a vast territory, leaving large expanses that are more or less unpopulated while the development of natural resources has caused discontinuous extensions of land use. The expansion of the Québec ecumene seems however to have stabilised in recent decades, new peripheral establishments are slightly compensated by the closing of certain remote centres, mainly decommissioned mining facilities (Proulx, 2003).

Québec is characterized by a centralisation and urbanisation movement in and around the large metropolitan areas (Montréal and Québec City mainly) fuelled over time by different concentration factors: the merging of various services (religion, health, education), the industrialisation process, the location of prime services (namely culture and higher education), and more recently the development of commercial centres (Proulx, 2003). Even the rural zones "undergo metropolisation" (Polèse and Shearmur, 2003) since a growing proportion of the Canadian rural population lives in proximity to a metropolis (from 38% to 42% between 1971 and 1996). Finally, significant territorial inoccupation and the concentration of the population around two large metropolitan areas have given rise to a distinctive political and administrative organization, divided into several levels of governance.

5 This figure excludes Cris and Naskapis in the North of Québec, which number 23.

6 For more information on the functions of these entitites, visit the MAMROT website (Ministère des Affaires Municipales, des Régions et de l'Occupation du territoire du Québec).

7 Some municipalities have also boroughs. For more information: ENAP (2010) *L'État québécois en perspective. Structure et taille de l'État. Les institutions, objets de décentralisation politique*. Online.

8 For more information: ENAP (2010) *L'État québécois en perspective. Structure et taille de l'État. Les institutions, objets de décentralisation politique*. Online.

Given these characteristics, it is useful to further examine the dynamics outlined in this section between metropolitan areas on the one hand and medium-sized cities on the other hand.

3. Metropolitan Areas and Medium-sized Cities: the Regional Dynamic the Cities

Apart from small towns, Québec has a network of medium-sized cities and metropolitan cities whose functions are quite varied, sometimes founded on the development of natural resources, sometimes on manufacturing or, for the most important among them, on services. This three-part section examines the dynamic thrust of large and medium-sized cities. In the first part, we discuss the metropolisation and urban decentralisation phenomena and employment breakdown. The two other parts deal successively with Montréal and Québec City.

3.1. Metropolisation, Urban Decentralisation and the Distribution of Employment

A growing part of the population and activities are now centralised in large centres, or in the proximity of large urban centres. The Canadian population living within one hour of access to one of the eight large Canadian metropolitan cities (Montréal and Québec City in Québec) is increasingly significant, at 72% in 1996 (Polèse and Shearmur, 2003: 32).

At the local level, as in numerous North American metropolitan areas, this movement of centralisation is characterized by a linked phenomenon, urban sprawl. Thus, we have witnessed in recent decades a deconcentration or dispersal of residential activities, then consumption services, manufacturing activities and even office rentals. This trend added to the rise of new peri-urban forms and a configuration of edge cities[9] (Garreau, 1991), edgeless cities[10] (Lang, 2003), of boomburbs[11] (Lang, 2007) and others.

In Québec, these types of urban forms are barely existent, even if sprawl can be witnessed. New attraction centres have been created, especially on the edges of large boulevards as well as in proximity to freeways (Proulx, 2003) but not to such an extent as can be observed in

9 Coined by Joël Garreau to designate a major concentration of office, commercial and leisure buildings and establishments at the periphery of urban centres, in proximity to freeways.

10 Neologism introduced by Robert Lang designating strings of scattered office buildings in "edgeless" suburban environments.

11 Cities with less than 100,000 inhabitants, not a regional capital but that have witnessed a growth of more than 10% over the course of recent years.

the US context. Québecois metropolitan cities (Montréal and Québec) thus retain their centre-driven attraction featuring their cultural character as well as major infrastructures such as their international airports.

Nevertheless, accessibility to the metropolis remains of crucial importance (particularly in the case of Montréal), as demonstrated by the geographic clustering in a number of economic sectors. The deconcentration observed in the manufacturing sector benefits mainly the metropolitan areas that offer good accessibility to and within the city (in particular high tech manufacturing which depends on information and knowledge), thus easing operations and transactions with suppliers and clients, providing access to transportation infrastructures for all, customised services and others. Tourism and entertainment as well benefit from being within or close to a metropolis, as does agriculture and resource development. Typically, resources are more readily available within rural zones said to be "central": the prime farmland is located near large cities, the summer season is also longer (relative to farmland further up north) and markets, suppliers, distributors and processing or manufacturing operations are located nearby (Polèse and Shearmur, 2003).

New information and communications technologies do not fundamentally change the situation in this respect: proximity to colleagues, airports, and the importance of face-to-face meetings remains essential. It was also observed that telework is used to offset commuting, especially in the vicinity of metropolitan areas (urban workers seeking a more natural living environment are attracted to nearby rural zones).

Economic and structural changes brought about by the phenomenon of metropolisation affect the cities of Québec very differently. The attractiveness of metropolitan cities contributes to the development of accessible medium-sized cities close to large urban centre; conversely, cities of mixed sizes located at the periphery or outside metropolitan areas may be subject to stagnation or to economic and demographic decline nearby (Polèse and Shearmur, 2003). It would seem therefore that "agglomeration economies progressively are progressively replaced by economies of accessibility" (Proulx, 2003).

As manufacturing activity gets thinner in the Montréal area, it is on the rise in medium-sized cities, sufficiently so to shape an identifiable "Québec Industrial Belt" (Polèse, 2009b: 12) or a "Québec Industrial Crescent" or bow (Proulx, 2003). This region, considered as the "seat of industrial thrust" (Polèse, 2009a) par excellence in Québec, extends from Saint-Jean-sur-Richelieu, near Montréal, up to Rivière-du-Loup approximately, and includes the cities of Granby, Victoriaville, Drummondville, Cowansville, Ste-Marie-de-Beauce and St-Georges-de-Beauce (Figure 4). According to Polèse (2009a; 2009b), this configuration confers a strategic advantage other regions in Québec do not have;

proximity to large cities and to the American market (compared to regions on the North Bank of the Saint Lawrence River) as well as the presence of a stable labour force, competitive salaries and a significant industrial pedigree. Thus, inter-city competitiveness is not played out between regional activity generators and Montréal, but involves medium-sized cities themselves, especially those that shape and structure the industrial belt (Polèse, 2009).

Montréal has had and increases its leadership in prime services employment. Montréal is considered as the city whose economy is the most diversified in Canada (Consortium de la Communauté Métropolitaine de Montréal – CCMM –, 2008). Montréal's economy is based *inter alia* on the information industry (telecommunications, broadcasting, digital media, etc.), professional services, distribution (apparel and food), financial intermediation and portfolio management, manufacturing, cultural events and performing arts, air and rail transportation, post-secondary education, hospitals, fashion design, and pharmaceuticals as well as power distribution. Montréal also shows noticeable progress on the Canadian scene with regard to several "metropolitan" functions, namely culture and the information sector as well as expert services such as architecture, engineering, electronic data processing, computing services, and graphic design (Polèse, 2009a).

The tertiary industry activity remains predominant and is thriving; between 2004 and 2008, Montréal's service sector share of total jobs rose from 80.5% to 84.5%. More specifically, trade and commerce, health care and social services as well as professional, scientific and technical services make up the main employment fields within the Montréal agglomeration.

The sectors displaying the most significant increases in recent years are lodging and food services, professional services, scientific and technical services, health care services and social assistance services. The later have provided close to 54,300 additional jobs between 2004 and 2008 (City of Montréal, 2009).

Running parallel to this tertiarisation of the economy, the City of Montréal recently suffered important losses in the manufacturing sector where staff cuts (especially in textiles and clothing) and the industrial deconcentration have affected several areas of the Montréal Island. While the manufacturing industry (mainly apparel, textiles, food and beverage and the chemical product industry) accounts for close to 11% of the employment, jobs in this sector have plummeted in recent years, dropping by 37.8% between 2004 and 2008 (City of Montréal, 2009). On the other hand, a number of city districts experience a vocational transformation, particularly Mont-Royal, Westmount and Outremont where structural adjustment is gradually occurring; the economy is changing from manufacturing and business services to an important increase in services to the population. The southwest, once industrial heartland of the country (CMM, 2008), tends toward a growing specialisation in the offer of professional (design, marketing and advertising, computing and data services) and cultural services (film and television, performing arts, cultural events, and entertainment) due to its proximity to the downtown core. The Plateau Mont-Royal is also experiencing a shift in employment. While professional services and the cultural industry underwent rapid growth during the 1990s, these sectors showed signs of saturation at the turn of 2003. A recent economic upswing is mainly due to a notable increase of employment in the education and health sectors, and the treatment and restoration of film.

Figure 4. Regional economic growth in 2006

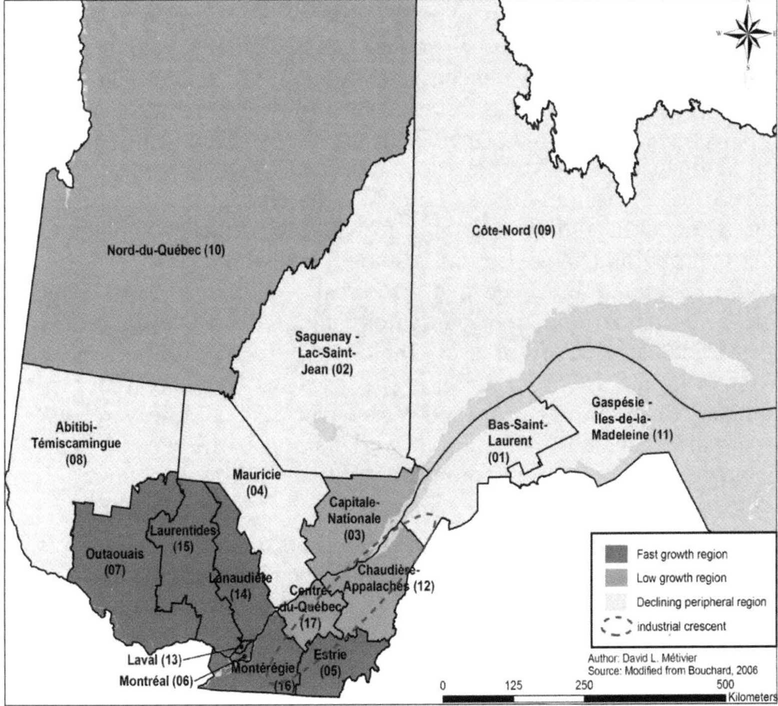

Source: Map by David L. Métivier.

Transformations occurring in Greater Montréal also benefit the suburban rings where the strong increase in employment observed in recent years endures, mainly in services to residents and construction services, but also, more recently, in professional services, transportation, wholesale trade, business support and financial services. The development of suburban belts is also buoyed up by the movement of industrial decentralisation in urban areas and suburbs welcome a growing number of industrial and manufacturing facilities.

The City of Québec also experienced a significant employment growth while the local economy diversifies and the manufacturing sector is buoyant. In addition to preserving its status as a regional service generator, Québec City is staging a vocational realignment with the emergence of technology centres often connected to the expansion of Montréal companies looking for a stable and well-trained workforce and savings on real estate expenditures.

3.2. Montréal among Large Canadian Cities

Formerly Canada's economic and demographic capital, Montréal has lost its national leadership over the recent decades. Since the 1970s especially, Montréal's rate of annual demographic growth was lower than that of Toronto and Vancouver (Martin, 1998) and the same applies to the economy. There has been, from 1980s and forward, a massive displacement of activities (mainly in the fields related to management and to finance) from Montréal towards Ontario's metropolis. From the mid-1960s to the mid-1990s, Toronto experienced an extremely rapid rate of economic and demographic growth, taking the lead among Canadian metropolitan areas, leaving Montréal far behind (Polèse, 2009a). It should be noted however that Montréal's economy never relented. Québec's number one metropolitan city simply lost its drive relative to other Canadian cities. With a 1% annual growth rate in employment between 1996 and 2005, Montréal lags behind the country's other metropolitan areas (Toronto 1.8% and Vancouver 2.3%). However, in 2005, Montréal accounted for 2 million workers and is the second Canadian labour market, second to Toronto with its 3 million workforce (Heisz, 2006).

This upheaval in the ranking of Canadian metropolitan cities is not due to Montréal's less valuable or central geographical position, but to cultural and socio-economic factors. Polèse and Shearmur (2004) suggested that Montréal illustrates the case of Québec's shrinking hinterland: the rise of a francophone elite (anglophones had been dominant in Québec prior to the 1960s) and the transformation in the social fabric (political and linguistic activism, etc.). Québec is surrounded by an Anglophone population almost as such, Toronto has a definite advantage over Montréal and benefits from the cultural and linguistic affinity with matching markets, especially with respect to the location of prime services (Polèse and Shearmur, 2004). In a context marked by disturbing political uncertainty, inefficient management in the public sector (construction of second airport facility in Mirabel, costly Olympic installations in 1976) and lip service in the recognition of Montréal's specificity by higher level governments (Martin, 1998), Montréal witnessed an out-migration of decision-makers, the relocation of head offices and financial activities, and a substantial drop in airport activity and traffic flow, by and large in favour of Toronto.

Figure 5. Population growth rate in Montréal, Toronto and Vancouver, 1986-2006

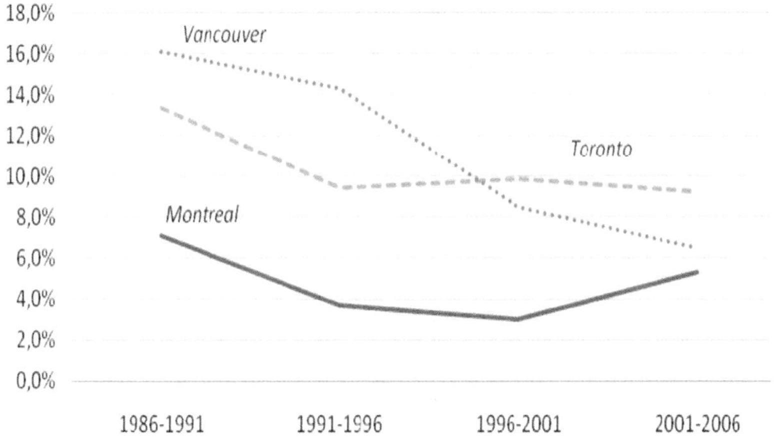

18,0%
16,0%
14,0%
12,0%
10,0%
8,0%
6,0%
4,0%
2,0%
0,0%
Vancouver
Toronto
Montreal
1986-1991
1991-1996
1996-2001
2001-2006

Source: Compilation and table by Anick Laforest (Statistics Canada 1991, 1996, 2001, 2006 censuses).

Recent indicators seem to show that Montréal is recovering while it adjusts progressively to its new hinterland and "reverses the losses relative to its Canadian standing" (Polèse, 2009a). For example, the evolution of the demographic growth rate in the three major metropolitan cities of the country shows that Montréal has been progressively recovering and approaches Toronto and Vancouver between 2001 and 2006 (Figure 5). What is more, as much as the rate of employment of Montréal has remained less than that of Toronto over the last twenty years, the difference between the two metropolitan areas has diminished in a significant way and is even tending to settle. Finally, not only does Montréal now exhibit good economic performance, but it seems less affected by the recent financial crisis (rising energy process and problems in the banking sector) because it is less dependent on the financial sector and the automobile industry than Toronto (Polèse, 2009a).

Montréal displays a number of typical metropolisation patterns. The observed large-scale concentration phenomenon is accompanied, at the regional level, by urban sprawl.

In addition to a movement of demographic concentration, it has been observed that Montréal's job market is increasingly specialized in fields that are typically metropolitan (finance, insurance, real estate, management, customised services), and particularly in the sector of information, culture and entertainment (Polèse, 2009a).

The metropolis is also gradually losing contact with its hinterland. The province of Québec is regarded as a quilt of peripheral resource-regions; their development is barely dependent of the Montréal market, but more of American markets. As Montréal is turning towards Canada and the rest of the world, it gradually frees itself from its hinterland, their respective markets presenting a low level of interdependence (Martin, 1998).

Figure 6. Population growth rate in Montréal agglomeration and metropolitan area, 1961-2006

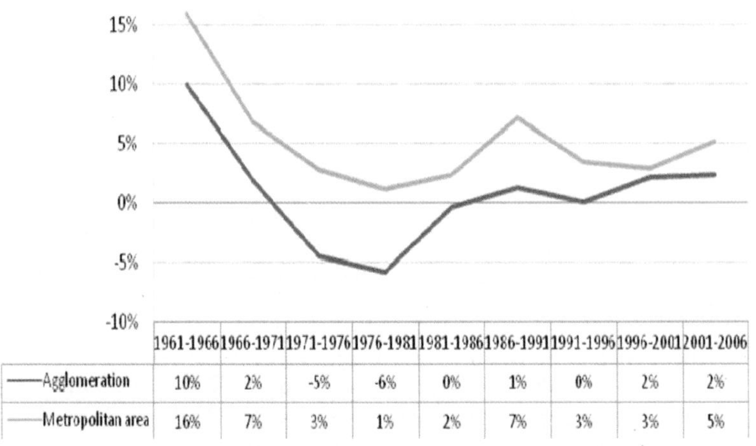

	1961-1966	1966-1971	1971-1976	1976-1981	1981-1986	1986-1991	1991-1996	1996-2001	2001-2006
Agglomeration	10%	2%	-5%	-6%	0%	1%	0%	2%	2%
Metropolitan area	16%	7%	3%	1%	2%	7%	3%	3%	5%

Source: Compilation and table by Anick Laforest (Observatoire du Grand Montréal 2009).

During the 1970s and 1980s especially, the population of the agglomeration of Montréal (i.e., municipalities on the Island of Montréal) decreased significantly while that of Greater Montréal continued to rise in the wake of the fabulous development of the suburbs – or boomburb (Figure 6). The suburban trend translates into a loss of the city's relative weight within the agglomeration and the metropolitan region over the recent decades; Montréal's population which was 78% of the metropolitan population in 1961 had dropped to 52.5% in 2006.

Moreover, the gap between the central city and its suburbs is not expected to toggle back in view of the Institut de la statistique du Québec's forecast predicting that, between 2009 and 2031, the population of the Island of Montréal will grow approximately 12%, while the South suburban ring could experience a 21.5% growth rate and Laval 28.6%. The larger suburban ring further to the North could experience a population growth rate nearly three times greater than Montréal's: Lanaudière 37.9%, Laurentides 34% (ISQ, 2009: 50).

The urban sprawl phenomenon around metropolitan Montréal is also reflected in the labour Employment attractors located outside the core, as they captured 57% of the metropolitan employment growth between 1996 and 2001 (Terral and Shearmur, 2008). These attractors now play a structuring role in Montréal's economy. Despite this tendency, the city centre is still the dominant job magnet in the metropolitan region, strong and dynamic, and still in command of the critical resources (Terral and Shearmur, 2008). In view of this phenomenon, it is suitable to qualify the ongoing process which takes place in the Montréal region as a "multipolarisation" and not a form of polycentrism; the latter involves a transfer of central and core functions to secondary or ancillary attractors (Terral and Shearmur, 2008). Therefore even where we observe a form of employment decentralisation which fosters the rise and subsequent development of secondary attractors in the city's peripheral areas, large scale enhancements such as edge city or edgeless city phenomena observed in the American context do not seem to typify Montréal (only Varennes and Mirabel have been identified as configurations reminding edgeless cities but the hypothesis remains to be substantiated) (Shearmur and Coffey, 2002).

The spatial distribution of employment is also an evidence of new centralities in the metropolitan region, largely and increasingly bundled or clumped on either side of freeways. Freeways are decisive location factors for all economic sectors and result in polarisation (Terral and Shearmur, 2008).

All the same, Montréal is still considered as one of the cities the least representative of the North American model in terms of urban sprawl (Coffey and Drolet, 1994; Collin and Mongeau, 1992; Filion, 1987). To take up the point of view held by Filion about the Québecois context, three factors could explain the lesser importance of exurbanisation (Filion, 1987): the relative [compactness of the agglomerations, which relates to a greater reliance on public transportation as well as restrictions on highway construction; the higher employment density and more retail outlets in central cities; and the presence of a middle class and an upper class in the town centre which puts off racial issues and crime. Finally, a fourth factor in Montréal relates to culture and quality of life which are typical of the Québecois metropolis.

In Montréal, quality of life is explained by different amenities. A network made up of 17 large parks covering close to 2,000 hectares is a major asset (City of Montréal, 2009). As well, several recent projects illustrate the different approaches that Montréal is taking as a sustainable city. The 1995 green neighbourhood program implemented in several boroughs intended to support environmental awareness and projects defined and managed by different local organisations. Moreover, Montréal offers several alternatives in terms of sustainable mobility: a well-developed public transportation system, a car sharing network (Communauto), and the latest improvement since the summer of 2009, convenient access to public bicycles or Bixi bikes (Figure 7).

Figure 7. Bixi bike station, Mount Royal Park

Source: Photo by Sandra Romaniuk.

In recent years, the efforts to revitalise several central districts have also multiplied. The Cité Multimedia (Figure 8), an international district and the Angus Science Park were developed successfully. More recently, the *Bassins* du Nouveau *Havre* project was approved, authorising the redevelopment of a section along the banks of the Lachine Canal.

Figure 8. Montréal's Cité Multimédia

Source: IFCS.

Montréal offers the image of a global metropolis. It is a multicultural and international city where large and highly visible cultural events are held annually – the Festival International de Jazz de Montréal (Figure 9), FrancoFolies de Montréal, the Just For Laughs Festival – and are main attractions. The recent creation of the Quartier des spectacles (cultural district) in downtown Montréal makes it the heart of today's cultural metropolis.

Figure 9. Montréal International Jazz Festival

Source: Photo by Anik Laforest.

3.3. Québec City: a Capital under Transformation

The time when Québec City outranked other cities in both the Province and Canada is now very much in the past. Nonetheless, the city is the second provider of employment in the province. Québec City is a very important regional attractor in eastern Québec. Even if the capital city does not encompass all the metropolitan functions, it displays specialisation in service sectors that brings Québec City apart from other urban centres in the province. Of course, public administration jobs are very important to the economy of the city for the reason that it is the provincial capital.

In 2006, employment in the service sector accounted for more than 85% of jobs in Québec City. More than half of the labour market (54%) is concentrated in 5 large sectors of activity. The most important is public administration, followed by health care and social services, retail trade, manufacturing as well as lodging and food services (CMQ, 2009). In Québec City, key growth sectors include applied technology – optics, photonics, electronics, software, data and computer services, geomatics, video games, defence and security –, health sciences, health and nutrition – biotechnologies, mainstream medicine, pharmaceutics, nutraceutics, functional foods, agro-food transformation –, as well as materials transformation, conversion or processing – plastics, composites, secondary wood processing – (Québec City, 2009).

In Québec City, the economy appears to evolve along several R&D lines such as the manufacturing of scientific equipment, optics, photonics, electronic data and computing, and biotechnologies (Polèse and Roy, 1999). It is also important to remind how significantly tourism contributes to the capital's regional economy. According to the Québec City Tourism Association, more than 4.5 million tourists visited the city in 2007, and that year, the local touristic industry provided close to 38,000 jobs. The richness of history and heritage makes Québec City as a very popular touristic destination. With more than 400 years of history, the City of Québec is considered the birthplace of French civilisation in North America. The city also provides a reminder of important steps in the colonisation of the continent by the Europeans. It was notably the capital of New France, the stage for the battle of the Plains of Abraham, and the capital of the new British colony. The historical district of Old Québec is classified as a remarkable and well conserved example of a fortified colonial city. Metropolisation phenomena are readily perceptible in the region of Québec City, especially exurbanisation. According to Statistics Canada (2006), population density in the Québec City CMA (Census Metropolitan Area) was close to four times less than in the Montréal CMA with 218.4 inhabitants/km^2 against 853.6. The strong development of highway systems between 1960 and 1980 certainly contributed to accelerate the suburbanisation process. During this period, urban expansion of the Québec City region was very unequal and was concentrated mainly along the new transportation vectors (Bolduc, 1999). At that time, road planning relied on much too optimistic demographic forecasts, as was later revealed. It was expected that the Québec City region would grow close to 2 million of population towards the end of the 20th century. Today, the Québec City CMA population is only 700,000 people. From the 1950s, population growth was very unequally distributed in the agglomeration. For several decades, most of the population growth occurred in the periphery while central districts generally experienced a decrease until the mid-1990s (Villeneuve and Trudelle, 2008), but figures indicate over the last several years a reversal of the situation. Downtown Québec City shows signs of a demographic and economic revival while several suburbs are stagnant or display negative growth rates.

Figure 10. Quartier St. Roch and Park St. Roch

Source: Photo by Simon Belisle 2009.

The borough La Cité, which includes the city's core sectors, shows a relatively good demographic growth rate between 1996 and 2001 (Table 4)[12].

La Cité also exhibits a sizeable increase in the number of jobs over the same period (Villeneuve and Trudelle, 2008). Apparently, the revitalization of the Saint-Roch sector (Figure 10) played a vital role in the revival of one of Québec City's core areas.

12 On November 1st, 2009, the number of boroughs in Québec City dropped from 8 to 6. The former boroughs *La Cité* and *Limoilou* were merged into a new entitiy: *La Cité – Limoilou*. The Laurentian borough was split in two; the south part was tied to the former *Sainte-Foy – Sillery* which became *Sainte-Foy – Sillery – Cap-Rouge* and the north was tied to *La Haute Saint-Charles*.

Table 4. Population and growth rate in Québec City boroughs

	1991	1996	2001	96/91	01/96
La Cité	61 970	60 015	62 110	-3,2 %	3,5 %
Limoilou	47 000	45 030	44 980	-4,2 %	-0,1 %
Sainte-Foy–Sillery	70 110	68 955	68 410	-1,6 %	-0,8 %
Beauport	69 160	72 920	72 810	5,4 %	-0,2 %
Charlesbourg	70 790	70 945	70 310	0,2 %	-0,9 %
Les Rivières	56 225	59 185	59 195	5,3 %	0,0 %
La Haute-Saint-Charles	41 625	46 605	47 215	12,0 %	1,3 %
Laurentien	72 930	80 545	82 965	10,4 %	3,0 %
Ville de Québec	489 810	504 200	507 995	2,9 %	0,7 %
RMR de Québec	645 550	671 890	682 755	4,1 %	1,6 %

Ref.: Villeneuve and Trudel, 2008. Source: Canada Census.

This core area saw a sharp expansion in the middle of the 19th century with the beginning of industrialisation in the city. It was the prime location for manufacturing industries such as textile and shoe factories. For close to a century and up until the middle of the 20th century, it was Québec City's business centre (Desormeaux and Collin, 2004). But in the second half of the 20th century, in the wake of sweeping suburban development and the relocation of many businesses, the area declined rapidly and deteriorated. It was the object of a revitalization program initiated by Mayor Jean-Paul L'Allier at the outset of the 1990s. The focal element of this revitalization project was the conversion of vacant lots into a large 8,000 square metres urban park, with an objective to embellish and improve the urban setting (Desormeaux and Collin, 2004). Since the early 1990s, more than 375 million dollars have been invested in the sector by various public and private players (Desormeaux and Collin, 2004).

As such, the revitalization of the neighbourhood attracted to St. Roch numerous companies, including higher learning institutions, computer and data corporations, as well as multimedia companies.

Despite the recent economic recovery in downtown Québec City, the rationale for employment decentralisation is at work and the phenomenon endures. Within the city, two major economic generators can be distinguished, the downtown area and the Boulevard Laurier sector (Figure 11) in St. Foy.

Figure 11. Boulevard Laurier, Sainte-Foy-Sillery borough

Source: Photo by Simon Bélisle, 2009.

For the longest time, downtown Québec City has been the regional employment attractor. While the core sector remains a well-established provider of jobs, the St. Foy sector is undoubtedly the second major employment attractor in the region and has been so over many years. The Sainte-Foy – Sillery borough sustains 65,150 jobs, of which 58,000 are along the Boulevard Laurier commercial strip, while La Cité, the core sector, offers 79,825 jobs (Québec City, 2009). St. Foy is an integral part of the regional metropolis (since 2002).

Summary

Apart from smaller towns, Québec includes an array of medium-sized regional cities, and two metropolitan cities; their functions are quite diverse and include the development of natural resources, or manufacturing or on services in larger places. Manufacturing activity has progressively concentrated in medium-sized cities, especially those that make up Québec's industrial crescent (Polèse, 2009b) while the Montréal and Québec metropolitan areas have tertiarised their economy.

The worldwide metropolisation phenomenon affects Québec; it causes the concentration of activities and of populations mainly in two metropolitan areas: Montréal and Québec City. Where Montréal displays an urban landscape that is relatively more compact than Québec City, both metropolitan areas have retained their chief central function. Urban sprawl is a major issue in the context of climate change. The Québecois metropolitan cities both implemented sustainable development policies in order to advance toward more viable urban systems, either through the rehabilitation of core areas or the development of assets, public transportation policies, etc. Nevertheless, demographic reality must not obscure regional issues and the assets; indeed regions are faced with problems of another nature.

4. Peripheral Regions: From Decline to Repositioning

Peripheral regions are quite different from metropolitan regions: they are low density regions located away from larger urban agglomerations. Different criteria can be used to characterise these regions. Coffey and Polèse (1988a; 1988b) define peripheral regions on the basis of two criteria: size (number of inhabitants) and distance (from large centres) whereby "all large metropolitan areas are viewed as central (more than 500,000 inhabitants in 1996), including all urban and rural sectors located less than one hour's drive approximately from these metropolitan cities. Beyond this distance, or travel time, regions are classified in the *peripheral* category" (Polèse *et al.*, 2002: 5).

The concept of peripheral region is a subject of debate[13], be it only that a "region is always peripheral relative to other places" (Polèse *et al.*, 2002, p. 4). A large part of the Québecois territory belongs to what we may call peripheral regions, regions that would have been designated as rural regions at the time of the Agricultural and Rural Development Act (ARDA). We call them resources' regions as well, simply because their economy is mainly based on the development of natural resources. As Polèse *et al.* explain, peripheral regions "are regions without a large urban centre and distant from important markets. They can be called non-metropolitan regions, remote regions, peripheral regions and so forth, but none of these terms was satisfactory. Traditionally in Québec, peripheral regions are called resource regions. We have discarded this label because it implies that the future of these regions is permanently and exclusively linked to natural resources; what is more, we are also interested in the Atlantic Provinces and to four Nordic countries. We therefore embraced the adjective 'peripheral', because it expresses both the ideas of distance and relative situation" (2002: 4).

---

13 We are also reminded that periphery is used to designate a zone of urban decay in metropolitan regions, especially in core sectors; this issue was one of the key messages of a 1989 report tabled by the Conseil des affaires sociales, *Deux Québec dans un*.

The reliance of peripheral regions on natural resources has nevertheless ensured their development for many years but more recently, overexploitation of resources has modified the potential of many regions. We have seen the status of regions in Québec; the next section analyses the situation in peripheral regions.

4.1. Regions of Québec: Contrasts in Demographic Evolution

The regions of Québec are not equal in terms of development. Some regions have clearly done better than others. The regional population increase clearly shows this. The following chart (Figure 12) shows the population growth rate in the 17 administrative regions of Québec and in the province over the last 10 years (two 5 year periods: 1998-2003 and 2003-2008). Upon analysing the chart, two elements clearly stand out. First, population growth increased from one period to the other. It was more rapid in the last 5 years (6.9 per 1,000) than in the previous 5-year period (5.1 per 1,000), as was also observed elsewhere in Canada.

Secondly, if population growth is positive in most administrative regions, even though sometimes weak, five regions saw their population shrink (Figure 12). The five regions are Abitibi-Témiscamingue, Bas Saint-Laurent, Côte-Nord, Saguenay – Lac-Saint-Jean, and Gaspésie-Îles-de-la-Madeleine. But, as noted by Girard and St-Amour: "Regions in demographic decline over 1998-2003 all saw their situation improve in 2003-2008. Three regions, Côte-Nord, Saguenay – Lac-Saint-Jean and Gaspésie – Îles-de-la-Madeleine still registered population losses, but they were of lesser range. Abitibi-Témiscamingue and Bas Saint-Laurent saw their population stabilise between 2003 and 2008. In Mauricie, population growth resumed" (2009: 26). At the other end of the spectrum, population growth is strong in Outaouais which benefits from its proximity to Ottawa, Canada's capital. Population growth was especially vibrant in Montréal's suburban ring, which concerns 4 regions: Lanaudière, Laurentides, Laval and Montérégie. There, population steadily increases while population growth has been fading in Montréal. Montréal is the only Québecois region where demographic evolution deteriorated in a significant way over the last ten years, dropping from 7.8 per thousand over 1998-2003 to 0.5 per thousand over 2003-2008.

It is somehow possible to recognize a basic organisation in Québec, which was identified in 1989 by the Conseil des affaires sociales: (i) a central Québec located to the South East and which includes Montréal, Gatineau and Québec City; and (ii) another Québec, peripheral, that covers a belt extending from Gaspésie – Îles-de-la-Madeleine to Abitibi-Témiscamingue, and includes the Bas Saint-Laurent, Côte-Nord and Saguenay – Lac-Saint-Jean. The belt regions contrast with metropolitan areas. They constitute the peripheral regions that we examine next.

A new balance is being pieced together among the regions of Québec. Polèse identified "two trends that bring with them the promise of a new equilibrium between Montréal and the other regions of Québec: that is, the industrial potency of the cities in south east Québec and, concurrently the emergence of residential economies based on lifestyle choices, notably for retired population" (2009b: 12). Both trends stimulate the regions outside the Island of Montréal although they depend on the metropolis to a certain extent.

Figure 12. Average annual population growth rate of growth, administrative regions and the whole of Québec, 1998-2003 and 2003-2008

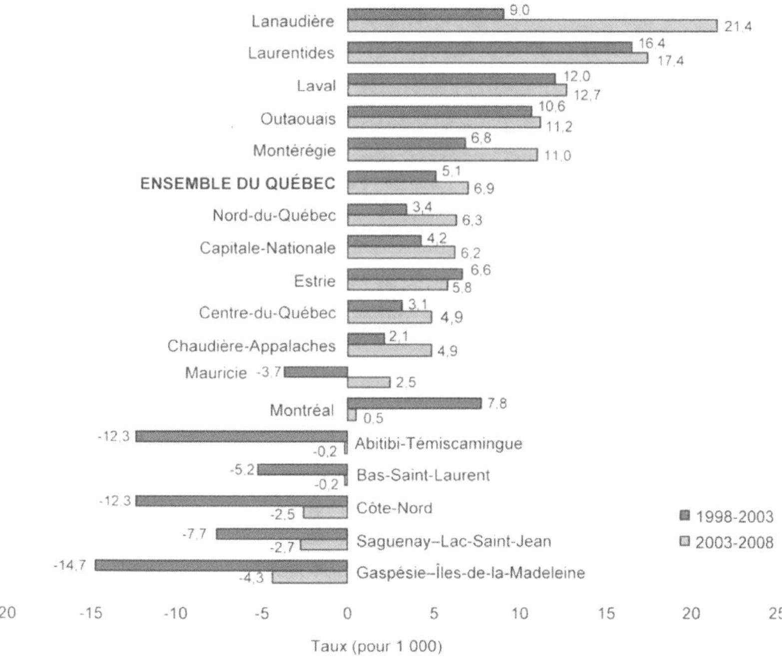

Source: Girard and St-Amour, 2009.

The first trend is related to the tertiarisation of Montréal's economy that increasingly specialises in prime services and in industries with value-added cultural and international content, as we have seen already. This specialisation results from the displacement of the more cost-sensitive manufacturing industries (labour, real estate, traffic congestion) towards smaller cities.

For the most part, cities of South East Québec benefited from this relocation. They form an industrial belt extending from Saint-Jean-sur-Richelieu to Rivière-du-Loup and include the Eastern Townships (Estrie), Centre-du-Québec and Beauce (Polèse, 2009b).

The second trend is more or less the upshot of an ageing population; segments of this population tend to relocate outside of metropolitan environments. Again, the successful regions are in the south of Québec, near the large cities. The phenomenon gives rise to residential economies as defined by Davezies' (2008). Regions that come out as residential economies are north of Montréal (Laurentides and Lanaudière) and south of Montréal (Estrie); another region, Centre-du-Québec, is also a winner in this respect.

Both trends are profitable for the regions located in the south east of Québec, even where they simply point to re-adjustments or shifts overall. At the same time, population shifts allow the determination of an array of peripheral regions which are now probed.

4.2. Peripheral Regions: Development Perspectives

The peripheral regions of Québec include five regions located in Northern and Eastern Québec: Gaspésie – Îles-de-la-Madeleine, Bas Saint-Laurent, Côte-Nord, Saguenay – Lac-Saint-Jean and Abitibi-Témiscamingue. Chenard *et al.* (2005) identified eight peripheral regions: to the five above mentioned regions, they suggest adding the western part of the Outaouais administrative region[14] (i.e., the regional municipalities of Pontiac County and La Vallée-de-la-Gatineau), and the northern part of Mauricie (La Tuque and Mékinac), whose economies are based on the development of natural resources, as well as Nord-du-Québec where development arises in a rather different context from that which is observed in the south. Overall, the eight regions aggregate a little more than 800,000 inhabitants (2001 data), but 25% of those are in Saguenay – Lac-Saint-Jean. Peripheral regions thus account for a little more than 10% of the population of Québec.

These regions are somewhat marginalised; their development capacity is clearly less than what is observed in metropolitan Québec, as we have seen earlier. Chenard *et al.* (2005) identified three characteristics shared among the eight peripheral regions of Québec and which account for their peripheral status:

---

14 As noted by Chenard *et al.* (2005: 12), "the definition of a peripheral region applies less and less to Outaouais (west)" chiefly because the population of the Outaouais region is expected to increase as a consequence of the expansion of the zone of influence or Umland of the Ottawa-Gatineau urban area: seasonal homes, proximity-based tourism, telecommuting, etc.

- Demographic vector: with the exception of Nord-du-Québec, all peripheral regions experience a decrease in population. The phenomenon is relatively recent, except perhaps in Gaspésie, and attends to the economic difficulties of the last several decades.
- Industrial vector: all peripheral regions are strongly dependent on resource-based industries or primary processing industries. This dependence "almost inevitable means future employment losses" (Chenard *et al.*, 2005: 14), especially where the natural resources have thinned off as a consequence of over-exploitation, as is the case for fisheries and forestry. Moreover, globalisation gives rise to a reorganisation of the international production system, as it can lead to more concentration in urban regions.
- Urban vector: all peripheral regions don't have a strong urban structure. Not only are these areas barely populated, but, in addition, their population is scattered over a vast territory (Côté et Proulx, 2002). As a result, there are no large towns and prime services are under developed in the local employment portfolio. There has been no evolution in this respect since the early 1980s, except possibly in Outaouais. This disadvantage is clearly a consequence of an ineffective urban structure; mainly, peripheral regions are not urbanised[15]. The absence of large cities is a legacy that is difficult to turn around; it also restricts the scope of development opportunities.

The challenge consists in successfully addressing the adverse demographic trends. All the peripheral regions are faced with a decrease in population with an exception for the Outaouais region which shows progressive integration with the economy of the Ottawa-Gatineau metropolitan region. From 2000 and forward, the population chart shows decreased losses, but it is too early to talk about sustained growth. The case of Saguenay – Lac-Saint-Jean is a concern in this respect. Not only is the region losing its young people, but also young adults at the launch of their career, which indicates "a labour market with distinctive problems" (Chenard *et al.*, 2005: 20). Moreover, it would seem that such migration is essentially permanent.

In these circumstances, job creation and company start-ups are probably a first step in meeting the challenge facing these peripheral. Development strategies should be adapted to the different regional profiles, given that their individual potential is basically different, be it in terms of the resources at hand and their relative location to larger systems. Likewise, the consolidation of urban attractors appears to be an influencing factor in view of developing the peripheral regions. As Chenard

---

15 We are reminded on the other hand that rampant destructuring affects local populations whose access to services is more than often tedious.

et al. have indicated: "The development of the upper tertiary sector and of more knowledge-based industries relies largely on the availability of an appropriate urban framework" (2005: 21). But, urban attractors in peripheral regions do not seem to be undergoing consolidation. The cities of Rivière-du-Loup and Rimouski, both in Bas Saint-Laurent, seem to be increasing their sway in their regions of influence. Their growth, however, draws partly on the exodus from the cities and towns in their own hinterland.

Over the course of the recent decades public authorities have paid great attention to peripheral regions in an attempt to promote their development; this question is now examined.

4.3. Countering the Decline

In order to tackle the problems that the peripheral regions have experienced and continue to experience, governments have designed measures to counter regional decline and to boost development in peripheral regions (Douay *et al.*, 2008). The Québecois approach to regional development is original, and yet it is similar to trends observed in other developed countries. It was in 1963, at the very beginning of the Quiet Revolution that the Government in Québec started to intervene to revive its peripheral regions. The Eastern Québec Planning Bureau was created and, based on a participant survey, was assigned to prepare a master plan for a region of the Lower Saint Lawrence and Gaspésie. This was a double break-through in the ways of addressing regional development. An ongoing experimentation was then launched to determine or outline regional development processes. In June 1966, the Planning Bureau released a substantial report with no less than 231 recommendations that supported six major objectives: modernising traditional sectors, creating new dynamic activities, appreciation and upgrading of the labour force, implementation of a planning and participation framework, stimulating regional awareness, and rational structuring of regional space. Fifty years after the publication of the report, the same goals are still sought after, which is significant.

The Québecois model of regional development has greatly evolved since, in keeping with evolving practices observed elsewhere. There are, as Bourque (2000) suggests, two main periods in the approaches to development in Québec. A first period covers the 1960s and 1970s; it was based on fordist principles and marked by State intervention where the provincial government played a leading role: "state intervention is perceived as necessary to dampen or curb social and regional inequalities" (Bouchard *et al.*, 2005: 3). For all intents and purposes, implementing regional systems and infrastructures was designed to provide equal opportunity to develop regional potential.

The second period began roughly in 1980: the thrust is now based on the mobilisation of local and regional players and stakeholders who take action. The first step was the creation of regional county municipalities (MRC), designed to include relatively small areas matching the zone of influence of small urban agglomerations. As Douay *et al.* (2008) explained: "This second period is essentially one of partnership. While, in the first period, only higher orders of government (federal or provincial) were responsible for planning, cities and regions have become major players." At the outset, the government of Québec entrusted regional county municipalities (MRC) with the responsibility for planning, which translated into the adoption of regional plans. The process did not lead to a rethinking of territorial organization but MRCs were nonetheless recognized as significant stakeholders. Thus the provincial government assigned regional municipalities the responsibility for planning, and namely to integrate local development centres. In 2002, the process was extended to include and eventually merge both planning and development. MRCs were then required to prepare "a strategic vision of cultural, economic, environmental and social development designed to foster coherent governance within their jurisdiction" (Loi sur l'aménagement et l'urbanisme (LAU), art. 5 [9]). MRCs were now expected to preside over planning and development, but these two responsibilities are more often parallel or concurrent than naturally convergent or integrated. The rationale behind this move was that the same designated representatives were at once responsible for planning and for development and therefore co-ordination would be possible. As it turned out, there were problems with this approach because planning is essentially thought out at the municipal level while development is first designed on a wider scale, at the level of administrative regions which is provincial.

Douay *et al.* (2008) explain:

> The shift from the first to the second approach was not simple to achieve nor is it completed. Changes made since 2003 call for the implementation of Regional Conferences of Elected Representatives (Conférence Régionale des Élus, CRÉ); the model is sometimes believed to be a progressive move toward a basically neoliberal prototype (Bouchard *et al.*, 2005). It is nevertheless possible to view this as a contemporary re-interpretation of the State – Market – Civil Society partnership attempting to evolve along existing institutional capacities.

The current governance model remains complex. The administrative capacity of numerous players is often blurred, jurisdictions overlap and are often entangled. "The reforms of recent years have brought little change even where they allowed to advance the establishment of regional authority" (Douay *et al.*, 2008).

The establishment of Regional Conferences of Elected Representatives (CRÉ) was designed to assign accountability but in doing this the partnership model was transformed. The provincial government chose to assign responsibility for development to elected officials only, thus bypassing all civil society organizations who are needed to apply the development strategies:

> Whether cities and regions have become important stakeholders, it must be recognized that upper level governments still play a decisive role, essentially because of their financial capability. Indeed cities, MRCs and regions do not always have the means to tip the balance in terms of attractivity relative to other territories. Nor do they have all the fiscal and regulatory tools required to nurture their development; those are still within the jurisdiction of higher order governments. We therefore need to recognize that the Québecois model is characterized by both a centralised and decentralised devolution of authority (Douay *et al.*, 2008).

Summary

Overall in Québec the status of peripheral regions remains extremely complicated, from all points of view. Nonetheless, the "indicators of economic renewal" are abundant: "Next to the constraints, often external, that shape their evolution, we also witness the expression of internal dynamism that contributes to change the mix of activities within" (Côté and Proulx, 2002: 92). Signs of administrative renewal are manifest in the activities concerned with the regions' traditional resources (for example, the extension of transformation streams and processing lines toward developing secondary and tertiary transformation or processing), including the design of new activities that could be described as innovative. If development hurdles are numerous for peripheral regions as we have outlined them, there is nonetheless a genuine potential, available for development despite the fact that 40 years of attempts have not allowed to do so.

Conclusion

Québec is a region under meaningful remodelling, but at the same time large consistencies can be observed over time, especially the cleft between the populated Québec – a small strip-like zone located in the South – and a Québec made of resources' regions for the most part uninhabited. As the Conseil des affaires sociales noted in 1989, there are two Québecs: one is fully developed while the other is slowly declining If this processes dwells with the resources' regions which are experiencing very serious difficulties, some neighbourhoods in urban agglomerations and metropolitan regions are not exempt. In a sense, the conflict is not so much between urban Québec and 'rural' Québec as one could think; rather, the variance is between an entity that is plainly attuned to globalisation and ensuing changes, and another part that seems unable to follow changes. It follows that Québec's first challenge is to bring about development across its territory. Several years ago, it was proposed to completely discount extensive regional territories on the premise that there is no future in them. The proposition was not put into force, yet large expanses of the Québecois territory struggle painfully as indicated by stagnant population growth. Development problems are not exclusively in resource regions. They also affect many small and medium-sized cities, sometimes located in the vicinity of large centres, as well as numerous urban core areas.

Two other challenges will need to be met by Québec in the years to come. One is to adopt a sustainable development approach. By and large, Québec is recognized as a model for environmental protection. Québec manages its waterways to generate power and this is a definite advantage that the Province was able to draw on toward its development. But at the same time, much more needs to be accomplished to shape Québec into a sustainable development model, especially since Québec lags behind other regions of the world in this respect. Recent debates on the improvement of transportation systems in Montréal and elsewhere in Québec have shown all the difficulties that Québec faces when attempting to turn sustainable development principles into action. The same applies to the development of natural resources and to urbanisation patterns

The third challenge concerns demography. Numerous regions have stagnating demographics or, worse, a shrinking and ageing population as a result. Ageing population is an issue throughout Québec. In many respects, it is the key challenge facing Québec in years to come, be it only personnel renewal in the wake of massive retirements. Immigration is not the solution. In the current international competitive context, Québec will need to innovate to ensure sustainable development.

Review Questions

- Explain the importance of the St. Lawrence River, for the establishment of the first settlements in Québec.
- How can we characterize the distribution of the population in Québec.
- Describe Québec's Industrial belt. What is its importance for Québec's economy?
- Why can Montréal be considered a metropolitan area? Describe the metropolitan patterns, trends and challenges.

- What are the main characteristics of peripheral regions? How can their economic situation be explained?
- How is the territorial and political organization in Québec unique and why?

Bibliography

Baccigalupo, A. (1990), *Système politique et administratif des municipalités québécoises. Une perspective comparative*, Montréal: Éditions Agence d'Arc.

Bolduc, P. (1999), *L'étalement urbain et le développement du réseau autoroutier dans l'agglomération de Québec: le cas des banlieues situées à l'Ouest et à l'Est de la RMR*, Québec: Département de géographie, Université Laval, p. 36.

Bouchard, R. (2006), *Y a-t-il un avenir pour les régions? Un projet d'occupation du territoire*, Écosociété, Montréal.

Bouchard, M. *et al.* (2005), *Modèle québécois de développement et gouvernance: entre le partenariat et le néolibéralisme?*, Montréal: CRISES, Collection Études théoriques, No. 0505.

Bourque, G.L. (2000), *Le modèle québécois de développement. De l'émergence au renouvellement*, Québec: Presses de l'Université du Québec.

Bruneau, P. (1989), *Les villes moyennes au Québec*, Québec: Presses de l'Université du Québec.

Chenard, P. *et al.* (2005), *L'évolution économique et démographique et les perspectives de développement des régions périphériques du Québec*, Montréal: INRS.

Coffey, W.J., Polèse, M. (1988a), "Locational Shifts in Canadian Employment, 1971-1981, Decentralization versus Decongestion", *The Canadian Geographer*, 32, 3: 248-255.

Coffey, W.J., Polèse, M. (1988b), "La transformation de l'espace économique canadien 1971-1981: assistons-nous à un mouvement centre-périphérie?", *Revue d'économie régionale et urbaine*, 1: 103-117.

Coffey, W.J., Drolet, R. (1994). "La dynamique intramétropolitaine des services supérieurs dans la région de Montréal, 1981-1991", *Cahiers de géographie du Québec*, 44(123): 325-339.

Collin, J.-P., Mongeau, J. (1992), "Quelques aspects démographiques de l'étalement urbain à Montréal, de 1971 à 1991 et leurs implications pour la gestion de l'agglomération", *Cahiers québécois de démographie*, 21(2): 5-30.

Collin, J.-P., Léveillée, J., with the collaboration of Rivard, M., Roberston, M. (2003), *L'organisation municipale au Canada. Un régime à géométrie variable, entre tradition et transformation*, Montréal: INRS Centre Urbanisation, Culture et Société, Online: http://vrm.ca/documents/ICPS_FR.pdf.

Courville, S. (2000), *Le Québec. Genèses et mutations du territoire. Synthèse de géographie historique*, Québec: Presses de l'Université Laval.

Communauté métropolitaine de Québec (2009), *Portrait statistique de la CMQ*, Online: http://www.cmQuébec.qc.ca/centre-documentation/documents/statistiques/01-Portrait-CMQ.pdf. Viewed 20 Nov. 2009.

Conseil des affaires sociales, Comité sur le développement (1989), *Deux Québec dans un: rapport sur le développement social et démographique*, Boucherville: Gaëtan Morin Éditeur.

Consortium de la Communauté métropolitain de Montréal (2008), *L'emploi local dans la région métropolitaine de Montréal 2008*, http://www.cmm.qc.ca/fileadmin/user_upload/periodique/emploi_local_2008.pdf. Viewed 18 November 2008.

Côté, S., Proulx, M.-U. (2002), *L'économie des régions périphériques du Québec et son renouvellement actuel*, Chicoutimi and Montréal: CRDT and INRS.

Davezies, L. (2008), *La République et ses territoires: la circulation invisible des richesses*, Paris: Seuil.

Desormeaux, R. et Collin, D. (2004), "Le quartier Saint-Roch à Québec: exemple et applications", *Urbanité*, Nov.: 18-22.

Douay, N., Lewis, P., Trépanier, M.-O. (2008), "Le modèle québécois d'aménagement du territoire à l'heure des bilans", in J.-P. Augustin, *Villes québécoises et renouvellement urbain depuis la Révolution tranquille*, Pessac: MSH.

ENAP (2010), *L'État québécois en perspective. Structure et taille de l'État. Les institutions, objets de décentralisation politique*, Online: http://netedit.enap.ca/etatQuébecois/docs/ste/organisation/a-territorial.pdf.

Filion, P. (1987), "Concepts of the inner city and recent trends in Canada", *Canadian Geographer*, 31(3), p. 223.

Garreau, J. (1991), *Edge City-Life on the New Frontier*, Doubleday.

Girard, C., St-Amour, M. (2009), "Démographie", in *Panorama des régions du Québec*, Québec: ISQ.

Heisz, A. (2006), *Le Canada et ses villes mondiales: conditions socio-économiques à Montréal, Toronto et Vancouver*, Document analytique, No. 89-613-MIF au catalogue – No. 010, Ottawa: Statistique Canada, http://www.stacan.gc.ca/bsolc/olc-cel/olc-cel?lang=fra&catno+89-613-M2006010.

Institut de la statistique du Québec (2009), *Le bilan démographique du Québec. Édition 2009*, http://www.stat.gouv.qc.ca/publications/demograp/pdf2009/bilan2009.pdf.

Institut de la statistique du Québec (2010), *Coup d'oeil sociodémographique. La population des municipalités du Québec au 1er juillet 2009: quelques constats*, http://www.stat.gouv.qc.ca/publications/demograp/pdf2010/coupdoeil_sociodemo_fev10.pdf.

Lang, L. (2003), *Edgeless Cities: Exploring the Elusive Metropolis*, Washington DC: Brookings Institution Press.

Lang, L., LeFurgy, J.B. (2007), *Boomburbs: The Rise of America's Accidental Cities*, Washington DC: Brookings Institution Press.

LAU – *Loi sur l'aménagement et l'urbanisme du Québec*, http://www2.publicationsduQuébec.gouv.qc.ca/dynamicSearch/telecharge.php?type=2&file=/A_19_1/A19_1.html.

Manzagol, C. (1996), "La région géographique", in M.U. Proulx, *Le phénomène régional au Québec*, Québec: PUQ, 91-112.

Martin, F. (1998), "Montréal: les forces économiques en jeu, vingt ans plus tard", *L'Actualité économique*, 74(1): 129-153.

Polèse, M. (2009a), "Les nouvelles dynamiques régionales de l'économie québécoise: cinq tendances", *Recherches sociographiques*, 50(1): 11-40.

Polèse, M. (2009b), "Vers un nouvel équilibre régional? La force industrielle des villes moyennes du sud-est québécois et l'essor des économies résidentielles non métropolitaines", in *Panorama des régions du Québec*, Québec: ISQ.

Polèse, M., Shearmur, R. (2003), "La métropolisation du Canada, ou pour-quoi la population se concentre-t-elle autour des plus grandes zones urbaines?", in F. Charbonneau, P. Lewis, C. Manzagol, *Villes moyennes et mondialisation: renouvellement de l'analyse et des stratégies*, Montréal: Trames.

Polèse, M., Shearmur, R. (2004), "Culture, Language, and the Location of High-Order Service Functions: The Case of Montréal and Toronto", *Economic Geography*, 80(4): 329-350.

Polèse, M. and Shearmur, R., with the collaboration of Pierre-Marcel Desjardins and Marc Johnson (2002), *La périphérie face à l'économie du savoir: La dynamique spatiale de l'économie canadienne et l'avenir des régions non métropolitaines du Québec et des provinces de l'Atlantique*, Montréal: INRS Urbanisation, Culture et Société and l'Institut canadien de recherche sur le développement régional.

Polèse, M. and M., Roy (1999), "La dynamique spatiale des activités économiques au Québec: analyse pour la période 1971-1991 fondée sur un découpage centre-périphérie", *Cahiers de Géographie du Québec*, 43: 43-75.

Proulx, M.-U. (2003), "Polarisation dans la géo-économie contemporaine du Québec", in F. Charbonneau *et al.*, *Villes moyennes et mondialisation: renouvellement de l'analyse et des stratégies*, Montréal: Trames.

Proulx, M.-U., Brochu, I., Leblanc, P., Robitaille, M., Chiasson, G., Geoffroy, D., Gauthier, É. and Doubi, A. (2005), *Les territoires du Québec et la décentralisation gouvernementale*, Québec: Ministère des Affaires municipales et des Régions, www.uqar.qc.ca/crdt/documents/CRDT%20-%20Rapport%20MAMR%20-%20Territoires%20et%20d%E9centralisation.pdf.

Rivard, M., Collin, J.-P. (2006), *Le système municipal au Canada en bref*, Montréal: INRS Centre Urbanisation, Culture et Société, Online: http://www.vrm.ca/documents/SystemeMunicipal_Canada_Enbref.pdf.

Shearmur, R., Coffey, W.J. (2002), "A tale of four cities: intrametropolitan employment distribution in Toronto, Montréal, Vancouver, and Ottawa – Hull, 1981-1996", *Environment and Planning A*, 34(4): 575-598.

Simard, M. (2003), "L'espace social des villes moyennes au Québec", in É. Trames (dir.), *Villes moyennes et mondialisation: renouvellement de l'analyse et des stratégies*, Montréal, p. 334.

Terral, L., Shearmur, R. (2008), "Vers une nouvelle forme urbaine? Desserrement et diffusion de l'emploi dans la région métropolitaine de Montréal", *L'Espace géographique*, 37(1): 16-31.

Trépanier, M.-O. (2008), "Les arrondissements dans les nouvelles grandes villes au Québec", in Daniel Pinson *et al.*, *Métropoles au Canada et en France. Dynamiques, politiques et culture*, Rennes: Presses universitaires de Rennes, pp. 89-102.

Villeneuve, P. et Trudelle, C. (2008), "Retour au centre à Québec: la renaissance de la Cité est-elle durable?", *Recherches sociographiques*.

Websites

Institut de la statistique du Québec : http://www.stat.gouv.qc.ca/default.htm

Gouvernement du Québec : http://www.gouv.qc.ca

Ministère des Affaires municipales, des régions et de l'occupation du territoire: http://www.mamrot.gouv.qc.ca

Statistics Canada: www.statcan.gc.ca

Villes-régions du monde: www.vrm.ca

City of Montréal: http://ville.Montréal.qc.ca/

Québec City: http://www.ville.Québec.qc.ca/

CHAPTER 9

Ontario

The Shield, Suburbs, and Restructuring

Nairne CAMERON

Algoma University

Introduction

Ontario has long been known as the 'heartland' (McCann, 1987) of Canada for geographical, historical, political, and economic reasons. Located in the centre of Canada, it is well connected by road, rail, air, and sea with other parts of the country and to the northeast and Midwest United States, as well as to international destinations. It is one of the founding provinces of Canada and is now home to 13 million residents representing 39% of Canada's population[1]. As such, Ontario wields considerable political power with over a third[2] of all seats in the federal Parliament, which itself is also situated in Ontario in the city of Ottawa. Immigration is strong with 47% of all immigrants to Canada in 2007 choosing Ontario as their place of residence[3]. Ontario also continues to be an economic power with manufacturing businesses and financial and management industries. The province's population also sustains a large consumer market. However, current challenges to Ontario's dominant role include the: economic recession in the United States, the rising and fluctuating Canadian dollar and its effects on Ontario's manufacturers, instability in the key auto-manufacturing sector, all leading to a mounting deficit.

With Ontario being the country's second largest province, there are marked differences in physical landscape and settlement across its territory. The province is often divided into north and south regions

1 Statistics Canada, 2009a.

2 Parliament of Canada, 2009.

3 Ontario Ministry of Citizenship and Immigration, 2008.

(Figure 1). Northern Ontario covers 800,000 square km – and nearly 90% of province's area, but only has 6% of the province's population[4].

Figure 1. Ontario is Often Divided into North and South Regions

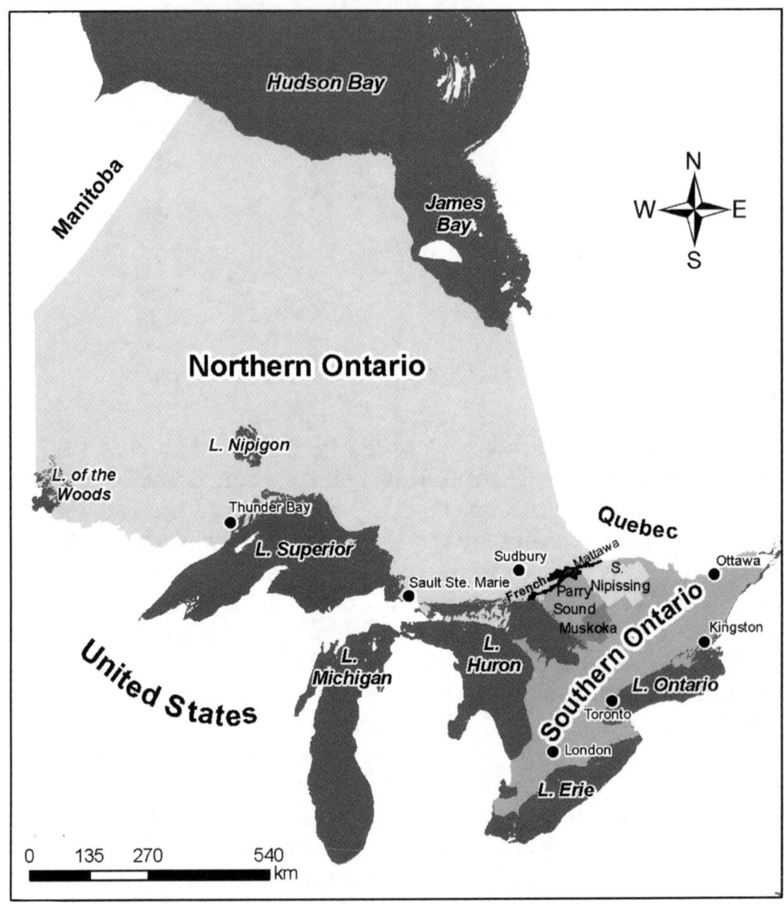

Source: Nairne Cameron, using spatial data from Natural Resources Canada, 2009a; Statistics Canada, 2006a; and Statistics Canada, 2010.

The actual boundaries of Northern Ontario are debated. Historically, Northern Ontario is the area north of Lake Huron, Georgian Bay, Lake Nipissing and the French and Mattawa Rivers (Woodrow, 2002). However, Parry Sound and Muskoka Census Divisions south of these boundaries are included in the Federal Economic Development Initiative of

4 Ontario Ministry of Northern Development, Mines and Forestry, 2009.

Northern Ontario (FedNor)[5]. Parry Sound is a designated area under the Province of Ontario Northern Ontario Heritage Fund, while the District of Muskoka became ineligible for Ontario northern development programs in 2004[6]. The area referred to as Northern Ontario in this chapter is pictured in Figure 2.

Figure 2. Northern Ontario

Source: Nairne Cameron using spatial data from Natural Resources Canada, 2009a; Statistics Canada, 2006a; and Statistics Canada, 2010.

The four main concepts woven through this chapter are: *density*, *diversity*, *division* and *conflict over space*, and *discontinuity* and *change*. While Ontario is Canada's most populous and highly urbanized province, there is variation in population *density* between sub-regions of the province and within cities. The dense population of Southern Ontario

5 FedNor, 2009.

6 Ontario Ministry of Northern Development, Mines, and Forestry, 2004.

(Figure 3) is contrasted with that of scarcely populated Northern Ontario reflecting the climate and physical characteristics of these two regions. The characteristics and size of urban areas in Northern and Southern Ontario are contrasted, and socio-economic differences between the two regions are explored. Sault Ste. Marie located on the border with the United States between Lakes Superior and Huron is highlighted as a northern city.

Figure 3. Southern Ontario

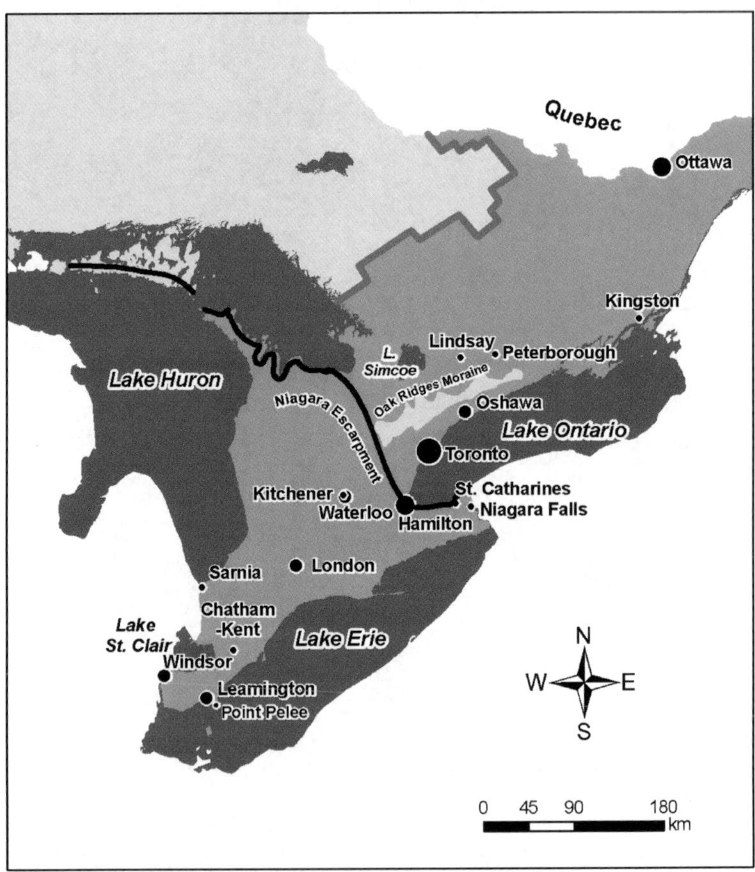

Source: Nairne Cameron using spatial data from Natural Resources Canada, 2009a; Statistics Canada, 2006a; and Statistics Canada, 2010. Feature boundaries also referenced from Eyles, 2002 and Oak Ridges Trail Association, 2010.

Next, the concept of *diversity* is examined from biological, economic, and population perspectives. The resource-based and generally less diversified economy of Northern Ontario is contrasted with the more diverse economy of Southern Ontario. By profiling two Northern Ontario enterprises, mining and forestry, a distinction is made between non-renewable and renewable resource industries. Then, Southern Ontario's diverse economy is highlighted, along with Canada's car industry. In relation to population, growth in the province has been propelled by immigration, but the spatial distribution of immigration has been uneven. Few immigrants move to Northern Ontario, whereas Toronto has been a magnet for new immigrants thereby creating a diverse city population.

Division refers to conflict over space, resource, wastes and land use. The chapter highlights the conflict and repercussions that have arisen in the province due to the conversion of agricultural lands to suburban areas. Finally, *discontinuity* and *change* encompasses both physical and economic events, breaks and shifts. For example, the increased accessibility with the construction of the St. Lawrence Seaway has brought in invasive species. The Ice Storm in Eastern Ontario in 1998 was a sudden event that interrupted regular life. Changes in trading agreements, international exchange rates, and globalization pressures have led to restructuring in Ontario's manufacturing sector. Also, the chapter explores the shift towards increased service sector employment in the province. Sudden economic change is illustrated by tracing the impact of the dot-com crash on Ottawa's high-technology industry and the decline of Ontario's auto-industry.

The supplementary concepts of *region*, *accessibility*, and *sustainability* are also addressed. Regions are geographic areas defined by common characteristics. They can help us to visualize patterns and compare areas. However, they must be treated cautiously as they may mask finer differences within each region. *Accessibility* or the ease of reaching a destination and transport modes has influenced the development of Ontario economy, settlement, patterns, and immigration. For example, the railway into Ontario's North opened up new mineral deposits.

While the economic slowdown of 2008-2009 reduced Ontario's energy demands, the province is a net importer of coal and crude oil to cover its energy needs. *Sustainability* is a particularly important concept in Ontario due to this energy challenge. Sustainability means the long-term capacity for both human and ecological systems to exist together and operate effectively. The province is a leader in Canada in terms of renewable energy. Ontario will need to continue to carefully manage its energy requirements for its future prosperity.

Physical Geography and Climate

Ontario is Canada's second largest province (following Québec)[7]. While the topography of the province is very flat with Ishpatina Ridge (north of Sudbury) being its highest point[89], the geology, soils, forests, water bodies and movement, and climate vary greatly over this vast territory.

Geology

Ontario is home to some of the world's oldest rocks with the Canadian Shield making up two-thirds of the province[10]. The Shield has created a landscape of rocks, lakes, and trees that inspired the sketches of the Group of Seven painters in the 1920s. Minerals mined from the Canadian Shield include: amethyst, cobalt, copper, gold, iron, nickel, platinum, silver, and zinc. The long history of mining in the Shield dating back to the 1800s has left several ghost towns (Brown, 2007).

The Precambrian Shield in Ontario consists of three provinces – the Superior, Southern, and Grenville, each formed during a different time period. The Superior found in Northern Ontario is the oldest of the three provinces and was formed 4.50-2.49 billion years ago[11]. Most of Ontario's mining activity takes place in this province (Eyles, 2002).

The Southern Province developed between 2.49 billion and 570 million years[12] ago and appears on the north shore of Lake Superior and the northwest shore of Lake Huron. This province includes the Sudbury Structure, mined for its nickel and copper, which was likely formed by a meteorite impact (Eyles, 2002). The youngest and southernmost Province is the Grenville which was formed one billion to 570 million years ago[13] and stretches eastwards from Georgian Bay to the Québec border and to the St. Lawrence River. The Shield areas of Southern Ontario are mainly used for recreational purposes such as summer cottages.

Later, between 570 and 63 million years ago[14], sedimentary rocks such as limestone, shale, and sandstone were deposited on the northern and southern edges of the Canadian Shield. These rocks are at the

---

7 Statistics Canada, 2005.

8 Government of Ontario, 2009.

9 It is located in the Temagami Region 95 km north of Sudbury (CanaTREK, the Summits of Canada, 2006).

10 Government of Ontario, 2009.

11 Ontario Ministry of Northern Development and Mines, 1994.

12 Ontario Ministry of Northern Development and Mines, 1994.

13 Ontario Ministry of Northern Development and Mines, 1994.

14 Ontario Ministry of Northern Development and Mines, 1994.

foundation of the Hudson Bay and James Bay Lowlands in Northern Ontario and the St. Lawrence and Great Lakes Lowlands in Southern Ontario. The Niagara Escarpment is a notable example of a sedimentary topographic feature, running 600 km from the Niagara Peninsula in a northerly direction to the Bruce Peninsula then onwards to Manitoulin Island. It has been preserved by a resistant top layer of Silurian aged (409-438 million years ago) dolostone overlaying softer shales. The Escarpment is the height of land over which the Niagara River flows forming Niagara Falls. It also surrounds the "Fruit Belt" (Krueger, 1959)[15] that borders Lake Ontario between Hamilton and Niagara Falls where vineyards for wine making are also located. The dolostone is suitable for aggregate and the escarpment for development creating land use conflict.

Glaciation has also influenced Ontario's landscape. Glaciers are slow moving, flowing ice bodies that form from a build-up of snow over several years which eventually turns to ice (Hambrey and Alean, 1992). During the Pleistocene epoch which began 1.8 million years ago, glaciers at the earth's two poles grew towards the equator. The Laurentide Ice Sheet moved over Ontario twice, completely covering Ontario the second time during the Wisconsinan Glacial Stage and reaching 3 km in thickness in some places. As the glaciers moved south, they scraped the soil off the rock surface of the Canadian Shield, leaving little soil cover.

Eventually, the temperatures warmed and the Laurentide Ice Sheet began melting, completely disappearing 6,000 years ago (Eyles, 2002). There are four main types of deposits that can be found in Ontario related to the retreat and melting of the glaciers[16]: till, glaciofluvial, glaciolacustrine, and glaciomarine. Till describes glacial deposits that have simply been released by glaciers, and have not been shaped by rivers or lakes. An example is *drumlins* which are teardrop shaped hills made of sand and silt that have their steeper end pointing towards the direction of glacial advance and tapered end leading in the direction of ice flow (McMaster University, 2010). There is a large drumlin field with more than a thousand drumlins (McMaster University, 2009) in the Peterborough area east of Toronto[17], making Peterborough the "Drumlin Capital of Canada" (Eyles, 2002).

Moraines are another type of till deposit. They comprise of sand and gravel dumped or moved in front of or between ice sheets as glaciers retreat. There are several moraines in Ontario. One of the largest is the

---

15 The late Dr. Ralph Krueger from Waterloo University published extensively on the Niagara Fruit Belt. http://library.wlu.ca/archives/collections/krueger.

16 Ontario Ministry of Northern Development and Mines, 1994.

17 Canada Centre for Remote Sensing, 2008.

Oak Ridges Moraine north of Toronto. It is a major source of aggregate (sand and gravel) for Southern Ontario. At the same time, it is a groundwater reservoir, creating a conflict between development uses and groundwater preservation.

Glaciofluvial deposits are created by rivers. *Eskers* are an example of such deposits which form when melting water flowing in streams underneath a glacier deposits sand and gravel. When the glacier melts, the eskers are left as winding ridges. The Munroe Esker is part of the province's Esker Lakes Park near Kirkland Lake and is one of Canada's largest eskers with a length of 250 km[18].

Glaciolacustrine deposits are formed in lakes filled with meltwater as glaciers retreat. The *clay belts* of Northern Ontario are glaciolacustrine in nature and formed when the glaciers retreated and large lakes formed in which fine sediments developed on the lake bottoms. Settlements in the clay belts include Hearst, New Liskeard, Timmins, and Kapuskasing. The soils in the clay belts are suitable for agriculture (Eyles, 2002). However, due to a short growing season, agriculture is challenging in this area (Kemp, 1993).

Deposits on sea floors related to glaciers, are termed glaciomarine. Glaciomarine clay deposits formed in the Ottawa area as the Laurentide Ice Sheet withdrew and the Champlain Sea expanded up the St. Lawrence Valley. One particular type of clay, called Leda, can become very unstable and liquefy upon physical disturbance or saturation with water. Leda clay deposits in the town of Lemieux, east of Ottawa, led the province to abandon the town in 1991 and resettle its residents. Two years later, a landslide devastated the old town's main street (Canadian Geographic, 2009).

Physiography and Soils

The term physiography encompasses both the combined topography and geology of an area. In Canada, the two main physiographic regions are the Shield and the Borderlands (Bostock, 1970). The Shield can be envisioned as a basin shaped core of ancient rocks. In the centre of the Shield lie younger Paleozoic sediments which form a subregion called the Hudson Bay Lowlands. Surrounding the Shield are the Borderlands, made up of younger sediments, with the portion in Ontario termed the Great-Lakes-St. Lawrence Lowlands (Figure 4). Within Ontario, the differing nature of the three main physiographic regions: The Hudson Bay Lowlands, The Canadian Shield, and the Great Lakes-St. Lawrence Lowlands (Trenhaile, 2007) combined with climatic and locational factors has influenced agricultural activity.

---

18 Ontario Parks, 2009.

Figure 4. Physiography of Ontario

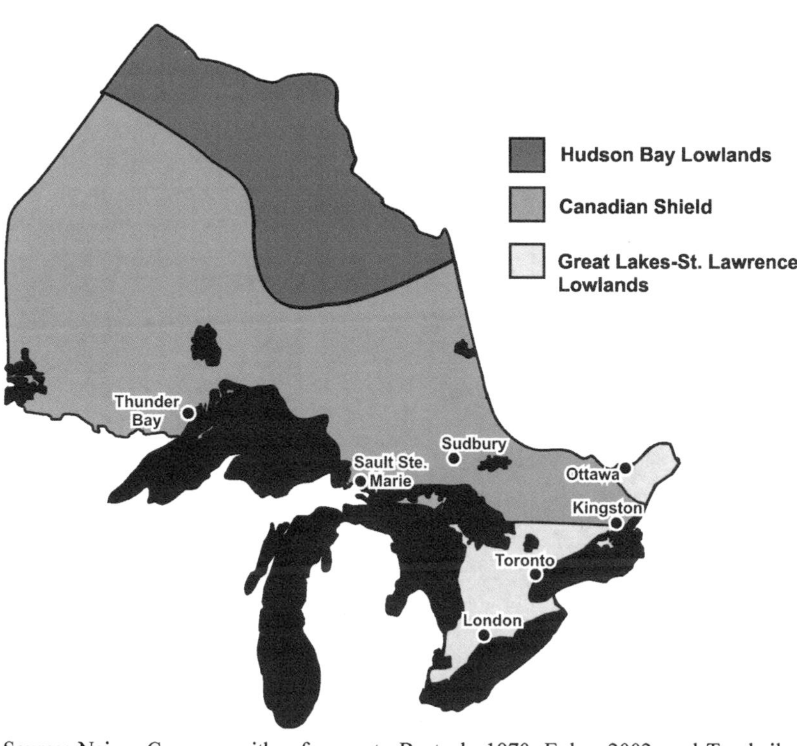

Source: Nairne Cameron with reference to Bostock, 1970; Eyles, 2002; and Trenhaile, 2007.

The poorly drained soils of slowly decaying organic matter limits agriculture in the northern Hudson Bay Lowlands have discouraged settlement. The soils overlaying the Canadian Shield are classed as mainly podzolic being thin, acidic and are poor for agriculture as their nutrients are easily lost when cultivated. The clay belts of Northern Ontario are an exception having luvisolic soils that are less acidic and can retain more nutrients (Kemp, 1993). By contrast, the sediments of the Great Lakes-St. Lawrence Lowlands are ideal for agriculture and historically Ontario's densest settlement has historically been focussed in this area.

Forests

Forests cover 66% of Ontario[19]. There are four main forest regions in Ontario (Armson, 2001) (Figure 5). In the far north lies the boreal forest and barrens. This forest is composed of white spruce, black spruce, and

[19] Ontario Ministry of Natural Resources, 2006.

tamarack. The second region, the boreal forest, covers much of Ontario and consists of white spruce, black spruce, balsam fir, jack pine, white birch, and aspen. Next, the Great-Lakes-St. Lawrence zone is a mixed forest with red pine, eastern white pine, eastern hemlock, yellow birch, maple, and oak. Finally, in the far south of the province a deciduous forest predominates made up of beech, maple, black walnut, hickory, and oak trees[20]. Generally the diversity of both deciduous and coniferous tree species declines moving north (Kemp, 1993). Forestry is an important industry in Northern Ontario.

Figure 5. Ontario's Forest Regions

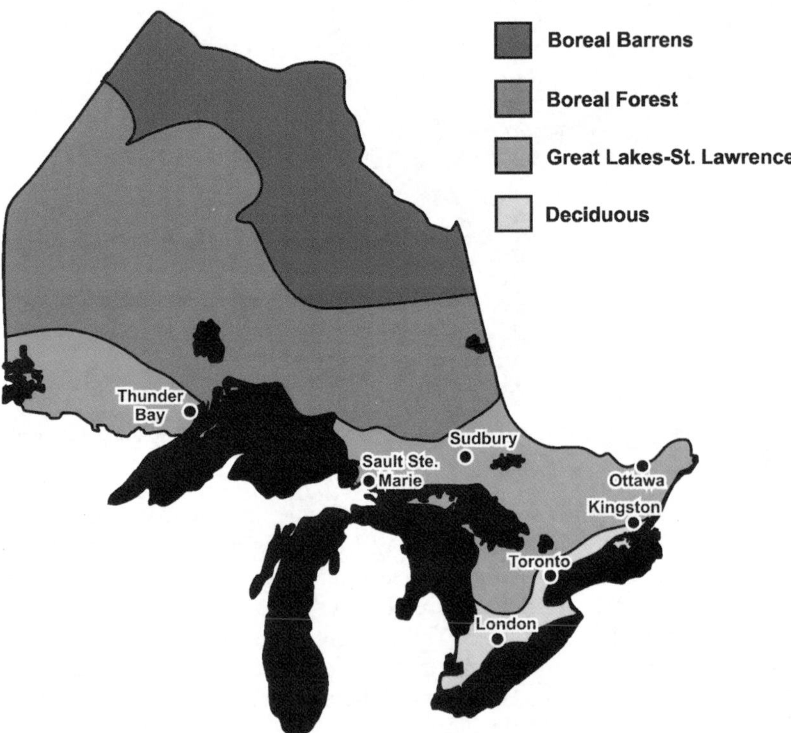

Source: Nairne Cameron with reference to Armson, 2001.

The province's vegetation sustains wildlife. Large mammals inhabiting Ontario include caribou, deer, moose, and bear. Lichen-eating caribou live in northern tundra areas of the Hudson Bay Lowlands, and in some areas north of Lake Superior. Birch, willow, and poplar eating

20 Natural Resources Canada. 2001.

moose are found throughout Ontario except in the Hudson Bay Lowlands. Areas with young deciduous trees are home to deer. Polar bears live in northerly areas near Hudson Bay[21], and black bears live throughout Ontario, although at lower densities in the far north[22]. Ontario also has walrus and beluga whale, and small mammals such as beaver, wolf, fox, bobcat, and lynx to name a few. Many birds also live throughout the province with less species diversity moving northwards. Some birds stay year-round, while others migrate south when colder temperatures curtail their food supply (Kemp, 1993).

Water Bodies and Movement

Ontario is thought to derive from an Iroquoian word for 'a large body of water'[23], and there are many lakes in the province. Watersheds demark the area drained by a waterbody or its tributaries. There are three primary watersheds in Ontario: Nelson River Basin (which flows into Hudson Bay), Great Lakes – St. Lawrence Basin, and the Hudson Bay Basin. Most of Ontario's land area drains into Hudson Bay to the north, while the remainder flows into the Great Lakes and St. Lawrence, which in turn drains into the Atlantic Ocean[24]. The Northern or Laurentian Divide (Gonzalez, 2003) separates water that flows to the Arctic and Hudson Bay from water flowing to other drainage basins. In Ontario, this Divide runs from west of Thunder Bay to north of Lake Nipigon (largest lake completely contained within Ontario)[25], then southeast to Chapleau, and finally just slightly north of Kirkland Lake.

Ontario's lakes comprise one-fifth of the world's fresh water supply[26]. Four of the five Great Lakes (Superior, Huron, Erie, and Ontario) border the province, whereas Lake Michigan is entirely located in the United States. Lake Superior is the largest freshwater lake by surface area in the world (Grady, 2007). Erosion by ice sheets 2.5 million years ago, created the Great Lake basins (Eyles, 2002). The lakes have been important historically as a transportation route, which led to the formation of settlements along their lakeshores.

The International Joint Commission (IJC) is a binational body originally created in 1909 to assist Canada and United States in resolving problems regarding the quality and usage of shared waters between the

---

21 Ontario Ministry of Natural Resources, 2008.

22 Ontario Ministry of Natural Resources, 2009a.

23 Could have originated from the Iroquoian dialects of Huron, Mohawk, and/or Seneca (Rayburn, 1997).

24 Natural Resources Canada, 2004.

25 Ontario Ministry of Natural Resources, 2009b.

26 Government of Ontario, 2009.

two countries[27]. These waters include the Great Lakes and St. Lawrence River system. With the Great Lakes being a major source of drinking water for surrounding residents, there is concern for water quality relating to industrial waste, contaminated substances, sewage, and farm run-off released into the lakes. The water levels in the Great Lakes are controlled to enable the generation of hydro-electric power and shipping through the Great Lakes St. Lawrence Seaway[28].

The Great Lakes St. Lawrence Seaway

Aboriginal people, explorers, and fur traders were early users of the St. Lawrence and Great Lakes. In the 1800s, canals were constructed on both sides of the Canadian/United States border to bypass obstacles such as rapids. The first Welland Canal was opened in 1829 to cross through the Niagara Escarpment and link Lake Ontario with Lake Erie as a detour around Niagara Falls. The canal has been reconstructed several times since then to straighten, shorten, and widen the passage (Jackson, 1997; *The St. Lawrence Seaway Management Corporation, 2003*). Large sea vessels were able to fully access the Great Lakes with the opening of the Seaway in 1959 (Jenish, 2009), a major engineering achievement which includes a series of seven locks between Montréal and Lake Ontario (Great Lakes St. Lawrence Seaway, 2009). Commodities shipped through the Seaway system include grains, iron and steel, iron ore, coal, coke, salt, stone, and manufactured products (Great Lakes St. Lawrence Seaway, 2009). Seaway transport has been challenged by other transportation modes such as railways and trucking.

With the increasing accessibility in the Seaway, there has been a rise in the number of invasive species including molluscs, fish, and plant life. Sea lampreys are an example of an invasive species from the Atlantic Ocean that first infiltrated the Great Lakes in the 1830s. They attack and feed on other fish using their suction cup mouth. In the 1940s and 1950s, they caused a severe decline in the Great Lakes' whitefish and lake trout fisheries[29]. However, with the introduction of the binational Sea Lamprey Control program[30] in 1955, their populations have been reduced by 90% (Bayfield Institute, 2001). The program entails applying lampricides to rivers and streams joining the Great Lakes where the adult lampreys travel to spawn (lay eggs) and their larvae grow. Geographic technologies such as global positioning systems and geographic information systems assist in targeting lampricides that kill the lamprey

---

[27] International Joint Commission, 2009.

[28] International Joint Commission, 2009.

[29] Fisheries and Oceans Canada, 2009.

[30] Fisheries and Oceans Canada, 2009.

larvae. Barriers have also been installed on streams to prevent the lampreys from travelling up streams to spawn, while allowing the passage of other fish (Bayfield Institute, 2001). Another strategy is to capture and sterilize male sea lampreys when they are migrating for spawning, thereby reducing the number of fertilized eggs. Canada's Sea Lamprey Control Centre is based in Sault Ste. Marie, Ontario near St. Marys River which is a breeding ground for sea lampreys that then return to feed on fish in Lake Huron and Lake Michigan.

Climate

First, a distinction, climate refers to average expected conditions based on historical data, whereas weather refers to day-to-day conditions. The climate varies across Ontario, but two main climatic zones can be recognized: Northern Ontario and Southern Ontario[31]. The province's climate is influenced by latitude, prevailing air masses, as well as proximity to large water bodies. Southern Ontario's climate is moderated by the Great Lakes, and as such has a modified continental climate with a warm, humid summer and a cold winter. Northern Ontario has a continental climate with a warm summer, and a longer and colder winter. Both Northern and Southern Ontario have four distinct seasons.

Latitude affects climate through the availability of sunlight. There is less sunlight at higher latitudes closer to the poles, compared to lower latitudes closer to the equator. Ontario, as a vast province, spans 15° of latitude from the central Hudson Bay coast at 56°51' N to 41°40' N at Middle Island near Point Pelee[32], the southernmost point in Canada. The northern part of the province is cooler overall as it receives less sunlight than the south. It can become quite cold in Northern Ontario, with the coldest ever Ontario temperature of -58.3 °C recorded at Iroquois Falls on Jan. 23, 1935 (Chadwick and Hume, 2009).

Climate is the most important factor in the formation of permafrost. Permafrost occurs at higher latitudes when the temperature of the ground remains under 0 °C for 2 years (Brown and Péwé, 1973). In the far north of Ontario, there is a small zone of continuous permafrost along Hudson Bay. Discontinuous permafrost (permafrost only forming in certain areas) appears below this continuous zone, bounded on the south by a line running east from Southern James Bay to the border of Manitoba.

Ontario's climate is affected by air masses from the Arctic, Pacific, as well as from the Gulf of Mexico. Arctic air descending from the north

---

31 Northern Ontario is known as the Northeastern Forest Zone and Southern Ontario is the Great Lakes/St. Lawrence Zone.

32 Middle Island and Pelee Island are actually south of the mainland Point Pelee.

tends to be cold and dry, while the air from the Gulf of Mexico is warm and moist. These two air masses are often competing, resulting in day-to-day variability in weather patterns. Pacific air masses from the west often create a wedge between the Arctic and Gulf of Mexico air masses (Kemp, 1993; and Chadwick and Hume, 2009). These air systems influence the type of vegetation cover across the province. For example, the tundra in Northern Ontario is subjected to Arctic Air for greater than 10 months/year, whereas the boreal forest grows in areas that are exposed to Arctic air 6-10 months/year. In Southern Ontario, the mixed forest grows where Arctic Air is present less than 6 months/year (Kemp, 1993).

The flatness of the Canadian Shield allows relatively free movement of weather systems across the northern part of the province (Kemp, 1993). The coast of Hudson Bay experiences high winds (University of Toronto, 1985). In the South, the Niagara escarpment in the Niagara Region is an example of a barrier to prevailing winds from the south, causing greater precipitation south of the escarpment, and creating a rainshadow (lower precipitation) north of the escarpment where fruit production occurs (Shaw, 1994).

Arctic Air is the main force during Ontario's winters creating cold, sunny, and clear weather, whereas there are more westerly air masses during the summer. The southern parts of the province more frequently experience hot air masses from the Gulf of Mexico leading to the need for air conditioning.

The length of the growing season influences agriculture as it dictates the type of crops that can be grown to maturity. The growing season (time when mean daily temperature exceeds 5 °C) in the North is shorter than in the South.

The Great Lakes have a strong influence on the climate in Southern Ontario by moderating temperatures and providing moisture to the air. Water bodies gain and lose heat more slowly than land masses, and water bodies can exchange heat with the air when there is a temperature difference, thereby providing a moderating influence. The Great Lakes add heat to the cold air in the autumn and winter, while they provide cooling in the spring and summer. Moisture added to the air by the Great Lakes leads to greater precipitation closer to the lakes. For example, there are higher snowfalls called 'lake-effect snow' or 'snowsqualls' east of the Great Lakes (Chadwick and Hume, 2009). The areas east of Lake Superior, Lake Huron, and Georgian Bay are known as 'snowbelts'. Precipitation in Northern Ontario increases from northwest to southeast due to exposure to the Gulf of Mexico air masses and the effect of the Great Lakes (Kemp, 1993).

Extreme weather events that can occur include cold spells in Northern Ontario, heat waves in Southern Ontario, and snow and freezing rain storms, as well as thunderstorms. Tornadoes also frequently strike Ontario at a higher rate than other provinces, particularly along a Tornado Alley extending in a northeast direction from London to Lake Simcoe (Chadwick and Hume, 2009). Finally, dwindling hurricanes can reach the province from time to time. Hurricane Hazel struck Southern Ontario in October 1954, killing 81 people and causing $100 million worth of damage (Toronto and Region Conservation, 2009).

The Ice Storm of 1998

The Ice Storm started in Eastern Ontario and Western Québec[33] suddenly on January 4, 1998, and lasted for 6 days (Soulard *et al.*, 1998). The storm was damaging due to its length and large coverage area, involving freezing rain, combined with ice pellets and snow falling for an extended period of time. For example, more than 85 mm of precipitation was recorded in Ottawa[34]. In the week before the storm, an Arctic air system filled Southern Ontario and Québec with cold air. At the same time, warm, moist air system moved northward from the Gulf of Mexico. Since the warm air was not able to displace the cold air and as warm air has a lower density compared to cold air, it flowed on top of the cold air. When rain began to fall from the warm layer, it cooled as it passed through the lower and colder Arctic air layer, and then froze on ground level surfaces (Risk Management Solutions, Inc., 2008).

Thousands of households were without electricity and as a result had to seek temporary accommodations with friends and family, while some sought refuge in shelters. Unfortunately, 25 Canadians died as a result of the storm. Transportation systems, especially air and rail, were interrupted. Farmers' operations were also impacted particularly due to the lack of electricity. Approximately 14,000 Canadian soldiers were dispatched to aid in the emergency[35]. Many trees, power lines, transmission towers, telephone cables, and hydro poles collapsed under the weight of the ice load[36]. Total damages in Ontario as a result of the storm were estimated to be $200 million (Soulard *et al.*, 1998).

---

33 The storm covered a wide area from Muskoka and Kitchener to Eastern Ontario and Québec, and to parts of Maritime Provinces and northern United States. However, Eastern Ontario and Western Québec were more severely affected (Soulard *et al.*, 1998).

34 Statistics Canada, 1998.

35 Statistics Canada, 1998.

36 Environment Canada, 2002.

The Economy

Ontario has long been considered the economic engine of Canada, particularly Southern Ontario. Southern Ontario's *dense* population has formed a pool of well-educated workers[37] as well as a consumer market for goods and services, and the region has also had a political power base and government support allowing a manufacturing sector to develop. The location of Southern Ontario relatively close to the US Manufacturing Belt allowed cross-border alliances to develop and was a convenient location for branch plants and joint ventures. The concentration of the US population in the northeast also represented a large market for Ontario exports. Northern Ontario's economic development has been slower than that of the South, but the North has historically played an important role supplying the South with resources. Together, Northern and Southern Ontario developed a *diversity* of businesses. As the Ontario economy has developed, the service sector has become more important, compared to primary (natural resource harvesting) and secondary (manufacturing) sectors. A wide range of resources is available in the province including half of all Canada's agricultural land in Southern Ontario, as well as mineral and forest resources in the North. *Sustainable* renewable energy such as hydroelectricity and increasingly wind and solar, has contributed to the success of the economy.

The three-sector hypothesis describes the change in the relative proportions of primary, secondary, and tertiary sectors of an economy as it becomes more developed (Fourastié, 1949). The *primary* sector involves harvesting activities such as fishing, agriculture, mining, and forestry, whereas the *secondary* sector further processes raw materials, for example, manufacturing. The *tertiary* sector includes services such as transport, retailing, hospitality, and maintenance. Two other categories *quaternary* and *quinary* can be recognized as specialized subcategories of the tertiary sector. *Quaternary* activities involve managing information, for example, financial services. The *quinary* sector generates knowledge through research activities, and applies the knowledge in high-level decision-making. The three-sector hypothesis states that as an economy develops, a greater portion of activities will be service-related. For example, in 1881 when less equipment and technology was employed in the Canadian economy, employment was categorized as: ~50% primary, ~30% secondary[38], and ~20% tertiary[39] (Figure 6). By

---

37 Ontario has the highest level (26%) of university degree holders (aged 25-64) in Canada (Statistics Canada, 2009b).

38 Includes manufacturing and construction.

39 Dominion Bureau of Statistics, 1921.

contrast, currently in Ontario employment is ~2% primary, ~20% secondary, and ~78% tertiary[40][41].

Looking more closely within the province, Northern and Southern Ontario have similar levels of employment in the tertiary sector (78%), but Southern Ontario has a higher proportion of its employment in the secondary sector[42] (20%), compared to Northern Ontario (16%), and a lower proportion of employment in the primary sector (2%) compared to Northern Ontario (6%)[43].

Figure 6. Comparison of Percentage Employment in Sectors: 1881 and 2006

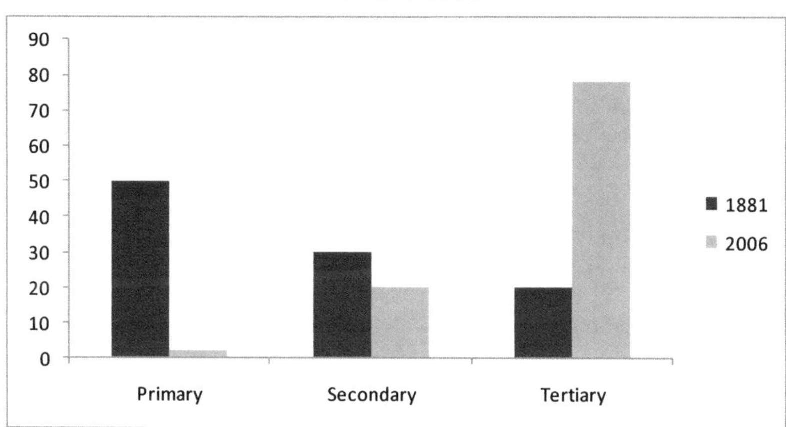

Source: Nairne Cameron using data from Dominion Bureau of Statistics, 1921 and Statistics Canada, 2006b.

But now *change* is confronting Ontario. Ontario became a 'have-not' province for the first time in 2009 receiving a $347 million equalization payment, followed by a $972 million payment in 2010-2011, and a $2,200 million disbursement in 2011-2012[44]. Equalization payments are distributed by the Federal Government to provinces with revenues below the average of all provinces, thereby allowing less wealthy or 'have-not' provinces to provide a parallel level of public services as the

---

40 Statistics Canada, 2006b.

41 These numbers based on employment differ from numbers by GDP: In 2008, the tertiary sector represented 73.7% of Ontario's GDP, while secondary represented 17.3%, primary 1.5% (utilities and construction 7.5%).

42 Includes construction.

43 Statistics Canada, 2006b.

44 Department of Finance Canada, 2011.

richer or 'have' provinces at similar tax rates[45]. Also, a $19.3 billion deficit was registered in 2009-2010, and a $16.7 billion deficit is forecast for 2010-2011[46]. Economic challenges facing Ontario include the US recession leading to a decline in demand for Ontario exports, a rising Canadian dollar undermining Ontario's exporters, and tightened post 9/11 security between Canada and the United States influencing tourist traffic and goods movement. Canada's leading export sector, the motor vehicle and parts industry, faltered with two major automakers: General Motors and Chrysler entering bankruptcy proceedings in 2009. However, following Ontario government emergency support for these two companies (along with federal support from the Canadian and US governments), the sector has stabilized as of early 2011[47]. Ontario's energy supply is also showing signs of strain with increasing demand, aging nuclear reactors, and the phasing out of coal sources. And, the growing economies in Western Canada are beginning to challenge Ontario's status as the core region.

Ontario's Economic History

Ontario's role as Canada's economic engine has historical roots which can be envisioned in the context of the core-periphery model of development (Friedmann, 1966). This core-periphery, or heartland-hinterland model as it is alternatively known, states that development is uneven. The core has higher levels of income, education, political influence, innovation, and sources raw materials from the periphery, which has lower incomes, educational levels, political influence, and innovation. Core-periphery relations can change over time. For example, when early Ontario was a British Colony, according to Mackintosh (1923) and Innis (1999), it was a peripheral region that sent 'staples' or natural resources such as fur, lumber, and wheat to Britain (the core) who in turn sold manufactured goods back to the colony (McCann, 1987).

As the new nation of Canada evolved, central Canada (Southern Ontario and Québec, and especially the Toronto and Montréal regions) became the country's core region with all other areas assuming a periphery role. Policies supporting Ontario's economy often came at the expense of other provinces. Following Confederation in 1867, Prime Minister John A. MacDonald aimed to strengthen Canada as an independent country. He developed a National Policy with three components: railways; land settlement; and, a manufacturing tariff[48]. MacDon-

---

45 Department of Finance Canada, 2009.

46 Ontario Ministry of Finance, 2011a.

47 Ontario Ministry of Finance, 2011a.

48 The tariff by itself is often referred to as the National Policy.

ald sought to build linkages within Canada by annexing the Western lands, then constructing a railway to bring new settlers to the West. The railway reinforced an east-west axis with immigrants travelling to the West and discouraged north-south trading with the United States by funnelling raw materials to central Canada for further processing (Eden and Molot, 1993), then returning manufactured products back to the periphery for retail. Construction of the railway required manufactured supplies which stimulated central Canada's manufacturing sector.

The manufacturing tariff was designed to shield Canada's developing industrial sector. Implemented in 1879, it levied taxes on imported manufactured items, but collected lower duties on raw products. The tariff was well received in Central Canada, especially among the business communities in Toronto and Montréal who politically supported MacDonald. The West, however, did not embrace the tax as it meant higher prices for agricultural equipment, and their grain, as a raw product, was not protected like manufactured goods and had to be priced competitively on the global market (Gibbins and Berdahl, 2003). In the east, trade between Atlantic Canada and American New England was curtailed. By the 1920s, a substantial Manufacturing Belt producing cars and household appliances among other goods developed in Southern Ontario in close proximity to the American Midwest industrial belt. Many of the businesses were branch plants of American companies that were created to avoid the tariffs on imported manufactured goods, and to take advantage of tax-free export to the British Empire.

World War II stimulated further industrialization in Ontario, with the manufacture of items for the war. Following the war, the province's energy system was boosted with an enhanced hydroelectricity network, the country's first nuclear plant, and a natural gas pipeline from Alberta. Also, the St. Lawrence Seaway was completed in 1961, enabling container shipping by sea vessels.

In addition to viewing the core-periphery model at a national scale, it can also be applied at a provincial scale. Historical events contributed to the development of a southern core and a northern periphery within Ontario's borders. Northern Ontario's periphery status can be traced back to the province's early history. The North was settled later than the South. In 1871 (shortly after Confederation), only 1% of Ontario's population lived in the North. Mowat, the first Premier of Ontario, recognized that since the government owned all the resource rights in the province, the Crown could benefit financially from natural resource development in the North. Mowat then began to actively encourage settlement, as well as mining and forestry activities in Northern Ontario (Baskerville, 2005). The raw products, however, were not generally processed in the North and the transportation network promoted movement of natural resources to the South, rather than good linkages and interaction within Northern Ontario. Furthermore, the profits from resource activities were not reinvested in the North, but instead were diverted southward (Martin, 1999).

Present-Day Ontario

Ontario is the wealthiest Canadian province as measured by Gross Domestic Product (GDP), and is the producer of over a third of Canada's GDP)[49]. The province has a highly diversified economy[50] with many different types of commercial enterprises, including a range of manufacturing businesses; agriculture, forestry and mining operations; tourism, and service sector businesses relating to areas such as finance, information and communications technology, and tourism. However, there are differences between the economies of Northern and Southern Ontario.

Northern Ontario's economy measured by GDP is in fact greater than four Atlantic Canadian provinces[51]. However, it lacks the diversity of Southern Ontario's economy with less manufacturing and more resource-dependent settlements. Also, some large resource-based businesses in the region that were previously Canadian owned are now under foreign ownership. For example, India based multinational Essar Global is the current owner of Essar Steel Algoma (Essar Steel Algoma, 2010), a major steel producer in Sault Ste. Marie. Also, Vale Inco which mines nickel (and other minerals) in Sudbury is under the control of Vale, a multinational based in Brazil (Vale Inco, 2010)[52]. Government programs have been implemented to support Northern Ontario. These include FedNor, a federal regional development agency. The provincial government's Northern Ontario Heritage Fund promotes job creation, infra-

---

49 Ontario's GDP was $578.2 billion and 37.9% of Canada's GDP in 2009 (Ontario Ministry of Finance, 2011b). Gross Domestic Product is the total value of all goods and services produced in a defined area over a set time period. The concept of GDP however has several limitations including an inability to describe the distribution of income among residents of the area. It is also a weak indicator of standard of living since the measure includes activities such as war, environmental degradation, and disasters, which may increase economic activity, but may have adverse effects on humans.

50 Provincial GDP for 2010 can be broken down as follows: 1.5% primary, 1.9% utilities, 5.3% construction, 15.4% manufacturing, 12.3% wholesale and retail trade, 4.0% transportation and warehousing, 4.1% information and cultural, 23.7% finance, insurance, and real estate and leasing, 11.8% health and education, and 20.0% other services (Ontario Ministry of Finance, 2011c).

51 GDP $25,602 million greater than Nova Scotia, New Brunswick, Newfoundland, and Prince Edward Island (Southcott, 2006).

52 In addition to the Sudbury site, the company has other Canadian operations.

structure and community development[53]. Additionally, several government offices were moved to the North such as offices of the Ontario Lottery Corporation[54] which were established in Sault Ste. Marie. The Ministry of Northern Development, Mines, and Forestry, Ontario Geological Survey, and also a federal Canada Revenue Agency Tax Services Office are located in Sudbury. And, the federal Department of Veteran Affairs regional client service centre is based in Kirkland Lake.

Ontario possesses a large (10.2 million) and well-educated (21.3% with a university degree) working age population compared to other Canadian provinces[55]. Again, however, there are differences between Northern and Southern Ontario labour forces. Northern Ontario has had a higher unemployment rate than the rest of Ontario[56], and starting in the early 1980s, there has been a trend towards less resource sector employment in the North (Dadgostar *et al.*, 1992 (Northwestern Ontario), Mulholland and Vincent, 1998). The northern population is aging, compared to the southern portion of the province, and there is a youth out-migration trend. Furthermore, the Northern Ontario workforce has lower average levels of education and professional skills (Southcott, 2006). The province is addressing the education and skills gap by expanding educational institutions in the region, including the newly independent Algoma University in Sault Ste. Marie.

Exports drive the Ontario economy, making up 53.4% of GDP in 2008[57]. In fact, Ontario's international trade is three times greater than its inter-provincial trade[58]. Most (78.7%) of Ontario's exports flow to the United States, with the next highest export destination (7.6%) being the United Kingdom[59]. The nearby state of Michigan is Ontario's largest trading partner representing 23.5% of the province's exports and 8.4% of imports[60]. The main international exports in 2010 were motor vehicles and parts 31.9%, precious metals and stones 11.2%, mechanical equipment 9.5%, electrical machinery 4.8%, and plastic products 3.6%[61]. In

---

53 Northern Ontario Heritage Fund, 2009.

54 Now Ontario Lottery and Gaming. The head office of the Ontario Lottery Corporation was established 1992-2000. In 2000, the head office was moved back to Toronto, but a corporate office remains in Sault Ste. Marie (Ontario Lottery and Gaming, 2009).

55 Data for 2006. Working age population is that aged 15 to 64 years. Provincial labour force differences by level of education (Statistics Canada, 2008a).

56 Data for 1986-2001. Southcott, 2006.

57 Invest in Ontario, 2009a.

58 Data for 2004. (Statistics Canada, 2006c).

59 Data for 2010. (Ontario Exports, 2011).

60 Data for 2010. (Ontario Exports, 2011).

61 Ontario Ministry of Finance, 2011b.

the next section, we will examine Ontario's exports and industries in more detail.

Manufacturing

Manufacturing is a key sector for Ontario making up 15.4% of GDP[62]. Auto manufacturing is particularly important yielding 4% of provincial GDP and for this reason is profiled separately. Other manufacturing products include: mechanical equipment, electrical appliances and machinery including computers, plastic products, iron and steel, and food products. Manufacturing is mainly based in Southern Ontario. The Ontario manufacturing sector has adapted to many trade rule changes over the last 20 years and is now facing further challenges with a rising Canadian dollar, increased energy costs, and global competition.

In 1988, Canada and the United States established the Free Trade Agreement (FTA) which removed trade barriers, including many tariffs between the two countries, stimulating more north-south trade. It also led to restructuring as many branch plants in Canada were no longer viable in the enlarged US-Canada market. In 1994, the North American Free Trade Agreement (NAFTA) brought Canada, United States, and Mexico into a common market. Again, this caused restructuring with some labour intensive companies moving their operations to Mexico, a source of cheaper labour. In 1995, the World Trade Organization agreement opened up new markets for Canadian companies, but also forced Canadian companies to compete on a global level. Compared to the nearby US Manufacturing Belt, the Canadian industrial belt has been thriving up until recently. Since the 1970s, the US belt, now known as the 'Rust Belt' has deindustrialized with plants shifting to the US South, Mexico, and offshore.

Starting in 2003, Ontario has experienced a decline in manufacturing employment, hitting all subsectors with the loss of 18% of its manufacturing jobs between 2004 and 2008[63]. This decline has been described as the 'China Syndrome' (MacDonald, 2007), caused mainly by the integration of China and other emerging countries into the global market, thereby increasing the world supply of manufactured goods, and raising the demand for raw materials to make those goods. This has led to cheaper prices for manufactured goods, and an increase in the prices of raw materials. As a major exporter of commodities, the Canadian dollar rose. While these changes caused a decline in manufacturing employment, the Ontario manufacturing sector has adjusted and there has been an increase in productivity.

---

[62] Data for 2010. Ontario Ministry of Finance, 2011c.

[63] Statistics Canada, 2009c.

In terms of geographic specialization in manufacturing, Windsor and Oshawa are focussed on motor vehicle assembly and parts. Sarnia, a port city near the southern tip of Lake Huron, is the 'petro-chemical capital of Canada' (Hill, 2002) with an oil refinery and plants producing rubber, glycol, and fibreglass, although it is now undergoing deindustrialization with job losses and high unemployment. Some manufacturing does occur in Northern Ontario. For example, Bombardier has a passenger train assembly plant in Thunder Bay. And, in addition to Southern Ontario's Hamilton steel centre, Sault Ste. Marie's Essar Steel Algoma is a northern steel manufacturer.

Agriculture

While agriculture makes up a small percentage of Ontario's GDP, the province has over half of Canada's prime agricultural land, and a relatively warm climate allowing the cultivation of a wide diversity of crops. The province's agricultural products include dairy foods, fruits and vegetables, beef cattle, corn, flowers and plants, tobacco, ginseng, soybeans, poultry, hogs, wheat, and eggs. Most of the province's fertile soils are located in Southern Ontario. The bulk of Northern Ontario's prime agricultural soils are in the clay belts, but farming has been difficult there in the past due to a colder climate and shorter frost-free period[64]. Between 1948 and 2008, Ontario's climate has warmed, especially in the west of the province, and is estimated to further warm 2.5 °C to 3.7 °C by 2050, potentially enabling agriculture to shift farther north where soil conditions permit[65]. Despite the conditions, Northern Ontario has 2,479 farms, including 89 fruit and vegetable farms[66]. Northern Ontario's top crops in terms of the percentage of provincial acreage under cultivation are: oats (15.0%), barley (11.5%), hay (10.9%). The region has 10.5% of the province's beef cattle[67], and also has a substantial dairy industry[68].

There is a historical trend towards fewer and larger farms in Ontario as economics dictate larger operations. Between 1996 and 2006 the

---

64 Ontario Ministry of Agriculture, Food and Rural Affairs, 2009a. Since the Temiskaming and Northern Railway passed through the northern clay belts which were known to have fertile soils, the government promoted agricultural settlement. However, many settlers abandoned the area as they discovered that farming was difficult due to the short growing season.

65 This annual average temperature estimation is based on a moderate greenhouse gas emission decrease scenario. Temperatures are compared to 1961-1990 temperatures (Ontario Ministry of the Environment, 2009).

66 Data for 2006. Ontario Ministry of Agriculture, Food and Rural Affairs, 2009b.

67 Data for 2006. Ontario Ministry of Agriculture, Food and Rural Affairs, 2009b.

68 Ontario Ministry of Northern Development and Mines, 2007.

number of farms in Ontario decreased by 15%, and the average farm size increased by 13%. There are many mixed farms (with multiple crops and livestock herds) in Southern Ontario that are concentrated along the Ottawa River near Ottawa, then west bordering the St. Lawrence River and along the north shore of Lake Ontario and Erie, and south of Lake Huron and Georgian Bay.

Certain areas of the province are more specialized in particular crops. The fruit and wine-growing area is situated on the Niagara Peninsula. Ontario's apple growing areas can be found beside the shores of Lake Ontario, Lake Erie, Lake Huron, and Georgian Bay where the lakes moderate year-round temperatures and lengthen the growing season. Ontario grows half of all Canada's vegetables[69] with the main vegetable growing region of the province in Holland Marsh, north of Toronto, and south of Lake Simcoe[70]. Leamington, near Point Pelee, is the 'Tomato Capital of Canada', raising tomatoes, and other vegetables under greenhouses in its warm, sunny location. Leamington also has an H.J. Heinz plant that makes ketchup among other items[71]. Ontario grows over half of Canada's corn[72] with two main corn-growing areas: southwest Ontario, west of Toronto and east of Kingston to the Québec border. The 'tobacco belt' is located in southwestern Ontario on the north shore of central Lake Erie. Ginseng is produced in the Norfolk area between London and Hamilton, representing 87% of all Canada's production[73].

Forestry and Mining

Forestry and mining have historically been important industries in Northern Ontario, although employment levels have declined over time. Forests are an example of a *renewable* resource, meaning that once lumbering has taken place a new forest can be grown. Mining, by contrast, is an example of a *non-renewable* industry. Once an ore body has been exhausted, it cannot be renewed and new sources must be explored and mined elsewhere, as evidenced by several former mining settlements that are now ghost towns dotted across Ontario (Brown, 2007)[74]. Elliot Lake, south of Sault Ste. Marie, is an example of a town that has extended its life beyond uranium mining by becoming a retirement community. Both renewable and non-renewable resources are commodities that are sold on the global market and are subject to forces of supply

---

[69] Data for 2006. Statistics Canada, 2008b.

[70] The Holland Marsh was hard hit by Hurricane Hazel (discussed earlier), but the flooding in 1954 actually nourished the soil.

[71] The Municipality of Leamington, Ontario, Canada, 2009.

[72] Statistics Canada, 2008b.

[73] Statistics Canada, 2008b.

[74] Examples of such settlements are: Gold Rock, Silver Centre, and Burchell Lake.

and demand. Northern Ontario has several settlements that depend on these two industries, rendering them vulnerable to swings in global commodity prices, and leading to boom and bust economies.

Forestry

Ontario's forestry industry is based in the northern part of the province which hosts nearly 97% of the province's forests[75]. The majority of forested lands (88%) are owned by the provincial government, and 1% belongs to the federal government, with the remainder being private[76]. Companies pay stumpage fees for the right to harvest crown lands. The forestry industry is made up of two subsectors: wood products and pulp and paper. Ontario operations produce mainly pulp and paper making up 61.6% of provincial revenues, with the remainder being forest products (31.6%), and the value of logging within the province (6.7%)[77]. Almost half (44.7%) of all production is exported[78], mostly to the United States.

Many challenges currently face the province's forestry industry. Between January 2003 and December 2009, Ontario had 52 permanent and indefinite mill closures, and 10,885 people were laid off[79]. This downturn has affected forest dependent communities, of which there are 63 in the province[80]. Two-thirds of Northwestern Ontario communities are forest dependent (Thunder Bay Community Economic Development Commission (CEDC), 2009). The downturn has particularly impacted Red Rock, Dryden, Terrace Bay, and Kapuskasing. Kenora and Thunder Bay have also been affected, but their diversified economies including regional service functions have minimized the effect.

There are several reasons for the downturn including: increased competition from around the world especially from China, Russia, and South America, and declining demand for newsprint due to electronic communications and a shift in manufacturing location (for example, to China) and subsequent drop in demand for paper packaging. Technological changes and consolidation in the industry have also reduced employment. The high Canadian dollar has made Canadian products more expensive for American customers, and there has been a slowdown in US housing construction. Also, many of the mills are small in size and are outdated. Energy prices, including Ontario electricity rates, have increased production and transportation costs. Forest supplies are also dwindling near mills, meaning that companies have to go further to

75 Ontario Ministry of Natural Resources, n.d.

76 Ontario Forest Industries Association, 2009.

77 Data for 2001. Invest in Ontario, 2009b.

78 44.7% $8.5 billion in exports, $19 billion in sales. Data for 2003.

79 Natural Resources Canada, 2011.

80 Natural Resources Canada, 2006.

source supplies increasing transport costs. Finally, the softwood lumber dispute with the United States has also impacted exports. Softwood lumber includes spruce, pine, and fir and is often used for construction. The dispute began with the US contention that for many years Canada did not impose high enough stumpage fees. The dispute escalated in May 2002 when the US imposed 27% duties on Canadian softwood lumber. In April 2006 the dispute ended with the US assenting to remove the tariff and return greater than $4.5 billion US dollars of duties it had collected as part of a seven-year agreement[81]. As of 2009, there are still ongoing frictions regarding the dispute.

The provincial and federal governments have offered support to the industry including provincial investment of $1 billion over 5 years to boost the industry[82]. In June 2009, the federal government announced a $1 billion aid package for the Canadian pulp industry for environmental and energy efficiency incentives to compete with American subsidies for a pulp byproduct called black liquor, an alternative fuel[83].

Mining Industry

Ontario is the top Canadian mineral producer with 19% of the country's production in 2010[84]. Minerals can be categorized as *metallic* and *non-metallic*. Most of the province's metallic minerals are found in Northern Ontario within the Canadian Shield, while the non-metallic mineral production occurs south of the Canadian Shield. Ontario is a producer of metallic minerals such as: gold, nickel, copper, zinc, platinum, palladium, cobalt and silver. In 2010, Ontario yielded the following percentages of Canadian production: gold (52%), copper (30%), nickel (34%), and platinum group metals (75%)[85]. In terms of geographic distribution of mining, Red Lake, Hemlo, Wawa, and Marathon have gold deposits. The Thunder Bay area has gold, platinum/palladium, amethyst and agate, while the Timmins area hosts gold, copper, and zinc deposits. In Sudbury there is mining for nickel, copper, platinum, and palladium. During the resource boom 2003-2008, the mining sector did very well. However, with the drop in commodity prices in 2008, the mining industry has contracted with decreased production and layoffs of employees. There has been some recent expansion in the province with the opening of Ontario's first diamond mine, the DeBeers Victor Mine,

---

81 An export tax is levied on Canadian producers if the price of lumber drops below $355 US per thousand board feet. (Foreign Affairs and International Trade Canada, 2008).

82 Ontario Ministry of Natural Resources, 2009c.

83 Natural Resources Canada, 2009b.

84 Ontario Ministry of Northern Development, Mines and Forestry, 2011.

85 Ontario Ministry of Northern Development, Mines and Forestry, 2011.

in July, 2008 in the James Bay Lowland 90 kilometres to the west of Attawapiskat (De Beers Canada, 2009)[86].

Services

Most jobs in Ontario are now in the service industry. Three industries are highlighted: financial services, information and communications technology, and tourism.

Toronto is Canada's financial centre. The Toronto Stock Exchange is the third largest exchange in North America. The city is also home to the headquarters of all five major Canadian banks (known as the "Big Five"), as well as many foreign banks, and mutual fund companies are based in Toronto. Bay Street in Downtown Toronto is the hub of the financial district[87].

The information and communication technology (ICT) or 'high-tech' industry is concentrated in the Toronto, Ottawa, and Waterloo. Toronto has the largest cluster with 3,300 companies specializing in digital media and internet. The Ottawa cluster, also known as 'Silicon Valley North' has 1,500 companies and focuses on telecommunications, software, and photonics. The smallest cluster in Waterloo has 400 companies producing software, microelectronics, and telecommunications[88]. The University of Waterloo is an anchor for this cluster. Research In Motion, the company that makes the Blackberry® is headquartered in Waterloo. After a boom during the 1990s, the high tech industry collapsed in 2001. This Dot-Com Crash resulted in layoffs with high tech employment being reduced to 4.2% of the country's labour force, compared to 4.6% in 2001. The Ottawa cluster was particularly hard hit (Frenette, 2007).

Tourism

Ontario draws half (49%) of Canadian overnight trips by United States travellers, and Ontario is the most frequently visited province by international travellers[89]. Ontario also attracts substantial internal tourism with 63% of the province's tourism revenues coming from Ontario residents, while 17% comes from the United States, 13% from overseas, and 8% from other Canadian provinces[90]. There was a dip in tourism in 2003 coinciding with the outbreak of Severe Acute Respiratory Syndrome (SARS). SARS was an epidemic that originated in China, and

---

86 Government of Ontario, 2009.

87 Toronto Financial Services Alliance, 2009.

88 Ontario Exports, 2009.

89 Statistics Canada, 2008c.

90 Ontario Ministry of Tourism, 2009.

affected Canadians mainly in the Greater Toronto Area resulting in 44 deaths[91]. Since 2003, visits by Ontarians, other Canadians, and international travellers have increased. However, visits by Americans have declined since 1998, owing to the stronger Canadian dollar (higher exchange rate for Americans), higher fuel prices, and tighter border security following 9/11[92]. Northern Ontario is more dependent on Canadian and American visitors, compared to Southern Ontario, as the overseas market represents only 2% of its tourism revenues. Visitors to Northern Ontario come mainly in the summer for outdoor activities such as fishing, boating, and parks (Rogers, 2004)[93].

Vignette: Ontario's Car Industry

The Canadian automotive industry started with individual entrepreneurs scattered across the country building experimental cars, but quickly centralized in Southern Ontario (Durnford and Baechler, 1973). While it was possible for entrepreneurs to build a car, it was a business, logistical, and mechanical engineering challenge to mass produce the vehicles for a profit. Henry Ford in the United States developed a system to produce cars using an assembly line. He also paid his workers a salary high enough that they could afford to purchase the cars they manufactured. This process involving mass production and consumption is termed *Fordism*. Successful Canadian entrepreneurs struck up partnerships with American companies to benefit from their expertise and capital. Thus, the industry became established in Southern Ontario due to its proximity to the US industry based in neighbouring Michigan. Another reason for the industry's initial location in Southern Ontario can be attributed to the population *density* and concentration of the country's population in the area (and in nearby southern Québec), and thus a consumer market was readily available.

Ford of Canada became the country's first major auto manufacturer in 1904 (Mays, 2003). This enterprise was based in Walkerville near Windsor, right across the border from Detroit, the centre of the US car industry. Gordon McGregor, the founder, proposed to Henry Ford that he assemble cars based on Ford's design and production methods at his family's carriage making business. Henry Ford accepted the proposal and entered into a partnership to gain greater access to both the Canadian and British Empire markets. At the time, there was a 35% tax on American-made carriages imported to Canada, and the same rate was levied on automobiles. By making cars in Canada this import tax could be by-

---

91 Public Health Agency of Canada, 2004.

92 Statistics Canada, 2008c.

93 Data for 2004.

passed. There was also a 20-30% tariff on automotive parts imported from the United States (White, 2007). This tariff stimulated the beginning of Canada's auto parts industry which provided substitutes for taxed US parts. Another advantage of Canadian assembly was that Canada could export cars tariff-free to the British Empire, unlike American automakers. A second prominent Canadian automotive pioneer was Sam McLaughlin whose family ran a thriving carriage-making business in Oshawa. He launched a car making business by partnering with US based Buick (Robertson, 1995).

After World War I, US owners and majority shareholders gained greater control over the Canadian industry. A pattern was set at this time with the majority of the car assembly being concentrated in Southern Ontario, particularly in Windsor, Oshawa, and the Toronto region (Anastakis, 2005). By the late 1920s, auto production was Ontario's leading industry, and Canada (in fact Ontario where the industry was located) was the world's second largest producer of cars (White, 2007).

Then, the Depression set in and decreased auto production severely impacted cities such as Oshawa and Windsor who relied on the auto industry (Baskerville, 2005; White, 2007). During World War II, the Canadian automakers increased their capabilities through the production of military vehicles, and adapted their factories with financial support from the Canadian government (White, 2007). The term 'The Big Three' was coined following the War as 'Chrysler Corporation, Ford Motor Company and General Motors' gained market dominance in Canada. The post-war demand for cars increased dramatically (Rea, 1985), and was accompanied by a highway building boom and increasing suburbanization of the Canadian population.

By the end of the 1950s, the ability for Canada to export cars to the British Empire tariff-free was withdrawn, and Canadian automakers faced more competition from British and European manufacturers. Imports of American auto parts resulted in a substantial auto trade deficit with the United States[94]. Growth in the Canadian auto industry had slowed. In 1965, the Automotive Products Trade Agreement or Auto Pact was signed between Canada and the United States allowing freer trading of automobile parts and vehicles by eliminating tariffs, yet with some provisions to protect Canada. Companies had to build the same number of cars that they sold in Canada, and all cars built in Canada were required to be made with 60% Canadian labour and parts (Anastakis, 2005). Instead of having branch plants assembling cars with US parts specifically for the Canadian market, the agreement allowed Canadian factories to become specialized in making particular models of

---

94 Government of Canada, 2009.

cars which were sold throughout Canada and the United States. The Auto Pact was largely advantageous for Canada, especially as it guaranteed a certain percentage of vehicle assembly and parts manufacture. This agreement resulted in more north-south cross-border trade in Southern Ontario, further solidifying Ontario's status as the core of Canada.

The 1979 oil crisis spurred concerns of car fuel efficiency among consumers who began to buy fewer and smaller cars. Japanese car makers identified a market niche and in the 1980s established three operations in Southern Ontario: Honda in Alliston, Toyota in Cambridge, and a Suzuki and General Motors joint venture in Ingersoll. The pattern of Southern Ontario production set earlier in the history of the Canadian auto industry continued to be reinforced. The Japanese automakers had initially anticipated that they would be able to access the US market from their base in Canada. However, the more liberal trade relations specified in the 1989 Free Trade Agreement were restricted to the original participants in the Auto Pact (White, 2007).

Despite this obstacle, the Japanese automakers were able to successfully compete in the US market through their flexible and efficient production, and largely non-unionized employees paid at lower wages. They were so successful that in the 1990s the Big Three were forced to close several plants. These closures mainly occurred in the United States, as it was more profitable to make cars in Canada due to the low dollar, well educated population, highly productive workers, lower labour costs, and investment by the provincial and federal governments, and universal health care that freed employers from private coverage. Canada's share of the Northern American market has increased since the 1990s.

In 2001, the Auto Pact agreement was nullified after the World Trade Organization determined that it contravened its global trading rules (Johnson, 2004). Then, in 2004, Ontario became the top vehicle-producing jurisdiction in North America, exceeding Michigan's production (Scotiabank Group, 2005) mainly due to increasing Japanese production in Ontario. As of 2005, Japanese production now accounts for over one-third of Canadian production (Roy and Kimanyi, 2007). Since 2004, oil prices have increased and there has been more demand for fuel-efficient cars. In the past, the Big Three have produced mainly larger cars and light trucks with higher profit margins. The consumers' preference for smaller, more fuel efficient, and higher quality vehicles are contributing to the declining production share of the Big Three in Ontario.

Ontario has 6 light vehicle manufacturers – Chrysler, Ford, General Motors, Honda, CAMI (a joint venture between Suzuki and GM), and

Toyota – and more than 450 auto parts companies[95] including Ontario-based multinational Magna International (Magna International Inc., 2009). In 2008, the Canadian auto industry had employment of 64,500 in vehicle assembly and 98,700 in parts, having declined 19.8% and 29.1% respectively since 2004 (Bernard, 2009). Approximately 90% of light duty vehicles produced in Canada are exported[96], mainly to the United States (Roy and Kimanyi, 2007).

Challenges currently facing the auto industry are numerous. Chrysler and General Motors (GM) filed for bankruptcy protection in 2009. The Canadian federal and provincial governments responded by providing loans of $3.78 billion to Chrysler[97] and $10.6 billion to GM[98][99] and taking ownership stakes in both companies[100][101]. The restructuring plans included a Canadian share of North American production of 20% in Chrysler[102] and 16% in GM[103]. Chrysler's Ontario plants were shut for a two-month period in the spring of 2009 during the restructuring. At the same time, GM curtailed production at its facilities and has permanently closed a truck plant in Oshawa. Since then though, GM has repaid its Ontario and Canadian government loans. Also, production in the sector rose 40% in 2010 compared to 2009[104].

Several factors are presently influencing car production in North America including: saturation of the North American market; increasing demand for vehicles in emerging markets; rising levels of imported parts from outside North America; environmental concerns; and increasing fuel prices. The market for cars in Canada and the United States is largely saturated, meaning that a high proportion of residents already have a vehicle and replacement vehicles are now the main market. By contrast, there is higher demand for vehicles in emerging markets such as China, India, and Brazil, which have rising vehicle ownership rates. Production may shift closer to these regions for political and transportation reasons (to avoid shipping costs). Already, the North American share (Canada, United States, and Mexico) of the international vehicle

---

95 Invest in Ontario, 2009c.

96 Industry Canada, 2007.

97 Ontario Office of the Premier, 2009a.

98 Ontario Newsroom, 2009.

99 These amounts are in addition to $4 billion in interim loans given to the two companies in December 2008.

100 Office of the Prime Minister, 2009a.

101 Office of the Prime Minister, 2009b.

102 Office of the Prime Minister, 2009a.

103 Ontario Office of the Premier, 2009b.

104 Ontario Ministry of Finance, 2011a.

market dropped from 33% to 25% between 1975 and 2005 (Sturgeon *et al.*, 2009). The North American parts industry is now competing globally with offshore parts manufacturing in locations such as China. There are also environmental concerns about cars and their emissions and associated air pollution, as well as their role in creating traffic congestion in urban areas. Finally, increasing fuel prices are affecting consumer preferences towards a trend to smaller more fuel-efficient vehicles.

Factors specific to Ontario vehicle production within the North American market are: the shifting production locations; exchange rate, border issues, and research and development. Within North America, production is shifting from the Manufacturing Belt to the US South (for vehicle assembly) and Mexico (for parts). Also, the rising Canadian dollar makes Canadian exports more expensive for Americans. Additionally, there are concerns about tightened security at US-Canada border crossings since the auto industry employs Just-in-Time (JIT) transportation for vehicle assembly. This means that components are shipped for assembly a short time before they are needed, reducing the need for warehousing. The concern is that increased security at borderpoints could slow the transportation of parts between Southern Ontario and its American suppliers. Finally, the Canadian automotive industry is made up of 'branch plants' and most of the research, development, and design is based in other countries such as the United States and Japan.

The automotive industry has long been part of Ontario. Throughout its history in the province several trends have persisted including the strong linkage of Ontario automakers with American businesses; continued government support and influence; and concentration of the industry in Southern Ontario. Next, we will look at the energy sources powering Ontario industry and population needs.

Energy Resources

At the beginning of the twentieth century, Ontario relied heavily on coal from the United States (Baskerville, 2005). The government decided to develop and control a hydroelectricity system to provide a greater and cheaper supply of power to its growing manufacturing sector. Ontario's first hydroelectricity plant harnessed the power of the Niagara River near Niagara Falls, and then additional power plants were built around the province. There are now 180 plants (58 are linked to the grid)[105] supplying 21% of Ontario's electricity. Major hydroelectricity plants require a reservoir from which water flow can be controlled, and a substantial change in elevation (vertical drop) for the flowing water to

[105] Ontario Ministry of Energy and Infrastructure, 2009a.

generate power. While most large-scale sites were developed by the late 1950s (ICF Consulting, 2005), there are still possibilities for smaller projects mainly in Northern Ontario (Ontario Power Authority, 2009).

Over half of Ontario's energy is produced by nuclear reactors at three power plants (Figure 7). The Pickering power plant is in the eastern Toronto region, Darlington is 70 km east of Toronto, and the Bruce power plant is located 250 km northeast of Toronto on Lake Huron[106]. Between 1995 and 1998 eight reactors were closed due to operational and safety issues. Since then, four have been refurbished and restarted, two will be opened in 2010 (IESO, 2009), and two more are closed indefinitely[107]. Ontario was planning to build two new nuclear reactors at the Darlington site, but the project has been suspended due to the high costs[108]. Nuclear reactors have a lifespan of 30-40 years[109,110] and for continuing use they must be renovated. Concerns remain about the safety of the reactors and disposal of nuclear wastes.

Figure 7. Sources of Ontario's Electricity (2007)

Nuclear	52%
Hydroelectricity	21%
Coal	18%
Natural Gas and Others	8%
Wind	1%

Source: Data from Ontario Ministry of Energy and Infrastructure, 2009b.

Four coal-fired power plants produce 18% of Ontario's power. They also release emissions that contribute to air pollution and smog. For this reason, the province closed a coal-fired plant west of Toronto in 2005, and is planning to phase out its other four facilities at Atikokan, Lambton, Nanticoke, and Thunder Bay by 2014 at the latest. Natural gas is also used to generate electricity in Ontario. There are 60 gas plants, with 19 of these connected to the grid[111]. Electricity generation through coal and natural gas releases greenhouse gas emissions while nuclear, hydroelectricity, wind, and solar energy sources release few if any greenhouse gas emissions.

Only 1% of Ontario's power in 2007 was produced from wind sources. The province is planning to generate greater volumes of renew-

---

[106] Ontario Ministry of Energy and Infrastructure, 2009c.

[107] Ontario Ministry of Energy and Infrastructure, 2009c.

[108] Ontario Ministry of Energy and Infrastructure, 2009d.

[109] Canadian Nuclear Association, 2009.

[110] International Atomic Energy Agency, 2009.

[111] Ontario Ministry of Energy and Infrastructure, 2009a.

able energy and implement conservation measures through the Green Energy Act[112]. Wind energy, along with bioenergy[113], hydroelectricity, solar energy, geothermal energy, and ocean energy are all *renewable* sources of energy, meaning that they can be naturally regenerated[114]. The province has launched several renewable energy projects[115]. For example, the Prince Wind Farm, located on the shores of Lake Superior near Sault Ste. Marie, was completed in 2006. It consists of 126 wind turbines generating 189 MW of electricity (Brookfield Power, 2006). The government has also set up a system called 'net metering' that allows residents to receive credit for generating electricity through wind, water, solar or agricultural biomass and contributing it to the grid[116].

In 2007, Ontario was a net exporter of power to both the United States and other Canadian provinces[117]. Ontario trades electricity with its neighbours: Manitoba, Québec, Michigan, New York, and Minnesota (Peerbocus and Melino, 2007). Power is imported to Ontario during peak demand periods, and exported in times of over-supply and to generate revenue.

Settlement and Population

First Peoples

Ontario was first inhabited by peoples who crossed the Bering Strait from Asia 11,000 years ago. At first, these people were nomadic and relied on hunting and fishing for survival and used stone tools, but later they became more settled and began to gather berries, nuts, and wild vegetables to supplement their hunting and fishing activities (Wright, 1972). Between 1000 BC and 1500 AD, they crafted pottery, and participated in long-distance trade. The pottery indicated a more complex social structure as certain individuals became specialized in craft skills and were less burdened with subsistence activities (Baskerville, 2005). Rice Lake near the present-day Peterborough shows archaeological evidence of burial mounds differentiated by social class with leaders (Baskerville, 2005) and participation in the Hopewellian exchange network that extended into the present-day United States (Wright, 1972).

---

112 Legislative Assembly of Ontario, 2009.

113 Bioenergy is biological matter (from plants or animals) that is used to generate energy. Biomass refers to plant or animal matter that can be converted into a fuel (OECD, 2002).

114 Natural Resources Canada, 2009c.

115 Ontario Ministry of Energy and Infrastructure, 2009e.

116 Ontario Ministry of Energy and Infrastructure, 2009f.

117 Statistics Canada, 2009d.

The first record of corn being grown in Southern Ontario dates from 260-660 AD (Warrick, 2007a), and later there was more extensive farming of squash, sunflowers, beans, and tobacco (Wright, 1972; Warrick, 2007a). By 1000-1300 AD, the population lived in large and more permanent villages. Agriculture enabled the establishment of these settlements, reducing the need for hunting and gathering. The population grew in Southern Ontario to 60-65,000 people living in villages of 2-3,000 within longhouses which held more than a hundred people by the mid-1400s (Baskerville, 2005).

Even prior to European contact, there were differences between Northern and Southern Ontario peoples, and their languages, settlement patterns, and social relations. Algonquin speaking people, called Anishinaabe (known as Ojibwa by Europeans) lived in the North near Lake Superior. Another group, the Wendat[118], lived to the south near Lake Huron. The Iroquois people were based further south in various groupings including the: Neutral (western Lake Ontario), Petun (North of Neutral), and the Five Nations Iroquois (Mohawk, Oneida, Onondaga, Cayuga, and Seneca) in Northern New York State south of Lake Ontario (Baskerville, 2005).

The Iroquois and the Wendat in the South spoke Iroquoian, while the northern Anishinaabe spoke Algonquin. There were larger, more permanent, agricultural-based settlements in the South. By contrast, the people in Northern Ontario relied more on fishing and hunting, congregating only temporarily. In trading, the Wendat exchanged corn and other food items for Anishinaabe beaver and deer pelts (Baskerville, 2005).

European Contact

The Europeans came to North America looking for the Northwest Passage, a route to China and India. Champlain founded Québec City in 1608, and sent Étienne Brûlé to live amongst the Wendat in present-day Ontario. Champlain himself made two exploratory trips to the province. Later, Récollet and Jesuit missionaries from France in the early 1600s were sent to live with and convert Wendat. The Europeans brought disease that devastated First Nations and often killed the very young and very old, disrupting the social structure (Warrick, 2007b).

Southern Ontario was part of New France, originally claimed by Champlain. Hudson Bay, having been discovered by Henry Hudson in the early 1600s, belonged to the British. In 1670, the British Hudson Bay Company was granted authority over Rupert's Land, a territory covering the Hudson Bay drainage basin and including most of Northern Ontario. The Seven Years War between France and Britain ended with

[118] The Wendat were called the Huron by Europeans.

Britain gaining control of New France in 1763. In 1774, Southern Ontario along with northern American states was named part of Québec through the Québec Act. Tensions created by the Act contributed to the Declaration of Independence by the former British colonies south of the border in 1776.

Loyalists and Upper Canada

After the British lost the War of Independence in 1782, approximately 100,000 Americans who were loyal to Britain fled to Upper Canada. The majority returned to Britain, but approximately 8,000 (plus 2-3,000 First Nations) (Wood, 1991) stayed in Upper Canada. The majority of loyalists were not well educated or rich, but many did arrive with some farming equipment and animals (Wood, 1991). They were a diverse group, with varying backgrounds (Baskerville, 2005). Crossing at three main border points (near and east of Kingston, Niagara, and Detroit), they formed initial settlements close to their entry.

The settlement of the United Empire Loyalists initiated a pattern of a diverse immigrant population, and a population concentration in Southern Ontario near the US border. Haldimand, the Governor of Québec assembled land along St. Lawrence River east to Detroit by making agreements with the First Nations. It was the First Nations who gave up their land for the Loyalists, rather than the British. The Six Nations Iroquois were given 400,000 hectares along the Grand River (near Brantford), while the Loyalists each received 200-300 acres of land along with compensation for property taken by Americans. Early settlers congregated in separate townships according to their religion and ethnicity (Gentilcore, 1972).

Following the Loyalists, more Americans immigrated to Canada, as they simply had to swear allegiance to the Crown to obtain land (Gentilcore, 1972). To accommodate the growing Southern Ontario population under British law, and at the same time allow the French to live under French law and practice Catholicism, the British Government introduced the Constitutional Act in 1791 that created Upper and Lower Canada with elected assemblies. A Lieutenant Governor appointed by Britain represented the Crown, and Britain retained control of the political and legal system. John Graves Simcoe, the first Lieutenant Governor, directed the military to survey Upper Canada into townships, lots, and concessions using a grid system. This grid approach did not consider the varying physical characteristics and quality of the soils for agriculture, and later some settlers abandoned their land grants when they discovered they were not suitable for farming (Gentilcore, 1972).

Americans attacked Upper Canada in the War of 1812. The British Empire won the war, and the victory had a sustained influence on Upper

Canada. Prior to the war, some loyalists still retained American beliefs emphasizing freedom of the individual. However, following Upper Canada's defeat of the Americans in the three-year war, Upper Canadians expressed more loyalty to the British Empire with less emphasis on individual rights, and accepted a stronger role for government. The War of 1812 was important in developing Canadian identity (Berton, 2006).

Canal-Building

Also, the war spurred the construction of the Rideau Canal, a 202 km system of canals and 45 locks between Ottawa and Kingston connecting Lake Ontario to the Ottawa River via Cataraqui River, Rideau Lakes, and Rideau River. It was built 1826-1832[119] by British Lieutenant Colonel John By as an alternate route (besides the St. Lawrence River) between Lake Ontario and Montréal (Watson, 2006). The canal was useful for over 15 years to move goods and people bypassing the St. Lawrence rapids. However, by the 1850s, locks on St. Lawrence enabled more direct routing from the St. Lawrence to Lake Ontario. The Rideau Canal has been adapted in modern times for recreational boating. The first Welland Canal was constructed in 1829 by private business interests, and was later bought by the Province of Canada[120]. Like the Rideau Canal in Ottawa, it boosted settlement at St. Catherines. It also helped in moving migrants to the west.

European Immigration

Starting in the early 1800s, various waves of immigrants arrived in Upper Canada from Europe. First the British came, next the Irish fleeing the Irish Potato Famine. Then, German immigrants came after 1850 with many settling in the twin cities of Kitchener (known as Berlin until 1916)[121] and Waterloo, west of Toronto. The most successful immigrants were those who arrived first as they obtained the best land for farming (Baskerville, 2005). By 1855, most of the reasonably priced and fertile farmland was already taken. So, until the turn of the century when Ontario's industrial sector became more developed, there was little immigration and in fact there was out-migration from Upper Canada to the western United States and later to western Canada and Northern Ontario (Wood, 1991). *Natural increase*, the difference between births and deaths, contributed to Ontario's population growth through the 1800s, especially during the 1830s and 1840s when there was an extremely high birth rate among Upper Canadians (Wood, 1991).

---

119 Parks Canada, 2009.

120 Welland Public Library, 1973.

121 City of Kitchener, 2009.

Canada West and Confederation

In 1841, Lord Durham joined Upper and Lower Canada. Upper Canada (Southern Ontario) became Canada West. Northern Ontario was still part of Rupert's Land. Then, in 1867, the British North America Act created the country of Canada. Ontario was one of the four founding provinces (along with Québec, New Brunswick, and Nova Scotia). At Confederation, over 40% of the country's population lived in Ontario[122]. In 1876, Ontario expanded north and west following the acquisition of Rupert's Land; however, the province did not achieve its current northern boundaries until 1912[123].

'New Ontario'

Around the time of Confederation, substantial settlement began in the North spurred by the development of railways and the ambitions of the provincial government and the logging industry. Up until this time, Upper Canada was still dominated by Britain and transportation systems reflected an export focus with less integration in Canada. Areas outside of the shield were settled first, and then later Shield settlement was promoted (Gentilcore, 1972). The conditions in the North differed from those in the South. Lumber and minerals were the main focus of economic activities rather than the agriculture in Southern Ontario. *Accessibility* was of prime importance. The construction of transportation facilities such as roads and railways was necessary to develop the natural resources as well as bring in settlers. The government promoted settlement by offering free land, but it had business objectives that conflicted with those of the settlers. While the settlers were allowed to clear their properties, they were not permitted to sell lumber from crown lands for profit. Only companies were authorized to lumber Crown lands, with the government receiving royalties. For this reason, there were many abandoned properties in the shield areas by the early 1900s (Gentilcore, 1972). This is a clear illustration of the early dominance of the Southern Ontario-based government over the northern periphery. At the turn of the century, Northern Ontario was known as 'New Ontario', while Southern Ontario was 'Old Ontario' (Baskerville, 2005).

Immigration in 20th Century

There was a major immigration wave pre-World War I (Wood, 1991). Then, foreign immigration slowed in the 1920s but was countered by movement to Ontario from other provinces as industry devel-

---

122 Statistics Canada, 2008d.

123 Archives of Ontario, 2009.

oped in the province (Baskerville, 2005). During the 1930s, Ontario's immigration and birth rates declined due to the Depression. Following World War II, there was another wave of foreign and domestic immigration until 1961 (Wood, 1991). Immigrants arriving between 1900 and 1961 were predominantly European (92.7%), with four countries: United Kingdom, Italy, Germany and the Netherlands making up nearly two-thirds of all immigrants to the province. After 1961, immigrants from Europe declined, and an increasing proportion of Ontario immigrants came from Asia[124] – a trend that has continued (until 2006). Ontario gained population from other provinces in the 1960s, then, in the mid-1970s to the early 1980s, there was out-migration from Ontario to Alberta and British Columbia for employment (Wood, 1991). This out-migration trend reversed itself in the 1980s when there were declines in the western resource sector. Again, in the 1990s, some Ontarians emigrated to other provinces (Foot, 2002).

Ontario's Population Today

Ontario is the most populous province in Canada with 13 million people making up 38.7% of the nation's population[125]. Ontario's population is very concentrated, with approximately 94% residing in Southern Ontario, while only 6% lives in Northern Ontario (Figure 8). This is despite Northern Ontario having almost 90% of the province's land area[126].

Ontario has consistently had the country's largest share of foreign-born population with 46.3% of Canada's total in 1901 increasing to 54.9% in 2006 (Chui *et al.*, 2007)[127]. India (15%), China (14%), Pakistan (7.5%), Philippines (6.5%), and Sri Lanka (3.2%) are the most common origins of recent immigrants. Southern Ontario is the destination for most of these new migrants. By contrast, Northern Ontario's population dropped in the 1990s and increased only slightly (less than 1%) between 2001 and 2006. Prior to the 1960s, Northern Ontario had a mobile population and attracted immigrants. However, since that time, Northern Ontario has attracted little inbound migration and there has been substantial youth out-migration. Southern locations such as Muskoka and Parry Sound have the highest in-migration rates in Northern Ontario. Of the few migrants in Northern Ontario, almost all come from Ontario itself and of those most are from within Northern Ontario. There is little international migration to the region (Southcott, 2006).

---

[124] Ontario Ministry of Finance, 2003a.

[125] Statistics Canada, 2009a.

[126] Ontario Ministry of Northern Development, Mines and Forestry, 2009.

[127] Statistics Canada, 2000.

The province's (median age) was older than the national average through the 20th century, due mainly to lower birth rates and more immigration of young to middle working aged immigrants, compared to all other provinces. This changed in 2001 when Ontario's population became younger than the national average. This can be attributed to a relatively small number of Ontarians born between the First and Second world wars who are now 65 years or older, resulting in relatively fewer seniors[128].

Figure 8. Comparison of Percentage of Population and Land Area for Northern and Southern Ontario

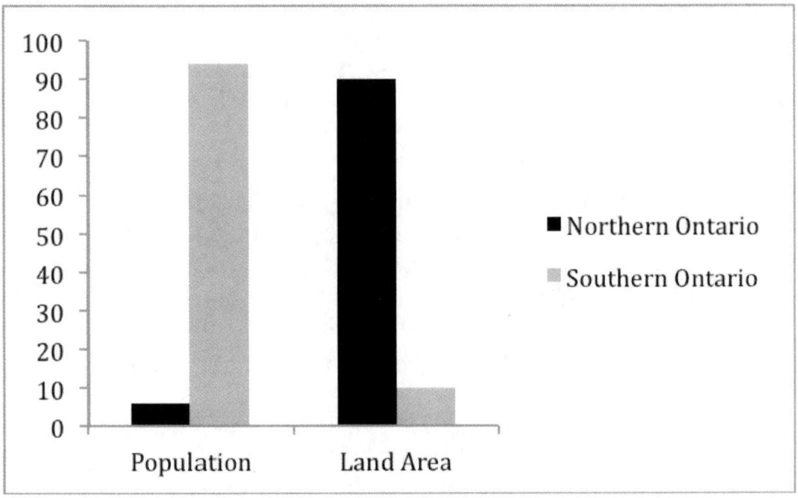

Source: Nairne Cameron using data from Ontario Ministry of Northern Development, Mines and Forestry, 2009.

Also, Ontario's high immigration rates of women of childbearing age contributed to more children being born and hence a younger population in 2006[129]. Still, Ontario is aging. The median age has risen over the province's history from 28.4 years in 1961 to 39.0 years in 2006[130]. The proportion of seniors (aged 65 years or older) registered 13.6% of the provincial population in 2006[131], and is forecast to make up 21.9% of the population by 2031[132].

---

[128] Statistics Canada, 2007a.

[129] Statistics Canada, 2007a.

[130] Statistics Canada, 2007a.

[131] Statistics Canada, 2007a.

[132] Ontario Ministry of Finance, 2009a.

Northern Ontario has an older population compared to Southern Ontario. This is due to out-migration of the region's youth and lack of immigration (Southcott, 2006). An exception to this trend is that the population in Northern Ontario aboriginal communities is much younger (60% under 30 years of age) than the population in Northern Ontario as a whole (37% under the age of 30 years) (Southcott, 2006). Elliot Lake, the former uranium mining town, turned retirement community, has the oldest population for a small city in Canada with almost a third of its population being seniors[133]. The older age structure in Northern Ontario and its health care demands poses a potential challenge as many communities in the region have been classified by the Ontario government as medically underserviced, meaning they lack an adequate ratio of medical staff relative to the population[134]. To help address this issue, the Northern Ontario School of Medicine was established in 2005 to train medical personnel in Northern Ontario as a partnership between Laurentian University in Sudbury and Lakehead University in Thunder Bay[135].

French speaking Ontarians make up 4.8% of the provincial population, with a particular concentration in Eastern Ontario[136]. Ontario is not officially bilingual, but since 1986 has made government services available in French in specific areas of the province[137]. Approximately 21% of all Canadian Aboriginal people live in Ontario, but make up only 2% of Ontario's population[138]. Two thirds are North American Indian, while nearly a third is Métis, and less than 1% is Inuit. The aboriginal population is a young population that is growing rapidly (almost 30% between 2001 and 2006). Nearly a third of Ontario Northern American Indians live on reserves, while most Métis and Inuit inhabit urban areas[139].

Urban Development

Ontario has historically had higher rates of urbanization compared to other Canadian provinces, and is partially related to increased industrialization in Ontario that attracted workers to its cities. Many of Southern Ontario's cities developed earlier than those in the western provinces and the North. Settlements that became established first possessed an

133 Elliot Lake was tied with Parksville, BC as the oldest small city (census agglomeration) in Canada (2006 Data). Statistics Canada, 2007b.

134 HealthForceOntario, 2009.

135 Northern Ontario School of Medicine, 2010.

136 Data for 2006. Ontario Office of Francophone Affairs, 2011a.

137 Ontario Office of Francophone Affairs, 2011b.

138 Statistics Canada, 2009e.

139 Ontario Ministry of Finance, 2009b.

initial advantage (Pred, 1965) in that they had skilled workers, built infrastructure, transportation, capital, and entrepreneurs, and therefore could more easily attract further growth. Transportation systems have shaped the pattern of Ontario's urban growth and their influence can be recognized in four eras[140]: 1) Water (pre-1850); 2) Railway (1850-1920); 3) Car (1920-1945); and, 4) Highway (1945-now).

The first two eras, water and railway, were important in forging the hierarchy of the urban system (network of cities). The initial locations of many current Southern Ontario settlements developed in the water transport era including Toronto, Hamilton, and Ottawa. Railways, along with accompanying industrialization drew and concentrated the population into settlements in Southern Ontario. In Northern Ontario, railways were a determining factor in the locations of new settlements. Automobiles and later highways had less impact on the overall structure of the urban system, but did enable city populations to suburbanize and disperse around urban centres.

Water Era

When Ontario was first explored by Europeans, water was the main mode of transport, supplemented by wagon roads. Thus, many settlements formed close to water and along major transportation routes such as the St. Lawrence River, Lake Ontario, and Hudson Bay. The settlements were frequently located beside natural harbours where resources such as furs, wheat, and timber could be assembled and shipped by boat. The habitations were often defensive in nature owing to competition and frictions between the British and French and the first peoples. For example, Southern Ontario was originally claimed by Champlain and was part of New France and controlled by the French. Hudson Bay, discovered by Henry Hudson in the early 1600s, belonged to the British. In 1670, the British Hudson Bay Company was granted authority over Rupert's Land, a territory covering the Hudson Bay drainage basin and including most of Northern Ontario. The Hudson Bay Company built forts and posts on the Hudson and James Bay coasts. France, in turn, constructed forts at two strategic locations with port facilities: Fort Frontenac (near Kingston), at the junction of the Cataraqui River and Lake Ontario, and Fort Niagara (now Youngstown, New York) where the Niagara River meets Lake Ontario.

Later, when the Loyalists fled to Ontario[141] they settled in three main areas: near and east of Kingston, Niagara, and across the river from Detroit. From these centres, smaller centres with ports and mills devel-

[140] Adapted from Borchert, 1967.

[141] Loyalists also settled in Lower Canada and the Maritimes.

oped along Lake Ontario and then there was further expansion along Lake Erie, and eventually northward into the interior (Mika and Mika, 1976).

John Graves Simcoe, Upper Canada's first Lieutenant Governor designed a plan for population distribution that promoted interior settlement to defend against the United States. First, he established Fort York, present-day Toronto, near the site of an abandoned fur trading post called Fort Rouillé. He moved the capital from Newark[142] (now Niagara-on-the-Lake) to York (The Friends of Fort York, 2009) on the north side of Lake Ontario, a greater distance from the United States. This led to Newark's decline, but York expanded. He envisioned population concentration in an east-west corridor between Windsor and the Ottawa River (along Lake Ontario and St. Lawrence River passing through Kingston, York, Oxford, London, and Chatham), and a second corridor encircling the west end of Lake Ontario to Niagara, and finally a third corridor north from Toronto to Lake Simcoe. Simcoe's plan has been enduring as Southern Ontario's current population largely reflects this distribution, except for the unpredicted development of the Waterloo and Ottawa regions (Wood, 1991).

Kingston was strategically located from a defence standpoint on Lake Ontario at the St. Lawrence River, and became a major processing and shipping point for Upper Canadian farm goods. In the 1800s, urban areas grew related to their shipping capacity of commodities such as wheat, flour and wood. York and Hamilton had ports for handling wheat from surrounding areas. Both ports also received imports and immigrants. In 1828, the Oswego Canal was completed linking Lake Ontario to the Erie Canal which in turn connected to New York City (Benn, 2008). This linkage greatly boosted the economies of York and Hamilton as direct trade with the States increased. With shipments no longer needing to proceed down the St. Lawrence through Montréal, this Kingston's trade volume decreased. Kingston expanded though when it became the capital of Canada West in 1841, but declined when the government seat was moved from the city in 1844. Kingston was deemed too small and too vulnerable to American attack, so the capital switched between Montréal and Toronto[143]. By 1851, Toronto had 31,775 residents and was considerably more populous than Kingston (Benn, 2008).

By the mid-1800s, settlement expanded into Ontario's interior and northward with habitations forming along roads. Interior service centres such as the twin cities of Berlin and Waterloo, London, and Peterbor-

---

[142] Newark was renamed Niagara in 1798, then Niagara-on-the-Lake in 1970.

[143] York was renamed Toronto in 1834 (Benn, 2008).

ough arose. Lumbering was an important industry at this time which accounted for the establishment of the cities of Peterborough and Lindsay along rivers. Ottawa (then called Bytown) was also a timber centre on the Ottawa and Rideau Rivers. Its population was boosted with the building of the Rideau Canal, and later in 1867 when it was named the new capital of Canada. St. Catherines, the terminus of the Welland Canal, expanded at the expense of Niagara which was situated along the former portage route. The Golden Horseshoe encompassing Toronto, Hamilton, St. Catherines, and Niagara began to form at this time. During the 1830s and 1840s, there was rural to urban migration. Toronto and Kingston had more women than men, as females moved to the cities looking for work (Wood, 1991). However, only 14% of upper Canadians lived in urban areas by 1851[144].

Railway

Upon the coming of the railways, a network of lines began to develop in the settled parts of Southern Ontario and the interior became more populated. Settlements with railway connections grew, while those without linkages declined. Hamilton's economy benefited from its location on a main line connecting Windsor and Toronto to the United States. Toronto, in particular, became a major railway centre handling wheat which led to the establishment of the Toronto Stock Exchange in 1855 (Gentilcore, 1972). With developing railway connections, Toronto increasingly benefited from the lumber and mining industries in the North.

A railway from the south to Gravenhurst, a timber centre on northern Lake Simcoe, was built in 1875, and later was extended northwards passing through the lumber towns of Bracebridge and Huntsville. Then, in the 1880s, the construction of the Canadian Pacific Railway (linking Canada from coast to coast) over the Canadian Shield was completed. The railway was pivotal in the development of Northern Ontario. The route passed through almost all[145] of Northern Ontario's now most populous cities: North Bay, Sudbury, Sault Ste. Marie, Thunder Bay, and Kenora. During construction, nickel and copper deposits were confirmed in the Sudbury area and mining began in 1886. Sault Ste. Marie, bordering the United States between Lake Superior and Lake Huron grew as a result of its location on a branch line of the CPR that originated from Sudbury and connected to the American railroad sys-

---

144 Statistics Canada, 2009f.

145 Timmins is not located on this route.

tem. Thunder Bay[146] was initially based on silver mining and lumbering, but expanded when it became a port handling wheat arriving by rail from the west. Thunder Bay is known as the 'Finnish Capital of Canada' due to the large Finnish population who began immigrating to the area in the 1870s to work on the railway (Multicultural Canada, 2009). Kenora, near the Manitoba border, also grew as a result of its location on the CPR and gold mining. Parry Sound, on eastern Georgian Bay, also became a centre due to railway development.

In the northeast of the province, the government began construction on the Temiskaming and Northern Railway in 1902 (Gentilcore, 1972) to stimulate development. Starting at North Bay, it connected to Lake Temiskaming and Cochrane, eventually reaching Moosonee (near James Bay) by 1932. Along the route, silver was discovered in Cobalt, and gold in Porcupine and Kirkland Lake.

Car and Highway Eras

By 1920, Ontario's current urban system was largely set with Toronto, Hamilton, London, and Ottawa as the largest cities. Ontario led other provinces in urbanization. The province was over half urban before World War I (1911), 10 years before the rest of Canada[147]. Migrants were attracted to cities by manufacturing jobs. The automobile industry led the manufacturing sector. They were mainly used for recreational purposes. Streetcar lines in major cities such as Toronto, Ottawa that had enabled the extension of suburban settlements gradually were phased out and replaced by bus and car transport. Following World War II, cars began to dominate cities and suburbanization forged ahead. And, Ontario continued to urbanize quickly after World War II, with over ¾ of the province's population city-dwellers by 1956, rising more slowly in the following years to 85% in 2006[148].

The 1950s and 1960s saw a surge in the development of highways. Urban sprawl began and facilitated by widespread car ownership, the population suburbanized. Simcoe's east-west corridor between Windsor and Ottawa (continuing to the east to Québec City) developed with considerable population, and has been termed the 'Main Street of Canada' (Yeates, 1975). More recently, the corridor has been considered as a route for high-speed rail due to its high population density.

In Southern Ontario where 94% of the province's residents reside, two-thirds of the provincial population is concentrated in the crescent-

---

146 Originally two separate settlements: Port Arthur and Fort William which joined in 1970 to form Thunder Bay.

147 Statistics Canada, 2009f.

148 Statistics Canada, 2009f.

shaped area – called the Golden Horseshoe[149] encircling the western end of Lake Ontario between Oshawa and Niagara Falls. Furthermore, within the Golden Horseshoe, 46% of Ontario's population lives in the Greater Toronto Area (GTA)[150] and growth in the GTA represented 63.2% of the province's growth between 2001 and 2006[151].

There is an inherent conflict between the population growth resulting in demand for housing and infrastructure and agricultural land use. Over half of all the highest quality farmland in Canada is located in Ontario, mainly in Southern Ontario, the same area where the province's population growth is occurring. Car dependent, low density suburban development that is taking place especially in the Golden Horseshoe area of Ontario not only removes agricultural land, but can also contribute to increased driving, rising vehicle emissions that cause air pollution and negative health effects including respiratory difficulties, cardiovascular problems, and lung cancer (Abelsohn *et al.*, 2005).

Toronto's Immigration and Suburbanization

The majority (68.3%) of the province's foreign-born residents live in Toronto (Census Metropolitan Area)[152]. Friends and family connections and employment draw immigrants to the Toronto region, and as of 2006, foreign-born persons make up 45.7% of the Census Metropolitan Area's population (Chui *et al.*, 2007). India and China are the origins of approximately a third of recent immigrants. The majority of recent immigrants (56.6%) are in the prime working ages of 25-54 (Chui *et al.*, 2007). In Toronto, most of the growth in the foreign-born population has taken place in Toronto's suburbs. The foreign born population in suburbs makes up around one half of population in Brampton, Markham, Mississauga, Vaughan, and approximately a quarter of the population in Ajax and Aurora[153].

Transportation

Southern Ontario now has a well-developed internal transportation system with road, air, rail, and sea as well as linkages to the States. Northern Ontario's linkages are more limited both internally and externally, with an incomplete passenger rail network that does not connect the major population centres, and threats to the freight rail network. For

---

149 Statistics Canada, 2009g.

150 The GTA is a provincial planning unit that includes the Toronto, York, Peel, Durham, and Halton census divisions. Ontario Ministry of Finance, 2003b.

151 Ontario Ministry of Finance, 2009c.

152 Statistics Canada, 2008e.

153 Statistics Canada, 2009g.

example, in mid-2009, the operators of Huron Central Railway announced the cancellation of their operations between Sault Ste. Marie and Sudbury providing freight service to customers such as Essar Steel Algoma. The rail line is in need of repair and upgrading to allow trains to move more safely and quickly. The City of Sault Ste. Marie, along with major customers, and the federal and provincial governments negotiated a one-year extension in service with a $15.9 million contribution for operating expenses (Mattia, 2009). However, the line's future is still very uncertain.

Northern Ontario's passenger bus service is also threatened with a recent announcement of cancellation of service in Northwestern Ontario. While a government report is being prepared, the company has agreed to continue operations, albeit at a lower frequency (Reuters Canada, 2009). Air travel in Northern Ontario is expensive due to market size, distances, and limited competition, but is an essential mode of travel, as some remote communities especially in the far north are inaccessible by road or water.

Conclusions

Ontario has historically dominated Canada geographically due to its central location and close proximity to major US centres; politically due to its large population; and economically owing to its industrial power, high level of urbanization, and strong consumer market. However, the province is struggling financially, now receiving equalization payments since 2009 for the first time in its history. This financial pressure is linked to the recession that is affecting trade especially with the US, the rising and fluctuating Canadian dollar that is challenging the province's manufacturers, and the downturn in Ontario's key auto-manufacturing sector. Many of Ontario's industries have already experienced substantial restructuring over the last 30 years due to changing trade arrangements and economic cycles. The province seeks to make itself more adaptable to future developments by investing in education and health care. Indirectly, the economic slowdown has helped Ontario by reducing energy demands, enabling the province to more easily abandon planned nuclear projects and commit to closing coal-fired generating plants. A sustainable energy focus has produced a Green Energy Act and further support for renewable energy projects.

This chapter has depicted the separate, yet interdependent histories of Northern and Southern Ontario, regions that are strongly defined by the imprint of the Canadian Shield on the province. The densely populated and growing, well connected, highly developed, economically diverse, and climatically warmer South is contrasted with the vast, sparsely populated, less well served and accessible, more resource-based, and climatically harsh North. Each Ontario has its own challenges. Southern Ontario's continued growth and suburban development continues to consume prime agricultural lands, create traffic congestion and contribute to air pollution. Northern Ontario with a more static, and in places declining population, does not experience the same growth pains. However, the North is adjusting to changes in its traditional resource-based economy and becoming more autonomous by expanding and strengthening its own educational and health service capacity.

Review Questions

Here are ten questions pertaining to the chapter listed below.

- Outline the general differences in population density across the province. What factors account for these differences? Cf. Natural Resources Canada, The Atlas of Canada, Population Density, 2006.
- How does Ontario's population distribution compare to the location of the province's forested areas? Cf. Natural Resources Canada, The Atlas of Canada, Population Distribution and Forested Areas.
- Describe the distribution of Ontario communities that are solely and highly reliant on the forestry industry. Cf. Natural Resources Canada, The Atlas of Canada, Forestry-reliant Communities, 2001.
- How is the foreign-born population of Ontario distributed across the province? Compare geographical differences between immigrants arriving between 2001 and 2006, and those arriving before 2001. What are the factors that might account for these geographical differences? Cf. Natural Resources Canada, The Atlas of Canada, Foreign-born Population, 2006.
- Check the current output of energy from wind power in Ontario and note the date and time. Then, examine the location of current and proposed wind projects in the province. Cf. IESO, Wind Power Generation in Ontario, http://www.ieso.ca. For comparison, examine a map of wind power at 80 magl (m.a.g.l. means metres above ground level, there are greater wind speeds at higher elevations). Are there any other potential areas for wind development in Ontario? Then, analyse these areas taking into account proximity to population centres and other factors portrayed in the other map layers. Cf. Ontario Ministry of Natural Resources, Renewable Energy Atlas.
- Explain the importance of the Great Lakes to the Province of Ontario, and then outline some of the potential impacts of climate change on the Great Lakes. Discuss some actions that the provincial government might consider taking in relation to the Great Lakes considering potential future climate change. Cf. Ontario Ministry of the

Environment, Adapting to Climate Change in Ontario, http://www.ene.gov.on.ca/.

- Contrast transportation accessibility by road, air, and rail across the province of Ontario. How does the breadth and connectivity of the networks vary? What are some possible reasons for the differences? Cf. Ontario Ministry of Economic Development and Trade, Invest in Ontario, Ontario Maps, http://www.sse.gov.on.ca/.
- Compare the Aboriginal population and population distribution in Ontario at the time of early European arrival (Eastern Native Population, Early 17th Century) with a later time period (Native Canada, circa 1820). You can check 'modern geography' to see the current boundaries of Ontario. By referring to the accompanying commentary, explain the differences between the two time periods. Cf. Historical Atlas of Canada Online Learning Project, Native Canada, http://www.historicalatlas.ca/.
- Examine the spread of Toronto's population over the period of 1971-2006. Describe the changes and note what factors have facilitated the expansion of the population. Cf. Statistics Canada, Toronto CMA, 35 Years of Change, http://www12.statcan.gc.ca/census-recensement/2006/.
- How important is the United States as a trading partner for Ontario, compared to other countries? Contrast the type and dollar value of Ontario's imports and exports to the United States over the time period given. What are the top 5 American states to which Ontario exports? Cf. Ontario Exports, Ontario Trade Fact Sheet, http://www.sse.gov.on.ca/medt/ontarioexports/.

References

Abelsohn, A., Riina Bray, R., Vakil, C., Elliott, D. (2005), *Report on Public Health and Urban Sprawl in Ontario: A Review of the Pertinent Literature*, Toronto: Ontario College of Family Physicians.

Anastakis, D. (2005), *Auto Pact: Creating a Borderless North American Auto Industry, 1960-1971*, Toronto; Buffalo: University of Toronto Press.

Armson, K.A. (2001), *Ontario Forests: A Historical Perspective*, Markham: Fitzhenry and Whiteside Limited.

Archives of Ontario (2009), *The Evolution of Ontario's Boundaries 1774-1912*, Retrieved September 26, 2009, from http://www.archives.gov.on.ca/english/on-line-exhibits/maps/ontario-boundaries.aspx#boundaries_1774.

Baskerville, P.A. (2005), *Sites of Power: A Concise History of Ontario*, Don Mills: Oxford University Press.

Bayfield Institute (2001), *Sea Lamprey Control*, Retrieved September 26, 2009, from http://www.dfo-mpo.gc.ca/regions/central/pub/bayfield/06-eng.htm.

Benn, C. (2008), "Colonial Transformations", in *Toronto: A Short Illustrated History of Its First 12,000 Years*, R.F. Williamson (ed.), Toronto: James Lorimer and Company Ltd.

Bernard, A. (2009), "Trends in Manufacturing Employment", *Perspectives on Labour and Income*, Ottawa: Statistics Canada.

Berton, P. (2006), *The Battles of the War of 1812: An Omnibus*, Calgary: Fifth House.

Bollman, R.D., Beshiri, R., Mitura, V. (2006), *Northern Ontario's Communities: Economic Diversification, Specialization and Growth*, Ottawa: Statistics Canada.

Borchert, J.R. (1967), "American Metropolitan Evolution", *Geographical Review*, 57(3), 301-332.

Bostock, H.J. (1970), "Physiographic Subdivisions of Canada", in *Geology and Economic Minerals of Canada*, R.J.W. Douglas (ed.), Ottawa: Dept. of Energy, Mines, and Resources, 10-30.

Brookfield Power (2006), *Brookfield Power Completes Canada's Largest Wind Farm*, November 22, 2006, Retrieved September 26, 2009, from http://www.brookfieldpower.com/content/2006/brookfield_power_completes_canadas_largest_wind_f-782.html.

Brown, R.J.E., Péwé, T.L. (1973), "Distribution of Permafrost in North America and its Relationship to the Environment: A Review, 1963-1973", in *Permafrost: The North American Contribution to the Second International Conference: 2nd Int. Conf. Permafrost, Yakutsk, Siberia, USSR*, Washington: National Academy of Sciences, 71-100.

Brown, R. (2007), *Ontario's Ghost Town Heritage*, Erin: Boston Mills Press.

Canada Centre for Remote Sensing (2008), *Tour Canada from Space: Peterborough, Ontario – Scene 1*, Retrieved September 26, 2009, from http://www.ccrs.nrcan.gc.ca/resource/tour/28/28scene1_e.php.

Canadian Geographic (2009), *In-Memorial: Lemieux, Ontario*, Retrieved September 26, 2009, from http://www.canadiangeographic.ca/magazine/so05/indepth/soc_lemieux.asp.

Canadian Nuclear Association (2009), *Major Reactor Types*, Retrieved September 26, 2009, from http://www.cna.ca/curriculum/cna_nuc_tech/reactor_types-eng.asp?bc=MajorReactorTypes&pid=MajorReactorTypes.

CanaTREK, the Summits of Canada (2006), *Ontario – Ishpatina Ridge*, Retrieved December 30, 2009, from http://www.summitsofcanada.ca/canatrek/summits/ontario.html.

Chadwick, P., Hume, B. (2009), *Weather of Ontario*, Edmonton: Lone Pine Publishing.

Chui, T., Tran, K., Maheux, H. (2007), *Immigration in Canada: A Portrait of the Foreign-born Population, 2006 Census*, Cat. No. 97-557-XIE, Ottawa: Statistics Canada, Retrieved September 26, 2009, from http://dsp-psd.pwgsc.gc.ca/collection_2007/statcan/97-557-X/97-557-XIE2006001.pdf.

City of Kitchener (2009), *History of Kitchener*, Retrieved January 7, 2010, from http://www.kitchener.ca/visiting_kitchener/history.html.

Dadgostar, B., Jankowski, W.B., Moazzami, B. (1992), *The Economy of Northwestern Ontario: Structure, Performance and Future Challenges*, Thunder Bay: Centre for Northern Studies, Lakehead University.

De Beers Canada (2009), *Victor Mine: Factsheet*, Retrieved September 26, 2009, from http://www.debeerscanada.com/files_2/victor_project/factsheet.html.

Department of Finance Canada (2009), *Equalization Program*, Retrieved January 7, 2010, from http://www.fin.gc.ca/fedprov/eqp-eng.asp.

Department of Finance Canada (2011), *Federal Support to Ontario*, Retrieved April 20, 2011, from http://www.fin.gc.ca/fedprov/mtp-eng.asp#Ontario.

Dominion Bureau of Statistics (1921), *The Canada Year Book 1920*, Ottawa: F.A. Acland, Retrieved September 26, 2009, from http://www66.statcan.gc.ca/cdm4/document.php?CISOROOT=/eng&CISOPTR=36697&REC=18.

Durnford, H., Baechler, G. (1973), *Cars of Canada*, Toronto: McClelland and Stewart.

Eden, L., Molot, M.A. (1993), "Canada's National Policies: Reflections on 125 Years", *Canadian Public Policy*, 19(3), 232-251.

Environment Canada (2002), *Ice Storm 1998*, Retrieved September 26, 2009, from http://www.msc-smc.ec.gc.ca/media/icestorm98/icestorm98_the_worst_e.cfm.

Essar Steel Algoma (2010), *Company*, Retrieved January 7, 2010, from http://www.algoma.com/company/.

Eyles, N. (2002), *Ontario Rocks: Three Billion Years of Environmental Change*, Markham: Fitzhenry and Whiteside Limited.

FedNor (2009), *FedNor Community Economic Development Officers*, Retrieved October 3, 2009, from http://www.ic.gc.ca/eic/site/fednor-fednor.nsf/eng/fn00805.html.

Fisheries and Oceans Canada (2009), *Aquatic Invasive Species*, Retrieved September 26, 2009, from http://www.dfo-mpo.gc.ca/science/enviro/ais-eae/index-eng.htm.

Foot, D.K. (2002), *Demographic Change and Public Policy in Ontario*, Research Report for the Panel on The Role of Government in Ontario, Retrieved December 31, 2009 from http://www.law-lib.utoronto.ca/investing/reports/rp3.pdf.

Foreign Affairs and International Trade Canada (2008), *The Canada-U.S. Softwood Lumber Agreement*, Retrieved December 31, 2009 from http://www.international.gc.ca/controls-controles/softwood-bois_oeuvre/notices-avis/agreement-accord.aspx?lang=eng.

Fourastié, J. (1949), *Le Grand Espoir du XXe Siècle: Progrès Technique, Progrès Économique, Progrès Social*, Troisième Édition Revue et Augmentée, Paris: Presses Universitaires de France.

Frenette, M. (2007), *Life After the High-tech Downturn: Permanent Layoffs and Earnings Losses of Displaced Workers*, Analytical Studies – Research Paper

Series, Cat. No. 11F0019MIE, No. 302, Ottawa: Statistics Canada, Retrieved September 26, 2009, from http://www.statcan.gc.ca/pub/11f0019m/11f0019 m2007302-eng.pdf.

Friedmann, J. (1966), *Regional Development Policy: A Case Study of Venezuela*, Cambridge, MA, and London: The M.I.T. Press.

Gentilcore, R.L. (ed.) (1972), "Settlement", in *Ontario*, R.L. Gentilcore (ed.), Toronto: University of Toronto Press.

Gibbins, R., Berdahl, L. (2003), *Western Visions, Western Futures: Perspectives on the West in Canada*, Peterborough: Broadview Press.

Gonzalez, M.A. (2003), "Continental Divides in North Dakota and North America", *North Dakota Geological Survey Newsletter*, 30(1), 1-7, Retrieved September 26, 2009, from https://www.dmr.nd.gov/ndgs/NEWSLETTER/ NLS03/pdf/Divide.pdf.

Government of Canada (2009), *Key Economy Events. 1965 – Canada – United States Auto Pact*, Retrieved September 26, 2009, from http://www. canadianeconomy.gc.ca/English/economy/1965canada_us_auto_pact.html.

Government of Ontario (2009), *About Ontario*, Retrieved September 26, 2009, from http://www.ontario.ca/en/about_ontario/ONT04_020887.

Grady, W. (2007), *The Great Lakes: The Natural History of a Changing Region*, Vancouver, Berkeley: Greystone Books.

Great Lakes St. Lawrence Seaway (2009), *The Seaway*, Retrieved September 26, 2009, from http://www.greatlakes-seaway.com/en/seaway/index.html.

Hambrey, M.J., Alean, J. (1992), *Glaciers*, Cambridge, New York: Cambridge University Press.

HealthForceOntario (2009), Ontario's New Physician Recruitment and Retention Programs Bringing the Underserviced Area Program into the 21st Century: A Consultation Paper, Retrieved January 7, 2010, from http://www. ontla.on.ca/library/repository/mon/23007/293996.pdf.

Hill, T.L. (2002), Canadian Politics, Riding by Riding: An In-Depth Analysis of Canada's 301 Federal Electoral Districts, Minneapolis: Prospect Park Press.

ICF Consulting (2005), *Electricity Demand in Ontario – A Retrospective Analysis*, Toronto, Ontario, Prepared for Chief Conservation Officer, Ontario Power Authority, Retrieved September 26, 2009, from http://www.conservation bureau.on.ca/Storage/14/1959_OPA_Report_FactorAnalysis_Final.pdf.

IESO (2009), *An Assessment of the Reliability and Operability of the Ontario Electricity System: 18 Month Outlook*, Retrieved September 26, 2009, from http://www.ieso.ca/imoweb/pubs/marketReports/18MonthOutlook_2009may .pdf.

Industry Canada (2007), *Cars on the Brain: Canada's Automotive Industry 2007*, Ottawa: Industry Canada, Retrieved September 26, 2009, from http://www.ic.gc.ca/eic/site/auto-auto.nsf/vwapj/2007_AutoStatisticsFlyer-ENG.pdf/$FILE/2007_AutoStatisticsFlyer-ENG.pdf.

Innis, H.A. (1999), *The Fur Trade in Canada: An Introduction to Canadian Economic History*, Toronto: University of Toronto Press.

International Atomic Energy Agency (2009), *Extending the Operational Life Span of Nuclear Plants*, Retrieved September 26, 2009, from http://www.iaea.org/NewsCenter/News/2007/npp_extension.html.

International Joint Commission (2009), *Who We Are*, Retrieved September 26, 2009, from http://www.ijc.org/en/background/ijc_cmi_nature.htm#role

Invest in Ontario (2009a), *Exports as a Share of GDP*, Retrieved September 26, 2009, from http://www.investinontario.com/siteselector/ooit_314.asp.

Invest in Ontario (2009b), *Forest Sector and Valued-added Products*, Retrieved September 26, 2009, from http://www.investinontario.com/north/industry_forest.asp?gonorth=y.

Invest in Ontario (2009c), *Ontario's Auto Industry: Driving the Future*, Retrieved September 26, 2009, from http://www.investinontario.com/automotive/default.asp.

Jackson, J.N. (1997), *The Welland Canals and Their Communities: Engineering, Industrial, and Urban Transformation*, Toronto: University of Toronto Press.

Jenish, D. (2009), *The St. Lawrence Seaway: Fifty Years and Counting*, Manotick: Penumbra Press, http://www.greatlakes-seaway.com/en/pdf/Jenish_en.pdf.

Johnson, J.R. (2004), "The WTO Decision – MFN, National Treatment, Trims and Export Subsidies", in *The Auto Pact: Investment, Labour and the WTO*, M. Irish (ed.), The Hague, London, New York: Kluwer Law International.

Kemp, D.D. (1993), *The Climate of Northern Ontario*, Lakehead University Centre for Northern Studies, Occasional Paper #11, Thunder Bay: Lakehead University.

Krueger, R.R. (1959), "Changing Land-Use Patterns in the Niagara Fruit Belt", *Transactions of the Royal Society of Canada*, 32(2), 67.

Legislative Assembly of Ontario (2009), *Bill 150, Green Energy and Green Economy Act, 2009*, Retrieved September 26, 2009, from http://www.ontla.on.ca/web/bills/bills_detail.do?locale=en&BillID=2145&detailPage=bills_detail_the_b.

MacDonald, R. (2007), *Not Dutch Disease, It's China Syndrome. Insights on the Canadian Economy*, Ottawa: Statistics Canada, Cat. No. 11-624-MIE, No. 017, Retrieved September 26, 2009, from http://www.statcan.gc.ca/pub/11-624-m/11-624-m2007017-eng.pdf.

Mackintosh, W.A. (1923), "Economic Factors in Canadian History", *The Canadian Historical Review*, 4, 12-25.

Magna International Inc. (2009), *Joint Press Release – Magna and Sberbank Offer Selected as the Preferred Solution for Opel*, September 10, 2009, Retrieved September 26, 2009, from http://www.magna.com/magna/en/media/pressreleases/default.aspx?i=229.

Martin, C. (1999), The Politics of Northern Ontario: An Analysis of the Political Divergences at the Provincial Periphery, M.A. Thesis, Dept. of Political Science, McGill University.

Mattia, E.D. (2009), "Railway saved – for now", *Sault Star*, Retrieved January 8, 2010, from http://saultstar.com/ArticleDisplay.aspx?archive=true&e=1699109.

Mays, J.C. (2003), *Ford and Canada: 100 Years Together*, Montréal: Syam Publishing.

McCann, L.D. (1987), "Heartland and Hinterland: A Framework for Regional Analysis", in *Heartland and Hinterland: A Geography of Canada*, L.D. McCann (ed.), Scarborough: Prentice-Hall Canada.

McMaster University (2009), *Case Study-Peterborough Drumlin Field*, Canadian Landform Inventory Project, Retrieved September 26, 2009, from http://libwiki.mcmaster.ca/clip/index.php/Main/CaseStudy-Peterborough DrumlinField.

McMaster University (2010), *Drumlins*, Retrieved January 8, 2010, from http://libwiki.mcmaster.ca/clip/index.php/Main/Drumlins.

Mika, N., Mika, H. (1976), *United Empire Loyalists: Pioneers of Upper Canada*, Belleville: Mika Publishing Company.

Mulholland, R., Vincent, C. (2007), *The State of Small- and Medium-sized Enterprises in Northern Ontario*, Sudbury: Laurentian University.

Multicultural Canada (2009), *Encyclopedia of Canada's People*, Retrieved September 26, 2009, from http://www.multiculturalcanada.ca/ecp/.

Natural Resources Canada (2001), *Forest Regions of Canada*, Retrieved September 26, 2009, from http://www.sof.eomf.on.ca/Introduction/Physical_Geography/Vegetation/Documents/forest%20regions%20of%20canada.pdf.

Natural Resources Canada (2004), "Drainage Basins", *The Atlas of Canada*, Retrieved September 26, 2009, from: http://atlas.nrcan.gc.ca/site/english/maps/environment/hydrology/drainagebasins.

Natural Resources Canada (2006), "Forestry-Reliant Communities, 2001", *The Atlas of Canada*, Retrieved September 26, 2009, from http://atlas.nrcan.gc.ca/site/english/maps/economic/rdc2001/rdcforest.

Natural Resources Canada (2009a), *Atlas of Canada 1,000,000 National Frameworks Data, Canadian Place Names*, Retrieved April 21, 2011, from http://www.geogratis.gc.ca/geogratis/en/download/framework.html.

Natural Resources Canada (2009b), *Government of Canada Announces $1 Billion to Support Environmental Improvements for Pulp and Paper Industry*, June 17 2009, Retrieved September 26, 2009, from http://www.nrcan-rncan.gc.ca/media/newcom/2009/200961-eng.php.

Natural Resources Canada (2009c), *About Renewable Energy*, Retrieved September 26, 2009, from http://www.nrcan.gc.ca/eneene/renren/aboaprren-eng.php.

Natural Resources Canada (2011), *Forest-Dependent Communities in Canada*, Retrieved December 30, 2009, from http://canadaforests.nrcan.gc.ca/indicator/communities.

Northern Ontario Heritage Fund (2009), *About the NOHFC – History and Overview*, Retrieved September 26, 2009, from http://www.mndm.gov.on.ca/nohfc/about_nohfc_e.asp.

Northern Ontario School of Medicine (2010), *About Us*, Retrieved January 7, 2010, from http://www.normed.ca/about_us/default.aspx?id=68.

Oak Ridges Trail Association (2010), *Oak Ridges Trail Map 2010*, Retrieved April 22, 2011, from http://www.oakridgestrail.org/images/Trail%20Map 2010.pdf.

OECD (2002), *Glossary of Statistical Terms*, Retrieved September 26, 2009, from http://stats.oecd.org/glossary/detail.asp?ID=4603.

Office of the Prime Minister (2009a), *PM and Premier Dalton McGuinty Announce Loans to Support the Restructuring of Chrysler*, April 30, 2009, Retrieved September 26, 2009, from http://pm.gc.ca/eng/media.asp? category=1&id=2547.

Office of the Prime Minister (2009b), *Backgrounder – Canada and Ontario: Joint support for General Motors Restructuring*, June 1, 2009, Retrieved September 26, 2009, from http://pm.gc.ca/eng/media.asp?id=2600.

Ontario Exports (2009), *Ontario's Information Technology Sector*, Retrieved September 26, 2009, from http://www.ontarioexports.com/resources/ sec_Infotech.asp.

Ontario Exports (2011), *Ontario Trade Fact Sheet*, Retrieved April 21, 2011, from http://www.sse.gov.on.ca/medt/ontarioexports/en/Pages/trade_fact_ sheets_ontario.aspx.

Ontario Forest Industries Association (2009), *Fast Facts*, Retrieved September 26, 2009, from http://www.ofia.com/about_the_industry/fast_facts.html.

Ontario Lottery and Gaming (2009), *History – 1990-1999*, Retrieved September 26, 2009, from http://www.olg.ca/about/who_we_are/history.jsp?contentID= about_history_90-99.

Ontario Ministry of Agriculture, Food and Rural Affairs (2009a), *Summary of Agricultural Statistics for Ontario*, Retrieved September 26, 2009, from http://www.omafra.gov.on.ca/english/stats/agriculture_summary.pdf.

Ontario Ministry of Agriculture, Food and Rural Affairs (2009b), *Northern Ontario Region at a Glance*, Retrieved September 26, 2009, from http://www.omafra.gov.on.ca/english/stats/county/northern_ontario.htm.

Ontario Ministry of Citizenship and Immigration (2008), *Immigrant Arrivals to Canada by Province (2002-2007) Percent of Total Canadian Landings*, Retrieved October 31, 2009, from http://www.ontarioimmigration.ca/ documents/arrivalsbyprovince-tables.pdf.

Ontario Ministry of Energy and Infrastructure (2009a), *Electricity Homepage*, Retrieved September 26, 2009, from http://www.mei.gov.on.ca/en/ energy/electricity/.

Ontario Ministry of Energy and Infrastructure (2009b), *Nuclear Energy: Electricity Supply*, Retrieved September 26, 2009, from http://www.mei.gov. on.ca/en/energy/electricity/?page=nuclear-electricity-supply.

Ontario Ministry of Energy and Infrastructure (2009c), *Nuclear Energy: Ontario's Nuclear Plants*, Retrieved September 26, 2009, from http://www. mei.gov.on.ca/en/energy/electricity/?page=nuclear-ontario-plants.

Ontario Ministry of Energy and Infrastructure (2009d), *Ontario Suspends Nuclear Procurement*, June 29, 2009, Retrieved September 26, 2009, from http://news.ontario.ca/mei/en/2009/06/ontario-suspends-nuclear-procurement.html.

Ontario Ministry of Energy and Infrastructure (2009e), *New Energy Projects*, Retrieved September 26, 2009, from http://www.mei.gov.on.ca/en/energy/electricity/?page=new-energy-projects.

Ontario Ministry of Energy and Infrastructure (2009f), *Net Metering*, Retrieved September 26, 2009, from http://www.mei.gov.on.ca.wsd6.korax.net/english/energy/renewable/index.cfm?page=net-metering.

Ontario Ministry of the Environment (2009), *Adapting to Climate Change in Ontario*, Retrieved December 30, 2009, from http://www.ene.gov.on.ca/en/publications/air/index.php#1.

Ontario Ministry of Finance (2003a), *Immigration to Ontario*, Retrieved January 8, 2010, from http://www.fin.gov.on.ca/en/economy/demographics/census/cenhi5.html.

Ontario Ministry of Finance (2003b), *Population Growth in Ontario's CMAs and the GTA*, Retrieved January 8, 2010, from http://www.fin.gov.on.ca/en/economy/demographics/census/cenhi1.html.

Ontario Ministry of Finance (2009a), *Demographic Quarterly: Highlights of the Second Quarter 2009*, Retrieved December 31, 2009, from http://www.fin.gov.on.ca/en/economy/demographics/quarterly/dhiq2.html.

Ontario Ministry of Finance (2009b), *Aboriginal Peoples of Ontario*, Retrieved December 31, 2009, from http://www.fin.gov.on.ca/en/economy/demographics/census/cenhi06-9.html.

Ontario Ministry of Finance (2009c), *Population Counts: Canada, Ontario and Regions*, Retrieved January 8, 2010, from http://www.fin.gov.on.ca/en/economy/demographics/census/cenhi06-1.html.

Ontario Ministry of Finance (2011a), *2011 Ontario Budget*, Retrieved April 20, 2011, from http://www.fin.gov.on.ca/en/budget/ontariobudgets/2011/.

Ontario Ministry of Finance (2011b), *Ontario Fact Sheet April 2011*, Retrieved April 21, 2011, from http://www.fin.gov.on.ca/english/economy/ecupdates/factsheet.html.

Ontario Ministry of Finance (2011c), *Ontario Economic Accounts – First Quarter of 2009*, Retrieved April 21, 2011, from http://www.fin.gov.on.ca/english/economy/ecaccts/.

Ontario Ministry of Natural Resources (2006), *Forest Resources of Ontario 2006. State of the Forest Report 2006*, Retrieved April 21, 2011, from http://www.mnr.gov.on.ca/en/Business/Forests/2ColumnSubPage/STEL02_179267.html.

Ontario Ministry of Natural Resources (n.d.), *Ontario's Forests Fact Sheets – Forest Facts*, Retrieved September 26, 2009, from http://www.mnr.gov.on.ca/en/Business/Forests/2ColumnSubPage/241205.html.

Ontario Ministry of Natural Resources (2008), *Polar Bears in Ontario*, State of Resources Reporting, Retrieved April 22, 2011, from http://www.mnr.gov.on.ca/stdprodconsume/groups/lr/@mnr/@sorr/documents/document/263710.pdf.

Ontario Ministry of Natural Resources (2009a), *Backgrounder on Black Bears in Ontario*, Retrieved April 20, 2011, from http://www.mnr.gov.on.ca/stdprodconsume/groups/lr/@mnr/@fw/documents/document/274503.pdf.

Ontario Ministry of Natural Resources (2009b), *Ontario's Crown Land Use Policy Atlas. 'Approved Land Use Strategy'*, Retrieved September 26, 2009, from http://crownlanduseatlas.mnr.gov.on.ca/supportingdocs/alus/landuse8.htm.

Ontario Ministry of Natural Resources (2009c), *Increased Energy Efficiency at Mill In Trenton Backgrounder*, Retrieved September 26, 2009, from http://www.mnr.gov.on.ca/en/Newsroom/LatestNews/273294.html.

Ontario Ministry of Northern Development and Mines (1994), *ROCK Ontario*, Retrieved September 26, 2009, from http://www.mndm.gov.on.ca/mines/rockon/rocontario_e.pdf.

Ontario Ministry of Northern Development, Mines and Forestry (2004), *Province Returning To Traditional Definition Of Northern Ontario*, Retrieved October 3, 2009, from http://www.mndm.gov.on.ca/news/NRView.asp?NRNUM=112&NRYear=2004&NRLAN=EN&NRID=3414.

Ontario Ministry of Northern Development and Mines and Forestry (2007), *Agriculture Sector*, Retrieved December 30, 2009, from http://www.mndm.gov.on.ca/nordev/documents/sector_profiles/agriculture_e.pdf.

Ontario Ministry of Northern Development, Mines and Forestry (2009), *Northern Ontario Overview*, Retrieved October 3, 2009, from http://www.mndm.gov.on.ca/nordev/documents/sector_profiles/northern_ontario_e.pdf.

Ontario Ministry of Northern Development, Mines and Forestry (2011), *Value of Ontario Mineral Production – 2010*, Retrieved April 21, 2010, from http://www.mndmf.gov.on.ca/mines/ogs/ims/documents/facts/production_facts_e.pdf.

Ontario Ministry of Tourism (2009), *Tourism Quick Facts*, 2007, Retrieved September 26, 2009, from http://www.tourism.gov.on.ca/english/research/quick_facts/index.html.

Ontario Newsroom (2009), *Ontario Helps General Motors Restructure*, June 1, 2009, Retrieved September 26, 2009, from http://www.news.ontario.ca/opo/en/2009/06/ontario-helps-general-motors-restructure.html.

Ontario Office of Francophone Affairs (2011a), *Portrait of the Francophone Community in Ontario*, Retrieved April 22, 2011, from http://www.ofa.gov.on.ca/en/franco.html.

Ontario Office of Francophone Affairs (2011b), *The French Language Services Act: An Overview*, Retrieved April 25, 2011, from http://www.ofa.gov.on.ca/en/flsa.html#.

Ontario Office of the Premier (2009a), *Ontario Helping To Put Chrysler On Sustainable Footing*, April 30, 2009, Retrieved September 26, 2009, from http://www.premier.gov.on.ca/news/event.php?ItemID=5788&Lang=EN.

Ontario Office of the Premier (2009b), *Ontario Helps General Motors Restructure*, June 1, 2009, Retrieved September 26, 2009, from http://www.premier.gov.on.ca/news/event.php?ItemID=6704&Lang=EN.

Ontario Parks (2009), *Parks Guide 2009*, Retrieved September 26, 2009, from http://www.ontarioparks.com/ENGLISH/pdf/parksguide2009.pdf.

Ontario Power Authority (2009), *Northern Ontario Transmission Resources*, Retrieved September 26, 2009, from http://www.powerauthority.on.ca/electron/Page.asp?PageID=1290&ContentID=675&SiteNodeID=142&BL_ExpandID=247.

Parks Canada (2009), *The History of the Rideau Canal*, Retrieved September 26, 2009, from http://www.pc.gc.ca/lhn-nhs/on/rideau/natcul/natcul2_e.asp.

Parliament of Canada (2009), *Party Standings in the House of Commons*, Retrieved October 31, 2009, from http://www2.parl.gc.ca/Parlinfo/lists/PartyStandings.aspx?Section=03d93c58-f843-49b3-9653-84275c23f3fb&Gender=.

Peerbocus, N., Melino, A. (2007), "Export Demand Response in the Ontario Electricity Market", *The Electricity Journal*, 20(9), 55-64.

Pred, A. (1965), "Industrialization, Initial Advantage, and American Metropolitan Growth", *Geographical Review*, 55(2), 158-185.

Public Health Agency of Canada (2004), *Learning from SARS – Renewal of Public Health in Canada*, Retrieved September 26, 2009, from http://www.phac-aspc.gc.ca/publicat/sars-sras/naylor/exec-eng.php.

Rayburn, A. (1997), *Dictionary of Canadian Place Names*, Toronto: Oxford University Press.

Rea, K.J. (1985), *The Prosperous Years: The Economic History of Ontario, 1939-1975*, Toronto: University of Toronto Press.

Reuters Canada (2009), *Greyhound Scales Back Bus Service in Ontario*, Retrieved January 8, 2010, from http://ca.reuters.com/article/domesticNews/idCATRE5B160L20091202.

Risk Management Solutions (2008), *The 1998 Ice Storm: 10-Year Retrospective*, RMS Special Report, Retrieved April 26, 2011, from http://www.rms.com/publications/1998_ice_storm_retrospective.pdf.

Robertson, H. (1995), *Driving Force: The McLaughlin Family and the Age of the Car*, Toronto: McClelland and Stewart.

Rogers, J. (2004), *An Overview of Tourism in Northern Ontario*, Sudbury: FedNor.

Roy, F., Kimanyi, C. (2007), "Canada's Changing Auto Industry", *Canadian Economic Observer*, 20(5), 3.1-3.11, Ottawa: Statistics Canada, Cat. No. 11-010-XIB, Retrieved September 26, 2009, from http://www.statcan.gc.ca/pub/11-010-x/11-010-x2007005-eng.pdf.

Scotiabank Group (2005), *Canadian Auto Report*, Retrieved September 26, 2009, from http://www.greatertoronto.org/pdf/ScotiabankAutoReport.pdf.

Soulard, F., Trant, D., Joe Filoso, J., Van Wesenbeeck, P. (1998), "Ice Storm '98!", *Canadian Social Trends*, Ottawa: Statistics Canada, Retrieved September 26, 2009, from http://www.statcan.gc.ca/pub/11-008-x/1998003/article/4006-eng.pdf.

Southcott, C. (2006), *The North in Numbers: a Demographic Analysis of Social and Economic Change in Northern Ontario*, Thunder Bay: Lakehead University, Centre for Northern Studies.

Shaw, T.B. (1994), "Climate of the Niagara Region", in H.J. Gayler (ed.), *Niagara's Changing Landscapes*, Ottawa: Carleton University Press, 111-138.

Statistics Canada (1998), *The St. Lawrence River Valley 1998 Ice Storm: Maps and Facts*, Retrieved September 26, 2009, from http://www.statcan.gc.ca/pub/16f0021x/16f0021x1998001-eng.htm.

Statistics Canada (2000), *Human Activity and the Environment*, Cat. No. 11-509-XPE, Retrieved September 26, 2009, from http://www.statcan.gc.ca/kits-trousses/hae-ahe2000/pdf/edu01f_0002c-eng.pdf.

Statistics Canada (2005), *Land and Freshwater Area, by Province and Territory*, Retrieved September 26, 2009, from http://www40.statcan.ca/l01/cst01/phys01-eng.htm.

Statistics Canada (2006a), *Boundary Files*, (Lakes and Rivers (polygons) and Coastal Waters (polygons), Retrieved April 21, 2011, from http://geodepot.statcan.gc.ca/2006/040120011618150421032019/021521140401182506091205 19_05-eng.jsp.

Statistics Canada (2006b), *Industry* – North American Industry Classification System 2002 (23), *Occupation* – National Occupational Classification for Statistics 2006 (60), Class of Worker (6) and Sex (3) for the Labour Force 15 Years and Over of Canada, Provinces, Territories, Census Divisions and Census Subdivisions, 2006 Census – 20% Sample Data, Cat. No. 97-559-X2006024, Ottawa: Statistics Canada.

Statistics Canada (2006c), *CYB Overview 2006*, Retrieved September 26, 2009, from http://www41.statcan.gc.ca/2006/1130/ceb1130_000-eng.htm.

Statistics Canada (2007a), *Portrait of the Canadian Population in 2006, by Age and Sex, 2006 Census*, Ottawa: Statistics Canada, Cat. No. 97-551-XIE, Retrieved September 26, 2009, from http://dsp-psd.pwgsc.gc.ca/collection_2007/statcan/97-551-X/97-551-XIE2006001.pdf.

Statistics Canada (2007b), "2006 Census: Age and Sex", *The Daily*, July 17, 2007, Retrieved January 8, 2010, from http://www.statcan.gc.ca/daily-quotidien/070717/dq070717a-eng.htm.

Statistics Canada (2008a), *Provincial Labour Force Differences by Level of Education*, Retrieved September 26, 2009, from http://www.statcan.gc.ca/pub/75-001-x/2008105/pdf/5215211-eng.pdf.

Statistics Canada (2008b), *2006 Census of Agriculture*, Retrieved September 26, 2009, from http://www.statcan.gc.ca/ca-ra2006/analysis-analyses/ont-eng.htm.

Statistics Canada (2008c), *International Travel 2007*, Retrieved September 26, 2009, from http://www.statcan.gc.ca/pub/66-201-x/2007000/part-partie1-eng.htm.

Statistics Canada (2008d), *Historical Statistics of Canada: Population and Migration*, Retrieved on January 8, 2010, from http://www.statcan.gc.ca/pub/11-516-x/sectiona/4147436-eng.htm.

Statistics Canada (2008e), "Census Snapshot – Immigration in Canada: A Portrait of the Foreign-born Population, 2006 Census", *Canadian Social Trends*, Ottawa: Statistics Canada, Retrieved January 8, 2010, from http://www.statcan.gc.ca/pub/11-008-x/2008001/article/10556-eng.htm.

Statistics Canada (2009a), *Population by Year, by Province and Territory (Number)*, Retrieved October 3, 2009, from http://www40.statcan.gc.ca/l01/cst01/demo02a-eng.htm?sdi=population%20province.

Statistics Canada (2009b), *Highest Level of Educational Attainment for the Population Aged 25 to 64*, 2006 Counts for Both Sexes, for Canada, Provinces and Territories – 20% Sample Data, Retrieved January 8, 2010, from http://www12.statcan.ca/census-recensement/2006/dp-pd/hlt/97-560/pages/page.cfm?Lang=E&Geo=PR&Code=01&Table=1&Data=Dist&Sex=1&StartRec=1&Sort=2&Display=Page.

Statistics Canada (2009c), *Trends in Manufacturing Employment*, Retrieved September 26, 2009, from http://www.statcan.gc.ca/pub/75-001-x/2009102/article/10788-eng.htm.

Statistics Canada (2009d), *Electric Power Generation, Transmission and Distribution*, Cat. No. 57-202-XWE, Retrieved September 26, 2009, from http://www.statcan.gc.ca/pub/57-202-x/2007000/t003-eng.htm.

Statistics Canada (2009e), *Aboriginal Peoples in Canada in 2006: Inuit, Métis and First Nations, 2006 Census*, Retrieved December 31, 2009, from http://www12.statcan.ca/census-recensement/2006/as-sa/97-558/p2-eng.cfm.

Statistics Canada (2009f), *Population Urban and Rural, by Province and Territory (Ontario)*, Retrieved January 1, 2010, from http://www40.statcan.gc.ca/l01/cst01/demo62g-eng.htm.

Statistics Canada (2009g), *2006 Census Analysis Series*, Retrieved September 26, 2009, from http://www12.statcan.gc.ca/census-recensement/2006/as-sa/index-eng.cfm.

Statistics Canada (2010), *Census Subdivision Boundary Files*, (2006 digital boundary file), Retrieved April 20, 2011, from http://geodepot.statcan.gc.ca/2006/040120011618150421032019/031904_05-eng.jsp.

Sturgeon, T.J., Van Biesebroeck, J., Gereffi, G. (2009), "The North American Automotive Value Chain: Canada's Role and Prospects", *International Journal of Technological Learning, Innovation and Development*, 2(1-2), 25-52.

The Friends of Fort York (2009), *Historic Fort York*, Retrieved September 26, 2009, from http://www.fortyork.ca/.

The Municipality of Leamington, Ontario, Canada (2009), *Community Overview*, Retrieved September 26, 2009, from http://www.leamington.ca/municipal/ edo_community.asp.

The St. Lawrence Seaway Management Corporation (2003). The Welland Canal Section of the St. Lawrence Seaway, Retrieved September 26, 2009, from http://www.greatlakes-seaway.com/en/pdf/welland.pdf.

Thunder Bay Community Economic Development Commission (CEDC) (2009), *Community Profile F – Forestry*, Retrieved September 26, 2009, from http://www.thunderbay.ca/Assets/CEDC/docs/Forestry.pdf.

Toronto and Region Conservation (2009), *Hurricane Hazel*, Retrieved September 26, 2009, from http://www.hurricanehazel.ca/.

Toronto Financial Services Alliance (2009), *Financial Services in Toronto*, Retrieved September 26, 2009, from http://www.tfsa.ca/financial/index.php.

Trenhaile, A.S. (2007), *Geomorphology: A Canadian Perspective*, Third Edition. Don Mills: Oxford University Press.

University of Toronto. Dept. of Geography (1985), *North of 50°: An Atlas of Far Northern Ontario*, Toronto: Royal Commission on the Northern Environment on behalf of the Government of Ontario by University of Toronto Press.

Vale Inco (2010), *Vale Inco*, Retrieved on January 7, 2010, from http://www.inco.com/default.aspx.

Warrick, G.A. (2007a), "The Precontact Iroquoian Occupation of Southern Ontario", in *Archaeology of the Iroquois: Selected Readings and Research Sources*, J.E. Kerber (ed.), Syracuse: Syracuse University Press.

Warrick, G.A. (2007b), "European Infectious Disease and Depopulation of the Wendat-Tionontate (Huron-Petun)", in *Archaeology of the Iroquois: Selected Readings and Research Sources*, J.E. Kerber (ed.), Syracuse: Syracuse University Press.

Watson, K.W. (2006), *Engineered Landscapes: the Rideau Canal's Transformation of a Wilderness Waterway*, Elgin: K.W. Watson.

Welland Public Library (1973), *A Brief History of the Welland Canal*, Retrieved December 31, 2009, from http://www.welland.library.on.ca/digital/history.htm.

White, R. (2007), *Making Cars in Canada: A Brief History of the Canadian Automobile Industry, 1900-1980*, Ottawa: Canada Science and Technology Museum.

Wood, J.D. (1991), "The Population of Ontario: A Study of the Foundation of a Social Geography", in *A Social Geography of Canada*, G.M. Robinson (ed.), Toronto: Dundurn Press Ltd.

Woodrow, M. (2002), *Challenges to Sustainability in Northern Ontario*, Toronto: Environmental Commissioner of Ontario.

Wright, J.V. (1972), *Ontario Prehistory: An Eleven-Thousand-Year Archaeological Outline*, Ottawa: National Museums of Canada.

Yeates, M. (1975), *Main Street: Windsor to Québec City*, Toronto and Ottawa: The Macmillan Company of Canada Limited and Ministry of State for Urban Affairs and Information Canada.

CHAPTER 10

The Prairie Provinces

A "Drive-thru" Region in Another Phase of Geographical Transition

Douglas C. MUNSKI

University of North Dakota

Introduction

When someone says the phrase, 'Prairie Provinces', what do you as an undergraduate geography student visualize about this part of Canada? For example, the traditional image of the provinces of Manitoba, Saskatchewan, and Alberta is that of people hurriedly travelling through the region by automobile or RV on the way elsewhere who remember the journey as taking place in a picturesque but monotonous landscape of a gently undulating topography of lowlands and plains upon which there is a rural countryside dotted with grain elevators overlooking small towns largely dependent upon a cash-grain economy. This is a settlement pattern often heralded as the epitome of representing nothing less than the successful immigration and acculturation of diverse European ethnic groups into the national Canadian economy between the 1880s through 1920s. It is proclaimed as the most positive consequence of late 19th century financial manipulations and engineering efforts to connect the Maritimes (Eastern Canada) and the Great Lakes-St. Lawrence settlements (Central Canada) with those in British Columbia. Were a traveller to follow the route of the Canadian Pacific Railway and the trackage of the rail lines that became the Canadian National Railroad, such an image most likely would have been close to the case. However, please examine the map of the Prairie Provinces (Figure 1). A sizable amount of the area of these three provinces actually is within the Canadian Shield and, in the case of Manitoba, a portion of that province is part of the Hudson Bay Lowlands.

Indeed, as noted by Everitt (2001) when considering the traditional image of the Prairie Provinces, the classic prairie grain elevator seemingly all but almost has disappeared from most small towns. Such losses in the built-environment definitely are helping to change the sense of place in this region. Increasingly, such 'cathedrals of the prairies' are being replaced only in strategic locations by the grain companies and the railways with 'inland terminals', some of which not even are associated with a specific community in terms of being a part of the small town streetscape (Heaver, 1993; Everitt, 1996). Such placement then in turn triggers a number of economic and quality of life issues for surrounding communities that are tied directly and indirectly to the agri-business infrastructure. From this perspective the prairie zone of Alberta, Saskatchewan, and Manitoba holds much in common with the Great Plains south of the 49th Parallel in terms of tremendous rapid agricultural restructuring, significant population out-migration of the young, increased aging-in-place of the elderly, and major negative alterations to a once dense, railway-oriented network of hamlets, villages, and small towns.

Yet, another image exists of the Prairie Provinces which is also known as the Western Interior (Lehr, 1982). This perception is mainly with respect to portions of Alberta and Saskatchewan but also a part of Manitoba as being a landscape dotted with oil rigs, natural gas operations, open pit coal mines, open pit potash mines, and economic activities associated with mineral exploitation. Indeed, the employment opportunities in 'The Oil Patch' that are perceived to exist have drawn workers from Atlantic Canada such that CBC television and radio broadcasts occasionally have featured the trials and tribulations of people from Atlantic Canada being long-distance commuters and temporary migrants in the Western Interior (CBC News March 14, 2006). This is a long way away from the offshore explorations of Hibernia and Sable Island.

Resource exploitation of metallic minerals and of forests also is an image that comes to mind. During the past several years, however, there has been a somewhat negative scene with various underground 'hard rock' mines closing, notably in Manitoba. This image is coupled to forestry corporations reducing or totally shutting-down operations in the zone of the Canadian Shield in which one finds the most commercially viable stands of boreal forest softwoods.

Another image that might be held is that of the Prairie Provinces being the home to isolated bands of First Nations People in remote parts of 'the bush' and who are living a subsistence-based existence of hunting, fishing, and gathering with trapping fur bearers as means to generate income for luxury items in a fashion similar to their ancestors who lived when HBCo (the Hudson's Bay Company) overwhelmingly dominated the economy in the mid-to-late 19th century. Yet, as noted by numerous authors, many of the Indians of Manitoba, Saskatchewan, and Alberta live in urban centers, including urban reserves (Peters, 1992; Barsh, 1994; Saku, 1999; Peters, 2001; Hanselmann, 2001; Deane, Morrisette, Bousquet, and Bruyere, 2004; Walker, 2005; Peters and Robillard, 2009). Also, there is a significant number of Metis (descendants of intermarriages of European fur traders and aboriginal women) (Harrison, 1985). It must be noted that the Metis have been exerting their historic rights to land in urban places in Western Canada such as Winnipeg.

However, how many people's images of the Prairie Provinces is that of dynamic, cosmopolitan urban centers? When one considers the total population of the region as being 5,597,400 in 2006 (Statistics Canada, 2006), 3,204,445 or 57.25% lived in Calgary (1,070,295), Edmonton (1,024,820), Winnipeg (686,040), Saskatoon (230,850), and Regina (192,440) (Figure 2). Indeed, Winnipeg is considered to be a primate city in central place hierarchy because with a total 2006 provincial population of 686,040, 57.94% lived within the boundaries of this census metropolitan area whereas the second largest city in Manitoba, Brandon, only had a 2006 census population of 41,511 people (Statistics Canada, 2006). Then, too, these main five census metropolitan areas are increasingly ethnically diverse, reflecting changes taking place in immigration to the Western Interior such that some of the historic Chinese neighborhoods are increasing in size and being joined by growing communities of people from South Asia, Southeast Asia, Southwest Asia and North Africa, the Caribbean, Latin America, and Africa South of the Sahara. While not as heterogeneous as Toronto, Montréal, or Vancouver, the citizens of Calgary, Edmonton, Winnipeg, Saskatoon, and Regina are less homogenous than in the mid-to-late 20th century.

As recently as the late 1960s and early 1970s, for example, the term 'immigrant' in reference to people in Winnipeg's North End meant generally a person of ethnic origin that most likely would be Slavic or possibly German or Scandinavian. This illustrated that the British were dominant overall but particularly west of the historic urban core a and south of the Assiniboine River with the Franco-Manitobans mainly located on the east side of the Red River of the North in the St. Boniface and St. Vital neighborhoods. Indeed, the presence of non-Europeans as new residents in the rural parts of the Prairie Provinces is sufficiently no longer that unique. Still, the stereotypes and problems of acculturation seemingly persist between 'Old Canadians' of European descent and 'New Canadians' of non-European source areas. As a case in point, a classroom discussion might be in order regarding the way 'Old Canadians' and newer immigrants who are of Middle Eastern background are portrayed within CBC television situation comedy shows such as *Little Mosque on the Prairie*.

Figure 1. Canadian Prairies

Source: Stefan Freelan, Western Washington University, 2/2011.

So, which image is right? Perhaps all of them in the sense of a kaleidoscope that changes with each twist of that optical device because the landscape does change through time depending upon the actions of people within and outside the region. Remember, please, that even the great grasslands along and to the south of what now is the Trans-Canada Highway once were viewed as an undesirable environment for agriculture. The semi-arid zone of today's southeastern Alberta and southwestern Saskatchewan was known in the mid-19th century as Palliser's Triangle, an area to avoid because of its lack of dependable moisture. During 'wet years', that same subregion became part of the 'Garden in the Grassland' (Emmons, 1971). Unfortunately during 'dry years', this subregion repeatedly has seen dreadful financial heartbreak and personal collapse of ranchers and farmers. There has been a cycle of wet and dry years with the early 21st century being part of a period of dry years such that some individuals would suggest that Palliser's Triangle ought not ever to have been developed for commercial agriculture.

Figure 2. Canadian Prairie Provinces

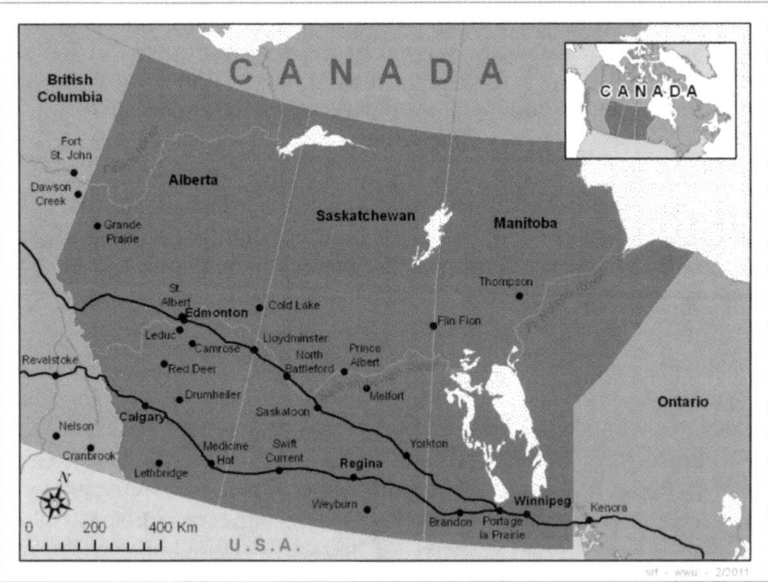

Source: Stefan Freelan, Western Washington University, 2/2011.

There are many images of the Prairie Provinces, so my purpose in introducing you to this part of Canada is to assist you to understand a bit better that these images reflect various interpretations of how specific locations, subregions, and the region itself demonstrate the interlocking nature of regional study which must have a temporal dimension to the spatial analysis of place.

The Prairie Provinces historically have been seen as part of the Canadian hinterland (McCann and Gunn, 1987) or as a semi-periphery zone within Canada but having cores, resource frontiers, downward transitional regions, and upward transitional regions (Bone, 2008). Although the three provinces overlap a number of different meta-regions in terms of geomorphology, hydrology, climatology, pedology, biogeography, economic activities, social geography, demography, settlement, and human-environment interaction, we group them together on political grounds which has become a convention of regional study for this part of Canada (Putnam and Putnam, 1979; Robinson, 1983; Boone, 2008).

Although the precedent is to view the Prairie Provinces as a major region of Canada, an area this large also can be subdivided into a number of interlocking and interconnected subregions in terms of regional concepts that underlie the notion of uniform (homogeneous) regions, nodal (functional) regions, and vernacular (perceptual) regions as determined according to the varying criteria agreed upon by geographers. Thus, we will be looking at the urban-rural dichotomy but recognizing that central place hierarchy helps tie together these places at the sub-provincial, provincial, and national scales. So, too, there will be consideration given of the spatial dimensions of the interplay of migration, ethnicity, and settlement associated with various economic activities that help to create a myriad set of neighborhoods, towns and cities, rural districts, and economic development zones. The focus of this chapter is more so upon people than the physical environment, but the physical environment is noted as the setting for people's assorted lifestyles.

When presenting the regional geography of the Prairie Provinces, I approach the topic in a fashion similar to the other authors in this text. First, there is a highly generalized overview to the physical geography of the Prairie Provinces. Second, the historical geography of the region is told in a way to highlight key economic and social processes which have transformed the area through the last several centuries to what it appears to be at the time of this publication. Third, the settlement geography of the three provinces is explained using broad themes of geographical analysis. Finally, there is a brief reflection about the future of the Prairie Provinces and speculation upon why this part of Canada continues to remain so ingrained in the Canadian psyche as home to 'the Canadian pioneer'. I make no pretence of being the definitive authority when dealing with these four aspects of the Prairie Provinces. Furthermore, I have come to the conclusion that the more one thinks one understands the regional geography of this part of Canada, the more one needs to explore it even more intensely. A scholar must do this constantly through additional readings, conversations with academic colleagues, and most importantly, visiting and talking with the people in the places which constitute today's cultural/physical landscape within Manitoba, Saskatchewan, and Alberta. The following presentation reflects a tendency to approach the regional geography of the Prairie Provinces using the core-periphery model, but the emphasis will be upon the local, sub-provincial, provincial, and national interconnections more so than the international relationship of the region in the global economy.

Contextualizing the Prairie Provinces into Mega-regions of Canadian Physical Geography

Approximately 19.8% of the area of Canada is within the Prairie Provinces (Bone, 2008). When one considers the mega-region of Canada's geomorphology, the provinces of Manitoba, Saskatchewan, and Alberta are within four different physiographic areas. A small part of the Canadian Rockies is to be found in southwestern Alberta although most of the Prairie Rose Province (a nickname for Alberta) is within the Interior Plains and only the northeastern portion of Alberta that is focused around the Lake Athabasca area is within the Canadian Shield. Slightly under half of Saskatchewan constitutes its northern area and which is in the Canadian Shield with the remaining southern area being overwhelmingly part of the Interior Plains. The province of Manitoba is about one-fourth within the Interior Plains south and west of Lake Winnipeg and the Winnipeg River with the vast majority of the province being part of the Canadian Shield and the northeastern part of the Keystone Province (a nickname for Manitoba) being within the Hudson Bay Lowlands. A considerable amount of the land in the Prairie Provinces that is located in the Canadian Shield is characterized by a variety of waterbodies. Found here is a mixture of small ponds, patches of muskeg, varying-sized streams and significant rivers such as the Saskatchewan River, and a set of mall-sized, mid-sized, and large-sized lakes including the internationally significant freshwater waterbody called Lake Winnipeg.

Generally speaking, the Hudson Bay Lowlands and the Interior Plains are gently undulating and limited in relief, particularly as compared to the Canadian Shield. However, there are three different levels of the Interior Plains trending from southeast to northwest that are prominent (Putnam and Putnam, 1979). The First Prairie Level is between the Canadian Shield on the east and the Manitoba Escarpment on the west; it basically is the terrain associated with glacial Lake Agassiz. The Second Prairie Level is between the Manitoba Escarpment on its east which is characterized by a series of places with local relief such as the Pembina Hills (southern Manitoba), the Porcupine Hills (Manitoba-Saskatchewan borderland), and the Pasquia Hills (east central Saskatchewan) to the Missouri Escarpment on its west. This area often is known as the Glaciated Plains or the Prairie Pothole Country because of the wetlands which are intermittent and semi-permanent water bodies. These resemble sloughs or marshes which, unless drained to convert the land into agricultural purposes, will expand and contract depending upon if there is a 'wet year' or a 'dry year' in that part of the countryside. The Third Prairie Level is between the Missouri Escarpment on its east and the Foothills of the Canadian Rockies on its west; this is the Canadian portion of the 'Northern Plains' region that stretches far south beyond the 49th Parallel into the United States. A geomorphic anomaly in this region is the area of the Cypress Hills, a part of the Great Plains that was spared certain aspects of glaciation. The Cypress Hills area represents a unique environment both in terms of being higher than the surrounding land and different relative to the ecology of the biogeography in that sub-region along the Saskatchewan-Alberta borderlands.

Because of the impact of glaciation in the past, the terrain of the Prairie Provinces is such that the regional hydrology is directed mainly toward drainage into Hudson Bay or the Beaufort Sea in the Arctic. The Saskatchewan River system with the Nelson-Hayes River system and the latter's connections with the Red River of the North, Assiniboine River, and Winnipeg River have been crucial transportation routes in the fur trade era and currently even more important *vis-à-vis* being parts of hydroelectricity projects. The Churchill River (yes, another Churchill River exists in the Labrador-Québec borderlands) is the other key river in the region's eastward flowing waters associated with the drainage into Hudson Bay. The Athabasca River-Peace River system of Alberta is a key component of the northward flowing McKenzie River system, and the Hay River is historically important as a tributary, notably relative to the fur trade. Exceedingly little of the Prairie Provinces is part of the Gulf of Mexico Basin, e.g., the Milk River of Alberta which crosses the 49th Parallel into the United States. Waterways were highly significant during the fur trade era (Huck *et al.*, 2002), but today their importance, as noted earlier, is related principally to hydroelectric production, locally significant fisheries, and to some extent ecotourism depending upon the particular location of the waterbody.

Depending upon your preferences, the Canadian mega-region of climate for the Prairie Provinces can be categorized according to the Köppen Classification or Canadian Climatic Zone (Bone, 2008) for climate regionalization. While the specific word choice is of interest principally to climatologists, the climatic types found in the region largely are reflective of the conditions of being deep in the interior of North America which increases the impact of continentality. Experiencing polar air masses more frequently than other forms of air masses, being at the latitudes between 49° North and 60° North, and existing largely in the rainshadow zone of the Canadian Rockies means that the more northerly parts of the Prairie Provinces are characterized by short, cool summers and long cold winters with low annual precipitation. The more southerly parts of the region are more a steppe-like environment with hot, dry summers and long cold winters plus low-to-moderate annual precipitation. The western part of the Prairie Provinces is semi-arid whereas the eastern zone is sub-humid with the line of demarcation fluctuating back and forth across the 100th Meridian partly depending upon if the area is in a wet cycle or a dry cycle. One could, as Bone (2008) notes, put the Hudson Bay Lowlands into the category of an Arctic climate, and this area is different than the other parts of the Prairie Provinces in additional ways, too.

There is no question that the extreme winter cold often is a factor in people's perception of the region as being highly challenging as a part of Canada to inhabit. However, the local humor emphasizes how during the long winters there are no mosquitoes, the bane of people and animals in the moderately short but generally warm summers. Climatic-related natural events such as chinooks and natural hazards such as blizzards, intense thunderstorms, dust storms, ice storms, flooding, and tornadoes can be experienced at various times during the year in a wide range of localities within the Western Interior.

The Prairie Provinces are within the Canadian mega-region of Canadian climatic zones such that the natural vegetation types and soil orders are best considered by looking at two west-east transects and a north-south transect. If we begin our first west-east transect immediately west of Calgary in Banff National Park and move to the east along that particular line of latitude, we go through an area with a cordillera style of climate, mountain forest, and mountain complex soil into a zone of prairie (steppe) climate, grassland and parkland vegetation, and chernozemic soil with the end stop being on the east in the subarctic climate that covers an area of boreal forest and podzolic soils. If we take our second west-east transect from the northwestern corner of the Wild Rose Province and go east, we are in the subarctic zone of climate and boreal forest with podzolic soils that then transition into a subarctic climate which is associated with a tundra-boreal transition in terms of vegetation with podzolic soils to a true tundra environment adjacent to Hudson Bay with its arctic climate, tundra vegetation, and cryosolic soils. Dealing successfully with permafrost is a significant issue in the more northerly parts of the Prairie Provinces when it comes to resource exploitation and ground transportation.

The problems associated with permafrost become most apparent if we take a south-north transect through the Prairie Provinces using a Manitoban example. If one tries to follow the line of longitude that runs through Winnipeg, one moves from a grassland and parkland with chernozemic soils into a boreal forest with podzolic soils that then becomes a tundra-boreal transition with podzolic soils and ends in a tundra vegetation with cryosolic soils. Climatically as one goes from south to north, the Köppen System most perfectly would be Dfb to Dfc to Dfd to ET, but one also would be crossing through areas of discontinuous and continuous permafrost once one goes north of 50° North. The location of the change from discontinuous to continuous permafrost seemingly has been moving farther north with the impact of global climate change in the subarctic and arctic areas of Canada (Smith, Burgess, Riseborough, and Nixon, 2005).

Such global climate change has ramification for placement of the Prairie Provinces within the Canadian mega-region for biogeography. Suffice it to say that ranges of various species from the southern portions of the Western Interior are extending northerly and that the habitat of the more northerly fur-bearers such as the polar bear is undergoing substantial environmental stress. Historically, the prairie grasslands and parklands were dominated by the plains bison, antelope, and mule deer with the boreal forest zone mainly the habitat of woodland bison, woodland caribou, white-tail deer, and moose with the Hudson Bay Lowlands being the range of the polar bear, arctic fox, and ptarmigan (Wishart, 2006). The importance of the furbearers, notably the beaver and the muskrat, cannot be underestimated for these were sources of sustenance and wealth long before the arrival of the Europeans. Consequently, it is appropriate to consider the historical geography of the Western Interior by beginning with a cursory overview to the First Nations People.

Historical Geography

The Prairie Provinces have had human habitation in an assortment of locations within what is now Manitoba, Saskatchewan, and Alberta since as least the end of the last Ice Age or even during that period of continental glaciation (Pettipas, 1984; Dickason, 2002). Archaeological evidence traces the ancestors of First Nations Peoples in the area adjacent to glacial Lake Agassiz to about 14,000 BP which would have been the Paleo Period, a time of the hunting of mega-fauna such as woolly mammoth and the mastodon by indigenous people using the Clovis point at the end of spears (Pettipas and Buchner, 1983). The Archaic Period, approximately from 8,000 BP to 2,000 BP, would have been the time first of the gradual end of the Ice Age and then into the start of a post-glacial environment. This was a period of marked continental climate change that was characterized by much warmer and dryer conditions. During this period of the regional chronology, the indigenous people would have been more likely found in the woodlands because of the comparable difficulty of living on the grasslands. Yet, this was a phase of cultural development characterized by adaptation by the hunters and gatherers to harvesting small mammals, waterfowl, fish, and plants to substitute for the mega-fauna that had been hunted to extinction as the Paleo Period came to a close (Nicholson, 1987). The formal culture of the Woodland Period of approximately 2,200 BP to 1,750 CE slowly was diffused into what is now the Prairie Provinces and has been noted as a time in which making of clay pottery, building of complex burial mounds including effigy mounds, introduction and adoption of the bow and arrow, and the cultivation of the 'Three Sisters' (maize, beans, and squash) would be significant cultural transformations (Pettipas, 1984; Pettipas, 1998).

No cultural innovations perhaps were more significant than the introduction during the fur trade period of the horse and the trade gun in terms of shifting the woodlands people into being also more able to survive successfully on the plains and in the parklands (Ray, 1974; Brown, 1980; Carter, 1999; Innis, 1999; Dickason, 2002). Yet, increased nomadic activity in the grasslands and the prairie parkland belt (an aspen savannah) was not necessarily voluntary when one takes into account the changes in balance of power among the various First Nations (Dickason, 2002). Some of this semi-forced migration can be viewed as a consequence of the struggles between the French and English prior to 1763, the impacts of competing Euro-Canadian fur traders between 1763 and 1821, the pressures of the rival American and Canadian business interests from 1821 through 1867, and the mixed results of Metis with the aboriginal people both as allies and enemies when dealing with the post-Confederation Canadian governments and the immigrants from Eastern Canada and Europe from 1867 through 1890s.

The historical geography of the fur trade in the Western Interior can be interpreted in many ways and seen from a variety of perspectives. The exploitation of the beaver, the bison, and other fur-bearers has been emphasized, notably by Innis (1999), as the basis of the staples approach to the economic geography of the Prairie Provinces. The fur trade laid the basis for the European and Euro-Canadian interests in developing the region later principally for agriculture in the south and for forestry and mining in the north. Yet, the fur trade also was a period of dislocation of the indigenous people with respect initially to controlling the resource base by external interests from outside Canada who were using Montréal as the key entrepot to the interior of North America (Ray, 1974; Gilman, Gilman, and Stultz, 1979). Furthermore, this was a time of population collapse in some instances for the First Nations relative to the effects of diseases and alcohol (Dickason, 2002). Such conditions were compounded in this era in which the First Nations and Metis lost land and were relegated onto small-sized, isolated reserves or were cheated out of land claims through unscrupulous speculators respectively (Flanagan, 1991). There is a special irony regarding Louis Riel, the leader of the Resistance of 1869-1870, and key figure in the Resistance of 1884-1885. He was vilified in helping to create the province of Manitoba in 1870 and executed in the late 19th century for what happened in what would become part of Saskatchewan and Alberta only for him to be rehabilitated in the late 20th and early 21st centuries such to have a the February Canadian ‘long weekend’ named in his honor by the Manitobans (CBC News September 25, 2007).

The tendency to think linearly about the historical geography of the Prairie Provinces and to compartmentalize it in terms of the fur trade being followed by a series of periods, i.e., taking the sequent occupance approach, is not as acceptable a model to follow for the professional geographer today as it was in the 1920s through 1960s when that was a dominant approach in studies of this sub-discipline. However, such a traditionalist model might be a reasonable approach for the neophyte geographer who is just beginning explorations in the historical geography of this part of Canada provided that the student recognizes a caveat. Remember, please, that there are overlapping interconnections between the periods of sequent occupance and continuations into the present of various themes within those periods of sequent occupance.

Having given the caveat above, what are those key themes within a sequence occupance approach to considering the historical geography of the Western Interior? First, the role of the railways in ethnic-oriented agricultural settlement of the region during the 1880s through 1910s is almost *de rigueur* when learning about the Western Interior (Studness, 1964; Stabler, 1973; Norrie, 1975; Richtik, 1975; Marr and Percy, 1978; Lehr, 1982; Friesen, 1984; Carter, 1999; Bone, 2008). Then, the rise of mining and forestry as economic lynchpins during the 1890s through 1920s is heralded as a major theme. Next is the theme of overcoming adversity when dealing with the impact of the Great Depression and drought years of the 1930s which is followed by the temporary resurgence of the rural centers and areas for producing foodstuffs for the Allies in World War II into the time of the Korean War. Then, the start of the oil booms of Alberta beginning in the 1950s and continuing through the latter 20th century with gradual development of the Athabasca tar sand is the basis for the theme of energy development across the three provinces which includes examining the roles of coal, hydroelectric, and new approaches alternative forms of energy. In a somewhat parallel time to this just-identified theme are the topics of the fluctuations of the mining and forestry sub-sectors of the primary sector from the 1950s through early 21st century as distinct themes overlapping with the energy development sub-theme of constructing hydroelectric megaprojects in the latter 20th century. Resource management is especially complex in the Western Interior (Timoney and Lee, 2001) and characterized with resulting pros/cons for First Nations people as an overarching issue for many places within the northern portion of the Prairie Provinces. Next comes the theme of the slow and steady decline of the rural centers and their trade hinterlands beginning in the 1960s through the present in terms of depopulation by the young people and aging-in-place of the older folks which seemingly was especially pronounced in Saskatchewan in the late 21st century although also occurring in certain parts of Manitoba and even Alberta. However, here is where a second caveat needs to be noted: certain rural areas that are part of the fossil fuel energy development boom are seeing some growth that, while it may be temporary, still is in contrast to earlier and ongoing declines occurring elsewhere in the rural zones of the Prairie Provinces. Finally, the theme of the increasing urbanization of the region, notably since the 1960s, is focused upon five major census metropolitan areas (Calgary, Edmonton, Winnipeg, Saskatoon, and Regina).

While such a simplification is somewhat useful, one cannot forget that frequently there have been interconnections between these themes and reflected in the cultural landscapes of the present as well as the past. Indeed, First Nation peoples now are to be found more often than not in the inner cities of the Western Interior's' Big Five' census metropolitan areas than on rural located reserves, and there is a steady push-pull of migration and counter migration of indigenous people between the reserves and these as well as small urban centers in the Western Interior as indicated in maps from the electronic *Atlas of Urban Aboriginal Peoples*.

It must not be forgotten that neither the First Nations Peoples nor the Metis disappeared from the Prairie Provinces when the fur trade ended (Harrison, 1985; Shore, 2001) Also, ethnic immigration continues, albeit in lesser numbers throughout the Prairie Provinces with the destinations of the 'New Canadians' from Asia, Africa, and Latin America being more often than not one of the major census metropolitan areas such as Calgary. Indeed, European-based ethnic immigration did not totally end with the Great Depression as one is inclined to think using a traditional periodization of the historical geography of the Prairie Provinces. For example, there even seems presently to be a small flow of Ukrainians and other Slavic groups into cities of the Western Interior, most often Edmonton and Winnipeg (Leshchyshen, 2009), with the 'face' of such movements experienced firsthand with some of the newest employees in eateries featuring perogies, holupchis, and borscht such as Alycia's Ukrainian Restaurant in Winnipeg's North End.

Because of the complexities of dealing in only a few pages of text with the interplay of the historical geography of the Prairie Provinces with the contemporary economic geography and social geography of this part of Canada, one does have to accept simplifying the general trends of modern economic activity and settlement in the Western Interior with the proviso that these three provinces have a wide range of sub-regions that often are highly unique when compared to the generalized regional geography of the Western Interior. Then, too, one always must recognize that any geographical description and spatial explanation of a place merely is an interpretation taken from a particular context of human-environment interaction and the core-periphery model of development.

Change is inevitable in terms of the geography of a place including even physical geography elements in some circumstances. For example, the cycle of flooding in the drainage basin of the Red River of the North has had profound effects upon agricultural activity, transportation infrastructure, settlement pattern, and political jurisdictions such that the 2009 flood season has caused some rethinking of flood mitigation that was based upon responding to the impact of the horrific 1997 flood and earlier floods such as the cataclysmic one in 1950. The theme of responding to natural hazards is in itself an element in trying to understand as much as possible the past, present, and future development of the Prairie Provinces because of the significance of such phenomena to the people who have, are, and will call the Western Interior home despite episodes of flood, forest fire, drought, thunderstorms, tornado, hail, blizzard, extreme cold, extreme heat, and even the occasional earthquake.

Thus, while not entirely satisfactory, perhaps the most useful approach to presenting the contemporary geography of the Prairie Provinces is to highlight the six themes which seem to be most important at the beginning of the 21st century in the context of the Western Interior's generalized historical geography and its potential for future economic development and spatial settlement pattern. These themes, however, are intertwined, and the result is more a 'patchwork quilt' than a 'Four Point Hudson Bay Blanket', but all regional geography is a 'work-in-progress'

A Thematic Perspective on the Contemporary Geography of the Prairie Provinces

Theme One: Transformation of Agriculture

Agriculture still is perceived as the backbone of the economy of the Prairie Provinces and while it remains a significant part of many communities in the region, as of 2005 only 4.7% of workers in the Western Interior were engaged directly in this sub-sector of the primary sector of the economy (Boone, 2008: 381). The key operational word in the previous sentence, however, is 'directly'. Indirectly, the needs of agriculturalists shape the construction and manufacturing activities in the region relative to the secondary sector of the economy, farmers and ranchers require a variety of services from the tertiary sector of the economy, and producers in the primary sector are influenced considerably by the research and development activities plus administrative decisions, both private and public, in the quaternary sector of the economy. Thus, we again see how the geography of a place is a set of intertwined strands of the core-periphery model. Why? Because the Western Interior is more so a part of Canada with a number of downward transitional regions, some upward transitional regions, a set of resource frontiers, and locally-important core regions all set within the context of the national and international framework of economic, cultural, and political influences.

Overlaying the core-periphery model for agriculture in the Prairie Provinces are cycles in terms of weather, costs for production of foodstuffs, and profit/loss margins for these foodstuffs. Being a farmer or a rancher is as much a lifestyle for some families as it is a source of income, so there are those factors to consider when people are faced with economic upheaval and 'wild card events' such as adverse public reaction to health-related issues directly or indirectly related to agricultural production practices, e.g., the occasional, isolated episode of 'mad cow' disease. Sometimes, a shift in cropping is a response to consumer demand such as the increased interest in more locally-produced products and more choices of organic foods which helps to explain the changes in some farming areas in the Prairie Provinces, notably in the zones of market gardening-style agricultural activity closest to the 'Big Five' census metropolitan areas of the Western Interior. Then, too, misinformation and panic amongst consumers concerning myriad health issues such as avian-related flu, often known as 'bird flu' and H1N1-type flu, commonly known as 'swine flu', respectively can wreak havoc for agriculturalists globally as seen in the early 21st century problems facing poultry producers and pork producers (English, 2006; Bremer, 2009; Bryson, 2009).

Let us reflect upon one dimension of such issues using hog production as the example. The shift to CAFO-style (concentrated animal feeding operations) hog raising in the Prairie Provinces initially was a response to growing demand for pork in East Asian markets in the 1980s and 1990s. Siting of such facilities was and remains highly controversial because of a number of factors but mainly over dealing with the frequently unpleasant hog lot smells and the manure generated by such massive concentrations of hogs. Genuine concern exists for protecting local and regional water quality, particularly if manure-holding ponds were to implode and contaminate both ground water and surface water (Tyrchniewicz, Carter, and Whitaker, 2000; CBC News March 4, 2008). Meanwhile, hog processing operations then were brought into communities such as Brandon, Manitoba, by corporations such as Maple Leaf Foods Inc. with mixed results. Historically, trade-offs have included relatively good-paying industrial jobs in food processing being created in addition to lower-paying meat processing employment. However, not all people or communities have benefited to the financial degree anticipated with some environmental issues, again related to worries over local water quality, being highly contentious (Tyrchniewicz, Carter, and Whitaker, 2000; CBC News March 4, 2008).

The early part of the 21st century has not proven all that good for pork producers. First, the economies of China, Japan, and South Korea have been affected by global recession with the consequence being that the demand has fluctuated considerably for pork products from the Prairie Provinces. Then, during the H1N1 flu season of 2009 in the Southern Hemisphere, ill-founded rumors regarding 'swine flu' being tied to pork consumption quickly reached the Northern Hemisphere in the Internet age of Facebook and twittering. Market prices for pork were affected adversely throughout the sub-continent of North America as noted by Bryson (2009), so there were resultant ripple-effects with especially economically-distressing impacts regionally and locally within the Prairie Provinces.

The irony is that the shift to CAFO-style hog operations and similar approaches to the production of cattle and poultry reflect the efforts of farmers and ranchers in the Prairie Provinces to become more efficient through increased mechanization, greater diversification of crops and livestock, and extensions of economies of scale. As noted by Statistics Canada (2007), the number of farms between 1971 and 2006 in the Western Interior has decreased by 35.4% with the greatest percentage changes occurring in Saskatchewan and Manitoba. Then, too, the average size of farms in the Prairie Provinces during 1971 to 2006 increased by 96%, i.e., the 'typical' farming operation nearly doubled in acreage going from 726 acres in 1971 to 1,425 acres in 2006 (Statistics Canada, 2007).

Meanwhile, although cash grain crops, notably spring wheat, are still significant in many parts of the Prairie Provinces, there is an increase in the production of oilseeds and even more growing of maize (corn) for ethanol-production in those and other sub-regions of the Western Interior (Olar, Romain, Bergeron, and Klein, 2004; Veeman and Gray, 2009). The key oilseeds are first canola followed by flaxseed, mustard seed, and sunflower seed (Small, 1999; Statistics Canada, 2009). Locally, other crops are important for national and nearby markets, particularly the organic and non-organic fruits and vegetables that are being sold in farmer's markets in the five key census metropolitan areas of the Prairie Provinces and at community markets in smaller cities in parts of the Western Interior.

Thus, there is an interesting dichotomy in some parts of the agricultural sub-sector of the primary sector of the economy in the Western Interior of seeing mid-sized farming operations in decline with the larger-sized operations becoming even larger in reaction to national and international factors but a number of smaller-sized operations surviving or actually coming into existence in response to more localized factors. This ongoing transformation of agriculture has implications for the central place hierarchy of the Prairie Provinces (Weida, 2002). The cultural landscape in many parts of the Western Interior now is partly the product of how hamlets and villages are disappearing, small towns are in decline, large towns are under economic stress, and the census metropolitan areas are seeing significant expansion into the rural-urban fringe.

Theme Two: Opportunities in Energy Production

When one considers energy production in the Prairie Provinces, one must be cognizant that there are both development opportunities in non-renewable fossil fuel (coal, oil, natural gas, and tar sands) and in renewable energy (hydroelectric power and alternative energy, notably wind power). Depending upon which sub-region of the Prairie Provinces is being examined, the types of energy development possible will vary considerably such that five vernacular regions might be argued to exist: 1) the Oil Patch (stretching from the Foothills of the Rockies in Alberta to southwestern Manitoba in the area around Virden); 2) the White Gold Zone (hydroelectric mega-projects of the Canadian Shield which mainly are in Saskatchewan and Manitoba); 3) the Alberta and Saskatchewan Coal Fields (spatially split with Alberta's sub-region in the more south-westerly portions of that province and Saskatchewan's sub-region focused upon Weyburn and Estevan); 4) the Athabasca Tar Sands (centered upon the projects in the area of Fort McMurray, Alberta), and 5) the Windy Grasslands (mainly those locations along key glacial beach ridges in the southern portions of these three provinces). The first four vernacular regions are far more in-grained into people's perceptions with the fifth one still to be recognized because of the relatively limited current amount of wind farm development compared to the other four forms of energy activity in the Western Interior.

The Oil Patch of the Prairie Provinces is found mostly within the province of Alberta with some operations in Saskatchewan and Manitoba being the consequence of earlier energy booms and busts which are once more part of an upward cycle of boom in the first part of the 21st century (Persaud and Kumar, 2001; Taylor, Bramley, and Winfield, 2005; Humphries, 2008). Alberta's production of petroleum, especially oil and natural gas, is consumed largely outside the Western Interior which is why pipelines such as the Alliance Pipeline, the Interprovincial Pipeline, and the TransCanada Gas Pipeline are crucial among an ever-growing number of pipeline projects in moving these energy products to the markets in central Canada and the Great Lakes region of the sub-continent of North America when looking at intercontinental commodity flows.

'White Gold' has been a vernacular term used in the past to describe any hydroelectric project, and such 'white gold' is to be found across the Canadian Shield in the Prairie Provinces (Ross, 1963; Lehr, 1982; Manitoba Hydro, 1997). The first such facility in the region actually was established in 1900 on the Minnedosa River (then known as the Little Saskatchewan River) close to its junction with the Assiniboine River near Brandon, Manitoba, but it was not a year-round operating facility (Ross, 1963). Hence the importance of using the Winnipeg River on the edge of the Canadian Shield in Manitoba as a year-round resource beginning in 1906 at Pinawa and followed in 1911 at Pointe du Bois (Bateman, 2005). The potential of the Nelson and Churchill Rivers as sources for hydroelectric production was studied in 1913, but it would not be until after World War II that the development of such water power resources would begin to be undertaken in northern Manitoba (Manitoba Hydro, 1997). Instead, the focus was upon the Winnipeg River with Great Falls Generating Station in operation in 1923, and despite the Great Depression, Seven Sisters Generating Station and the Slave Falls Generating Station came online in 1931 (Ross, 1963). Stepping back for a moment, the Island Falls Hydroelectric Station was begun in 1929 on the Churchill River in Saskatchewan near the Manitoba border, but the greater emphasis at this time was upon the hydroelectric potential of the Canadian Shield northwest of Lake of the Woods and southeast of Lake Winnipeg. Starting in 1936, hydroelectric power was being exported to the United States from the Winnipeg River area (Bateman, 2005).

Yet, not just export markets were important to stimulating hydroelectric production in the Prairie Provinces. Starting in 1945, rural electrification became a significant domestic market. Pine Falls Generating Station opened on the Winnipeg River in 1951 and the Pinawa Generating Station was dismantled in 1951-1952 so to make more efficient water flows possible for the Seven Sisters Generating Station (Manitoba Hydro, 1997). By 1955, the capacity of the Winnipeg River for hydroelectric production had been reached with the McArthur Generating Station coming online (Ross, 1963).

Post-World War II urbanization of Winnipeg and other parts of the Western Interior helped stimulate the need to look for resources for hydroelectric power elsewhere in the Canadian Shield. In 1960 the Kelsey Generating Station was online, the first such facility to be developed on the Nelson River. The Squaw Rapids Hydroelectric Station in northern Saskatchewan came online in 1963, and the Grand Rapids Generating Station was opened in 1965 on the Saskatchewan River close to its entry into Lake Winnipeg in Manitoba. Then came the completion in 1970 of the Kettle Generating Station on the Nelson near Gillam, Manitoba (Manitoba Hydro, 1997).

Additional facilities were created and improvements made between 1970 and 1990 so that the mega-project at Limestone on the Nelson River would be starting-up by 1990 (Bateman, 2005). Importantly, compensation had to be arranged for the Southern Indian Lake Commercial Fisherman's Association in 1984 as part of dealing with the damage cause by the Churchill River Diversion Project and raising South Indian Lake' s level as part of the terra-engineering to make such a mega-project even possible. Ongoing environmental consequences which adversely affected the indigenous people were contentious then and still are in this part of Manitoba and elsewhere in the Canadian bush (Rosenberg, Bodaly, and Usher, 1995; Hoffman, 2002; and Martin and Hoffman, 2008). However, compensation agreements were signed in 1992 with the Split Lake Cree First Nation, Cross Lake First Nation, and Nelson House First Nation as well as indigenous people of South Indian Lake in response to the losses suffered by the native peoples to have Limestone Generating Station fully operational by that year (Manitoba Hydro, 1997). Between the mid-1990s to the early part of the 21st century, the emphasis of hydroelectric development in the Canadian Shield within the Prairie Provinces has emphasized improvement in transmission lines, upgrading of facilities, and working with First Nations People to ameliorate the negative initial impacts of these developments plus to improve access of the various reserves to such energy so to have better quality of life in remote northern location of the Western Interior.

The Alberta and Saskatchewan Coal Fields are energy developments much in contrast to the hydroelectric projects in the Canadian Shield part of the Prairie Provinces. Environmental controversy surrounds the renewed exploitation of coal in places such as Canmore, Alberta, in no small measure because of its closeness to Banff National Park. This is an issue that can be traced to the early 1930s (Jones, 1933). When coal mines were abandoned in this part of the eastern slopes of the Canadian Rockies, subsidence associated with disused mine shafts and tunnels affected the potential for expansion for tourism in the area (Hartshorn *et al.*, 2005), an activity which was questionable in general according to some authors (Schindler, 2002). As of 2008, there were nine mines operating in Alberta and three mines being worked in Saskatchewan (Stone, 2008: 14.2) Demands for increased energy have revived interest in coal fields in many parts of the eastern slopes of the Canadian Rockies, but there are potentially easier and more profitable fields to be mined. Coal mining in the Prairie Provinces has changed from underground mining to surface mining. Thus, the coal production in southeastern Saskatchewan currently is focused upon this surface type of mining in the area between Weyburn and Estevan. Coal production here is essential for thermal-generated electricity for consumption within and outside that province.

The Athabasca Tar Sands is perhaps better known as the Alberta Oil Sands region, although the bitumen in this part of northeastern Alberta was recognized as having some resource potential even to the fur traders such as Mackenzie during the early days of the Athabasca fur trade in the late 18th century (Humphries, 2008). The problem with the oil sands is the projects are dealing with the immense difficulty and costs of converting bitumen (tar sand or oil sand depending upon whose parlance is used) into oil, an extremely intensive chemical process requiring substantial amounts of heat and water with long-term costs that possibly will exceed the short-term profits from exploiting such oil reserves (Kennett and Wenig, 2005). The Alberta Oil Sands are focused historically upon the Athabasca Tar Sands deposits near Fort McMurray, but other developments are being undertaken in the sub-regions of Cold Lake and Peace River (Humphries, 2008). During the early 21st century, the boom in the oil sands area drew workers from across Canada (Bone, 2008).

The Windy Grasslands is perhaps a vernacular region still to be recognized with respect to opportunities in energy development in the Western Interior. Much of the southern area of the Prairie Provinces is noted for a climate characterized by nearly steady, rapidly moving winds. The type of alternative energy capacity of Manitoba, Saskatchewan, and Alberta is of interest to commercial energy corporations as well as to environmentalists, politicians, and the general public, so the overall feasibility and profitable level of modern wind power is being taken more into account as a power option (Cuddihy, Kennedy, and Byer, 2005). Meanwhile, exploitations of solar power as another form of alternative energy are still in the early stages of successful development, but windmills are not new to the Prairie Provinces. During the late 19th century through mid-20th century and the start of rural electrification, many farmsteads used windmills for pumping water plus numerous ranches had windmills for the same purpose for the stock tanks needed to keep cattle well-watered in the drylands, drought-prone environment that is so typical of the prairie grasslands. Indeed, some ranchers still are using the older, functioning stock tank windmills as well as more modern versions of this invention which historically has been important for the settlement of this steppe environment.

Theme Three: Forestry Woes

When considering Canadian forestry, the images generally are of the activity in British Columbia or perhaps in parts of Atlantic Canada and 'somewhere' in the Canadian Shield, more likely than not Québec or Ontario because it is hard to conceptualize that the Prairie Provinces are part of this sub-sector of the primary economic sector. Indeed, only about 15% of Canada's commercial forestry products came from Prairie Provinces with about three-quarters of that originating in Alberta

(Boone, 2008: 397). Most of that activity is along the more southerly edge of the boreal forest or along the Foothills of the Canadian Rockies and is geared toward the harvesting of trees for pulp and paper milling with limited processing for lumber and construction material. While only a limited number of individuals remain in forestry jobs, some communities continue to be highly dependent upon such employment as single resource communities which is why the biological health of even the aspen trees is of importance while various tree diseases are expanding in the region (Brandt *et al.*, 2003). Such extreme functional specialization is to the disadvantage of these places when the booms and busts of North American housing construction has had such a disastrous implosion in the early 21st century as to adversely affect all the forestry sub-sector, particularly at times when pulp and paper prices have plummeted and the Canadian dollar's strength actually dissuades foreign buyers of these goods. The conditions are ripe for a 'firestorm' of trouble in the boreal forest zone.

Indeed, we need to take into account that fire is a crucial part of the ecology of the Canadian Shield (Johnson, Miyanishi, and Weir, 1998; Bergeron, *et al.*, 2001). The boreal forest's major tree species rely upon fire to help in regeneration of the entire ecozone and to maintain the present ecological succession regime. Most fires in the boreal forest are the product of lightening strikes, and a particularly dry summer can make for a long and tiring fire season for fire fighters. Sometimes, the various fires 'in the bush' produce sufficient smoke to affect air quality negatively beyond the boreal forest. Occasionally in Manitoba the smoke drifts as far south as Winnipeg even when the conflagrations originated as far north as Thompson.

Fire is a key concern in 'cottage country' because recreational pursuits are a major form of diversification of economy in this part of the Prairie Provinces. It must be noted, however, that while cottaging most often is associated with the major lakes to be found in the boreal forest zone of Manitoba such as Lake of the Woods, Lake Winnipeg, Lake Manitoba, and Lake Winnipegosis, it is not a new phenomenon and is found elsewhere in the region (Jaakson, 1986; Selwood, 1996), Some parts of Alberta, notably the Calgary area, have had second home development in the forested areas of the Canadian Rockies and the Foothills (Draper and McNicol, 1997) and continue so as one spinoff of the 1988 Calgary Winter Olympics (Kariel and Kariel, 1988). Then, too, the aspen parkland belt of Alberta and Saskatchewan is readily accessible cottage country respectively for Edmonton and Saskatoon before one reaches the sport fishing and hunting areas of 'the bush' in these parts of the Western Interior. Ecotourism and other forms of recreational activity in the boreal forests of Alberta, Saskatchewan, and Manitoba are fickle sources of supplemental or alternative revenue (Bogdanski, 2008; Scott, Jones, and Konopek, 2008). This especially is seemingly seen in the economic downtown of 2008-09 which appears to be continuing into 2010 relative to second home development and people's use of discretionary income for recreational activities with 'staycations', i.e., remaining close to home or at home rather than travelling to a vacation destination.

Theme Four: Not All Hard Rock and Hard Times

Often when the topic of minerals is discussed regarding the Prairie Provinces, the emphasis is upon fossil fuels principally in Alberta and less so metallic, non-metallic, and construction stone to be found in Saskatchewan and Manitoba. Yet, there are areas of those latter two provinces where one finds these non-fossil fuels minerals in significant quantity. There even is gold in Saskatchewan and Manitoba (Wagner, 2008: 18.2) as well as sulphur in Saskatchewan (Stone, 2008: 47.2) When examining the distribution of uranium and potash, one must look initially at the geology of northern Saskatchewan for the former and the geology of central and southern Saskatchewan mainly for the latter (Stanford, 2003). Interestingly, the hydroelectric development of part of northern Saskatchewan in the vicinity of Uranium City during the 1950s was tied to efforts in uranium mining; such mining has shifted more toward McArthur River and Rabbit Lake since the 1970s (Calvert, 2008). Canadian potash is concentrated in Saskatchewan and centered upon what is known as the Prairie Evaporite Deposit (Stone, 2008: 36.1). These deposits are especially of high quality and crucial in making Canada the world leader at present in potash production and exporting. As of 2008, there were 11 potash mining/processing operations in Canada with only one outside of Saskatchewan and that site being in New Brunswick (Stone, 2008: 36.2).

Generally-speaking, Manitoba has been the focus of this region's hard rock mining, i.e., the underground copper-zinc and nickel mining that is part of the overall economy of the Canadian Shield (Panagapko, 2008: 56.1). The Hudson Bay Mining and Smelting Company was instrumental in developing the copper-zinc mining in the area centered upon Flin Flon, Manitoba in the late 1920s with production commencing in 1930 (Morton, 1957). Again, the need for cheap power and the connection between sub-sectors of the economy are illustrated by how a subsidiary of this corporation underwrote initial hydroelectric power development on the Churchill River in Saskatchewan (Ross, 1963; Manitoba Hydro, 1997). Additional copper-zinc deposits would be exploited in the Swan Lake area of Manitoba, but mining there and around Flin Flon went into decline in the late 20th century when other sources of copper and zinc were mined, processed, and exported more cheaply elsewhere. By the early 21st century, active mining essentially had ceased temporarily or totally stopped in some mines with drastic ramifications for these two single-resource communities. Yet, in 2007 there was slightly renewed interest in re-starting such mineral exploitation in this sub-region of the Western Interior because once again, the cycle of fluctuations in ore prices was swinging upward in terms of global market demand (CBC News May 3, 2007). However, as Coulas (2008: 16.4) noted in the next year that the downward swing, coupled to tight credit market, resulted in a halt to such copper mining activities.

Similarly, the single-resource community of Thompson has been having its ups and downs with nickel production. Originally established in 1957 and linked at that time to southern Manitoba initially with the rail line of the former Hudson Bay Railway, Thompson was a prominent mining center for Inco starting in 1961 and lasting well into the latter part of the 20th century (Gill and Smith, 2008). However, Inco's overseas mines and its interests in the Voisey Bay nickel development on the Labradoran Coast of eastern Canada increasingly have been crucial factors in maintaining the Thompson smelting operation in the early 21st century (Minerals and Metals Sector Natural Resources Canada, 2008: 32.1). The mine at Thompson does survive at present because of improvements in productivity, but such effort is the result of increased mechanization and less labor to offset overseas competition. Consequently, the shrinking labor pool in Thompson has had negative spin-off effects for the community and environs. Relative to current economic development, Thompson has been re-making itself into a social services (education and medical) center for Manitoba north of 50° North latitude in response to population decline with changes in nickel mining employment (Gill and Smith, 2008).

Theme Five: Rural Central Place Hierarchy in Flux

As noted earlier in this chapter, agriculture remains seen by many people as the backbone of the economy of the Prairie Provinces even when so many of the citizens of Manitoba, Saskatchewan, and Alberta no longer are tied directly to it. As noted at the beginning of this chapter, more often than not the region's population is to be found living in or close-by to one of the 'Big Five' census metropolitan areas of the Western Interior. There often seems to be a near denial of negative consequences to agriculture from the general population out-migration, the radical restructuring of the railway infrastructure relative to grain handling with inland terminals, the aging-in-place in so many communities which is not being counter-balanced by in-migration of younger people in child-bearing and child-rearing stage of life, and consolidation of social services as well as efforts at restructuring telecommunications and transportation throughout the region by both governmental and private entities. There is no question that hamlets and villages have been more often than not wiped off the landscape of the Prairie Provinces since the mid-20th century with the advent of modern transportation making it easier to reach larger communities (Carlyle, 1994). Unfortunately, those surviving small towns and many medium-sized cities remain being at a disadvantage competing against each other for declining market share of trade hinterlands that are shifting more so to regionally-significant 'Big Five' census metropolitan areas of the Western Interior (Ramsey, Annis, and Everitt, 2002). Yet, various rural non-farm bedroom communities are increasing in popularity for residential relocation specific for commuting into Edmonton, Calgary, Winnipeg, Saskatoon, and Regina by the central city dwellers respectively from those cities.

Theme Six: Cosmopolitanization of Census Metropolitan Areas

While images of the Calgary Stampede and its emphasis upon the ranching roots of Alberta are popular lures for tourists to come on bus tours or individual travel to this gateway location to the Canadian Rockies, Calgary is anything but a 'cow town' and increasingly is a world-class city relative to culture and amenities as a consequence of the connection the place has to the international dimensions of the energy corporations based in this particular Prairie Provinces city. Similar cases could be made in turn for Edmonton, Winnipeg, Saskatoon, and Regina but in the context of more localized economic conditions, population composition, and ties to the world community relative to the economic geography and ethnicity of these other four members of the 'Big Five' of the Prairie Provinces.

A case in point can be made with Winnipeg. It is sited in a low-lying, flood-prone, mosquito-infested, extreme continental climate area where two key rivers join together (the eastward-flowing Assiniboine River is a tributary of the northward-flowing Red River of the North). This location has had some form of human habitation dating back over 10,000 years but only really became significant as a trading center in the era of the fur trade beginning with the post established by LaVerendrye in 1738 which was not maintained for any length of time and consequently abandoned (Ray, 1974; Pettipas, 1984). During the late 18th and early 19th centuries, rival fur trade companies, notably the last of the iterations of the mercantile ventures known as the Northwest Company and the omnipresent Hudson Bay Company (established in 1670), contested each other for control of the area first for fur, mainly beaver and muskrat, and then for bison-related products, most importantly pemmican (Moodie, 1987). The introduction of subsistence agriculture by Scottish immigrants as part of the Selkirk Settlement in 1811-1812 at Point Douglas (north of the confluence of the two rivers) eventually lead to increased friction between the Hudson's Bay Company and the Northwest Company. Thus, the Pemmican War of 1814-1816 was fought and followed with ongoing hostilities that lead to a legal struggle and the eventual merger of the Northwest Company into the Hudson's Bay Company (Morton, 1965; Bellan, 1978; Kienetz, 1983). Gradually, the French Canadians, the Metis, and the Selkirkers established a series of parishes in the vicinity of Upper Fort Garry (site close to the long-abandoned LaVerendrye fur trade post) that was administered overall in the interests of the Selkirk Estate until 1834 and then as part of the trading monopoly of the Hudson's Bay Company until free trade was the norm starting in 1849 (Gilman, Gilman, and Stultz, 1979; Pannekoek, 1981; Friesen, 1984). By the late 1860s, these parishes were on the verge of a settlement boom. That development was contingent upon how what was then part of the Northwest Territory would enter the Confederation of Canada which had acquired by 1869 what had been the lands of the Hudson's Bay Company (Richtik, 1975; Marr and Percy, 1978).

The Riel Resistance of 1869-1870 was a defining moment in the historical geography of all the Prairie Provinces, not just Manitoba (Morton, 1965; Friesen, 1984). The ultimate defeat of the French Canadian and Metis over land rights in Manitoba was based upon the replacement of the Québecois long lot land division model and traditional heritage of subsistence farming with an innovation of land division from Eastern Canada. Introduction and implementation of the Ontarian-style of range and township system of land holding and land sales for speculative purposes would become the manner in which the territories would be handled for settlement emphasizing commercial cash-grain farming in what eventually would become the rest of Manitoba, Saskatchewan, and Alberta (Morton, 1957; Friesen, 1984).

Promoting such a settlement pattern was the creation of a railway network that was centered upon what then finally was known as Winnipeg, the name given in the late 1860s to the agglomeration of parishes until then referred to as the Red River Settlements (Artibise, 1977). A branch line of the Canadian Pacific Railway between St. Boniface and Emerson to connect with American railways at Noyes, Manitoba, was made operational in 1878, but it would not be until 1881-1883 that the direct eastern connection with the Lakehead and onto Montréal would be solidified and finally constructed (Morton, 1951; Sharp, 1952; Studness, 1964; Artibise, 1975) Because of the situation of Winnipeg, the Canadian Pacific Railway's mainline and related regional branch lines funnelled into the city from the east and then funnelled out of the city to the west as eventually did the trackage of rival Canadian rail firms (Stabler, 1973; Artibise, 1981; Lewis and MacKinnon, 1987). Thus, the 'Belt Buckle of the Dominion' became a popular nickname during the rise of Winnipeg's prominence as the wholesaling and retailing center for Canada between the Great Lakes and the Pacific Ocean in the late 19th and early 20th centuries (Artibise, 1977). Winnipeg evolved as the chief point of access for the immigrants coming from Europe to settle what would become the Prairie Provinces with the city becoming the third most important urban center of Canada just before World War I (Fowke, 1956; Burghardt, 1971; Artibise, 1975).

Yet, Winnipeg's control of its western hinterland began to decline during the 1910s through 1960s. First, the opening of the Panama Canal in 1914 gave Vancouver a comparative advantage such that it could penetrate the Alberta portion of the Western Interior as well as dominate the province of British Columbia (Artibise, 1975). Then, the Winnipeg General Strike of 1919, which pitted the lower income, mainly Slavic working class against the higher income, mainly British commercial interests, stifled growth temporarily but the city rebounded in the 1920s with its suburban developments as well as central business district's improvements (Artibise, 1977). The Great Depression adversely would affect Winnipeg more so than other Prairie Province cities because of Winnipeg's substantial number of firms associated with the grain trade, commercial banking, and life insurance as well as companies producing agricultural equipment, transportation equipment, clothing, and goods associated with printing and publishing (Bellan, 1978). This economic downturn coupled to a delay in major building construction during World War II and the shift away from the inner city to suburban locations for wholesaling facilities actually would have positive turnaround in the 1970s through 1990s with the gradual gentrification-oriented revival of certain neighborhoods closest to the historic central business district.

Meanwhile, Winnipeg's prominence among Prairie Province cities suffered another major blow with the rise of the oil centers in Calgary and Edmonton in the 1950s and 1960s through 1970s which both surpassed the Manitoban primate city in terms of population, employment opportunity, retailing structure, transportation importance, and perception in terms of urban desirability (Friesen, 1984; Artibise, 1988) Winnipeg's creation and development of a Unicity government in the early 1960s did result in merging the central city and 11 of its suburbs taking effect in 1972 (Lightbody, 1978), but the urban renaissance anticipated did not take place until somewhat later. The inner city decline, notably in the North End (Broadway and Jesty, 1998; Deane, 2004), and other problems ultimately forced the issue such that significant urban renewal has taken place since the mid-1980s, especially in the Exchange District and Osborne Village but also extending out into other central city neighborhoods such as Chinatown and across former suburbs such as St. Boniface. These changes were contested by the various stakeholders and users of the public space of these neighborhoods, as in the case of Osborne Village (Ashford, 1999) and the North End (Deane, 2004).

Perhaps the redevelopment of the Forks is the crucial part to the cosmopolitanization of the Winnipeg census metropolitan area although the development of Portage Place Mall starting in 1985, located farther west in what is considered the true central business district of Winnipeg, often is considered as a similar trigger (Dafoe, 1998). Once the railyard of the Canadian National Railway adjacent to the eastern edge of the central business district, the land known as the Forks has been transformed into a destination center not just for local people but visitors from across Canada and the world (Dafoe, 1998; Leo and Pyl, 2007). Initially, existing buildings were joined together to create an indoor market effect and then other properties were renovated to serve as a children's museum, a children's theatre, and a retailing center. A major destination hotel was developed close to the Parks Canada Forks Historic Site and a series of entertainment stages as well as skateboard park have been added to the landscape.

Approximately during the same time period, a substantial minor league baseball stadium was constructed as part of the deal-making for bringing the Pan-American Games to Winnipeg in 1999 which stimulated more re-investment in the central city (Taylor and Row, 2005). Increased gentrification of the Exchange District was undertaken that also included new construction. Efforts even have been extended along Main Street to develop a cultural corridor that would include the Exchange District, key museums, and theatres from Portage Avenue into Point Douglas and adjacent to Chinatown.

Unfortunately, cosmopolitanization has not been equally beneficial to all the citizens of Winnipeg. Poverty and violent crime, notably among the newest Canadian immigrants and the First Nation Peoples, has been an ongoing problem and particularly hard to solve when alcoholism, illicit drugs, prostitution, and illegal gambling also are fuelling gang-related activity. Quality of life is not equitably distributed, especially when the disparity between gentrified neighborhoods and non-gentrified neighborhoods is so evident in part because of the propinquity of such places (Owen and Watters, 2005; Silver, 2006; Tiwari, 2008). Then, too, the influence of the globalization of retailing is becoming as evident in Winnipeg with the emergence of 'big box' stores (Lorch, 2004). Yet, there is a level of optimism that is reminiscent of the boosterism of Winnipeg's leadership of the 1910s. Now, nearly one hundred years later at the time of writing this chapter, one might see Winnipeg as being reflective of the influences of the core-periphery model as it affects the Prairie Provinces in general.

Conclusion: Still a "Drive-thru" Region in Another Phase of Geographical Transition or Not?

We have come to the end of this journey through the Prairie Provinces, and perhaps you are wondering still about the use of the term, 'drive-thru' region as part of the title to this particular chapter. All too often, people simply 'drive-thru' Manitoba, Saskatchewan, and Alberta along the route of the Trans-Canada Highway and believe that route is all there is to knowing about and understanding the Western Interior. Worse yet, there are those individuals who think only of Alberta, Saskatchewan, and Manitoba as part of that almost neglected and overlooked 'fly-over' region of the Great Plains of North America that stretches from the Mexican-American border to the Canadian Shield and fail to consider how the Prairie Provinces is a complex region of Canada. The Western Interior really is a set of interconnected, sometimes disparate but sometimes overlapping sub-regions.

What does the future portend for the Prairie Provinces? It depends upon a number of variables, including some 'wild card' factors and events that we only can speculate upon at the moment, e.g., any changes in the federal government as to which political party (or parties in a coalition government) would be controlling Parliament, a geopolitical crisis of international magnitude, or some new meltdown of the global economy or continuing melting of the polar ice caps! All that can be said with any certainty about the Western Interior is that is a region undergoing transition to whatever will be its next phase of development within the Dominion of Canada and as part of our global environment. So, you have a wonderful opportunity as a neophyte geographer to watch and to participate in that geographical transformation for the Prairie Provinces.

Review Questions

1. Based upon your reading of this chapter and reflecting upon your personal experiences, how has your perception of the Canadian Prairies changed, if any? Why is your response what it is concerning this part of Canada?

2. Human habitation can be traced well into the past for the Canadian Prairies with the foundation of the modern era rooted along the waterways and activities of the fur trade. Why have changes in transportation and industrial production been such important aspects in the develop-

ment and transformation of the economy and settlement pattern of Manitoba, Saskatchewan, and Alberta?

3. Different forms of mining have spatial as well as economic significance throughout various parts of the Canadian Prairies. How would you characterize the spatial distribution of mining in this region? Furthermore, what might be various economic and non-economic impacts of mining within and outside these three provinces?

4. Hydroelectric production is highly localized within the Canadian Prairies but its consequences are important to the overall electrical grid for North America. What issues need to be considered in terms of any future developments for increasing hydroelectricity in this part of Canada for regional and international use?

5. Urbanization and migration are changing the nature of settlement in the Canadian Prairies. How does the intra-provincial and inter-provincial migration of First Nations people coupled to the immigration of people from outside the Canadian Prairies impact the towns, cities, and metropolitan centers of Manitoba, Saskatchewan, and Alberta? What do you think the region's central place hierarchy will be in the future, particularly when taking into account the globalization aspects of the economy of the Canadian Prairies?

References

Artibise, Alan F.J. (1975), *Winnipeg: A Social Hisotry of Urban Growth, 1874-1914*, Montréal: McGill-Queen's University Press.

Artibise, Alan F.J. (1977), *Winnipeg: An Illustrated History*, Toronto: J. Lorimer and National Museum of Man, National Museums of Canada.

Artibise, Alan F.J. (1981), "Town and City: Aspects of Western Canadian Urban Development", *Canadian Plains Studies*, 10, Regina, Saskatchewan, Canadian Plains Research Center.

Artibise, Alan F.J. (1988), "Canada as an Urban Nation", in *Daedalus*, 117(4), pp. 237-264.

Ashford, Theresa K. (1999), Punk, Power and Place: Contesting Public Space in Winnipeg's Osborne Village, M.A. thesis, Simon Fraser University.

Atlas of Urban Aboriginal Peoples, http://gismap.usask.ca/website/Web_atlas/AOUAP/.

Barsh, Russel L. (1994), "Canada's Aboriginal Peoples: Social Integration or Disintegration?", in *The Canadian Journal of Native Studies*, XIV(1), pp. 1-46.

Bateman, Leonard (2005), "A History of Electric Power Development in Manitoba", *IEEE Canadian Review*, pp. 22-25.

Bellan, Ruben (1978), *Winnipg, First Century: An Economic History*, Winnipeg: Queenston House.

Bergeron, Yves, Gauthier, Sylvie, Kafka, Victor, Lefort, Patrick, and Lesieur, Daniel (2001), "Natural Fire Frequency for the Eastern Canadian Boreal Forest: Consequences for Sustainable Forestry", *Canadian Journal of Forest Research*, 31(3), pp. 384-391.

Bogdanski Bryan E.C. (2008), *Canada's Boreal Forest Economy: Economic and Socioeconomic Issues and Research Opportunities*, Information Report BC-X-414, Victoria, British Columbia, Natural Resources Canada.

Brandt, J.P., Cerezke, H.F., Mallett, K.I., Volney, W.J.A., and Weber, J.D. (2003), "Factors Affecting Trembling Aspen (Populus tremuloides Michx.) Health in the Boreal Forests of Alberta, Saskatchewan, and Manitoba, Canada", *Forest Ecology and Management*, 178(3), pp. 287-300.

Bremer, Jennifer (2009), "Inaccurate Flu Reports Affect Pork Industry", *High Plains/Midwest Ag Journal*, http://www.hpj.com/archives/2009/may09/may11/Inaccurateflureportsaffectp.cfm.

Bryson, Jay H., *Economic Effects of Swine Flu: Mexico and Beyond*, Wachovia Economics Group, April 28, 2009. http://www.docstoc.com/docs/5678750/Economic-Effects-Of-Swine-Flu.

Brown, Jennifer S.H. (1980), *Strangers in Blood: Fur Trade Company Families in Indian Country*, Vancouver: University of British Columbia.

Burghardt, A.F. (1971), "A Hypothesis about Gateway Cities", in *Annals of the Association of American Geographers*, 61(2), pp. 269-285.

Bone, Robert (2008), *The Regional Geography of Canada*, Fourth Edition, Don Mills: Oxford University Press.

Broadway, Michael J. and Jesty, Gillian (1998), "Are Canadian Inner Cities Becoming More Dissimilar? An Analysis of Urban Deprivation Indicators", in *Urban Studies*, 35, pp. 1423-1438.

CBC News, *New Rules Bring about 'Dark Day' for Producers: Pork Council*, March 4, 2008, www.cbc.ca/canada/manitoba/story/2008/03/04/hog-reax.html?ref=rss.

CBC News, *Manitoba's New Holiday*, September 25, 2007, http://www.cbc.ca/canada/manitoba/story/2007/09/25/stat-holiday.html.

CBC News, *Mining Scene Hot in Lynn Lake, Man.*, May 3, 2007, http://www.cbc.ca/canada/manitoba/story/2007/05/03/lynn-lake-mining.html.

CBC News, *More Flights Planned from Oilpattch to Island*, March 14, 2006, http://www.cbc.ca/canada/prince-edward-island/story/2006/03/14/pe-oil-flight-20060314.html.

Calvert, H. Thomas (2008), "Uranium", in Evelyn Godin (ed.), *Canadian Minerals Yearbook 2008*, Ottawa, Government of Canada, pp. 53.1-53.12.

Carlyle, William J. (1994), "Rural Population Change on the Canadian Prairies", in *Great Plains Research*, 4, pp. 65-87.

Carter, Sarah (1999), *Aboriginal People and Colonizers of Western Canada to 1900*, Toronto: University of Toronto.

Coulas, Maureen (2008), "Copper", in Evelyn Godin (ed.), *Canadian Minerals Yearbook 2008*, Ottawa: Government of Canada.

Cuddihy, John, Kennedy, Christopher, and Byer, Philip (2005), "Energy Use in Canada: Environmental Impacts and Opportunities in Relationship to Infrastructure Systems", in *Canadian Journal of Civil Engineering*, 32, pp. 1-15.

Dafoe, Christopher (1998), *Winnipeg: Heart of the Continent*, Winnipeg: Great Plains Publications.

Deane, L., Morrissette, L., Bousquet, J., and Bruyere, S. (2004), "Explorations in Urban Aboriginal Neighbourhood Development", *The Canadian Journal of Native Studies*, XXIV(2), pp. 227-252.

Deane, Lawrence (2004), Community Economic Development in Winnipeg's North End: Social, Cultural, Economic and Policy Aspects of a Housing Intervention, Ph.D. Dissertation, University of Manitoba.

Dickason, Olive P. (2002), *Canada's First Nations: A History of Founding Peoples from Earliest Times*, 3rd ed., Don Mills: Ontario, Oxford.

Draper, D. and McNicol, B. (1997), "Coping with Touristic and Residential Developmental Pressures: The Growth Management Process in Canmore, Alberta", *Challenge and Opportunity*, Waterloo: University of Waterloo Press, pp. 159-196.

Emmons, David M. (1971), *Garden in the Grasslands: Boomer Literature of the Central Great Plains*, Lincoln: University of Nebraska Press.

English, Ed, Federal Reserve Bank of Atlanta (2006), "Avian Flu Has Poultry Industry in a Flap", in *EconSouth*, 8(1), http://www.frbatlanta.org/pubs/econsouth/econsouth_vol_8_no_1_avian_flu_has_poultry_industry_in_a_flap.cfm?redirected=true.

Everitt, John (2001), "A Closer Look: The Canadian Grain Elevator", in Tom L. McKnight (ed.), *Regional Geography of the United States and Canada*, Third Edition, Upper Saddle River, NJ, Prentice-Hall.

Everitt, John (1996), "The Development of the Grain Trade in Manitoba", in John Welsted, John Everitt, and Christoph Stadel (eds.), *The Geography of Manitoba: Its Land and its People*, Winnipeg: The University of Manitoba Press.

Flanagan, Thomas (1991), *Metis Lands in Manitoba*, Calgary: University of Calgary Press.

Fowke, Vernon C. (1956), "National Policy and Western Development in North America", *The Journal of Economic History*, 16(4), pp. 461-479.

Friesen, Gerald (1984), *The Canadian Prairies: A History*, Lincoln: University of Nebraska Press.

Gill, Alison M. and Smith, Geoffrey C. (2008), "'Residents' Evaluative Structures of Northern Manitoba Mining Communities", *The Canadian Geographer*, 29(1), pp. 17-29.

Gilman, Rhoda R., Gilman, Carolyn, and Stultz, Deborah (1979), *The Red River Trails: Oxcart Routes between St. Paul and the Selkirk Settlement, 1820-1870*, St. Paul, Minnesota: Minnesota Historical Society.

Hanselmann, Calvin (2001), *Urban Aboriginal People in Western Canada: Realities and Policies*, Calgary: Canada West Foundation.

Harrison, Julia D. (1985), *Metis: People Between Two Worlds*, Vancouver: Douglas and McIntyre.

Hartshorn, J., Maher, M., Crooks, J., Stahl, R., and Bond, Z. (2005), "Creative Destruction: Building toward Sustainability", in *Canadian Journal of Civil Engineering*, 32(1), pp. 170-180.

Heaver, Trevor D. (1993), "Rail Freight Service in Canada: Restructuring for the North American Market", in *Journal of Transport Geography*, 1(3), pp. 156-166.

Hoffman, Steven M. (2002), "Powering Injustice: Hydroelectric Dvelopment in Northern Manitoba", in J. Byrne, L. Glover, and C. Martinez (eds.), *Environmental Justice: Discourses in International Political Economy, Energy and Environmental Policy*, Volume 8, New Brunswick, NJ: Transaction Publishers, pp. 147-170.

Huck, Barbara et al. (2002), *Exploring the Fur Trade Routes of North America*, Winnipeg: Heartland.

Humphries, Marc (2008), *North American Oil Sands: History of Development, Prospects for the Future*, Washington, DC: Congressional Research Service.

Innis, Harold A. (1995), "The Importance of Staples Products in Canadian Development", in Daniel Drache (ed.), *Staples, Markets, and Cultural Change*, Montréal: McGill-Queen's Press, pp. 3-23.

Innis, Harold A. (1999), *The Fur Trade in Canada,* Toronto: University of Toronto.

Jaakson, Reiner (1986), "Second-home Domestic Tourism", in *Annals of Tourism Research*, 13(3), pp. 367-391.

Johnson, E.A., Miyanishi, K., and Weir, J.M.H. (1998), "Wildfires in the Western Canadian Boreal Forest: Landscape Patterns and Ecosystem Management", in *Journal of Vegetation Science*, 9, pp. 603-610.

Jones, Stephen B. (1933), "Mining and Tourist Towns in the Canadian Rockies", in *Economic Geography*, 9(4), pp. 368-378.

Kariel, Herbert G. and Kariel, Patricia E. (1988), "Tourist Developments in the Kananaskis Valley Area, Alberta, Canada, and the Impact of the 1988 Winter Olympic Games", in *Mount Research and Development*, pp. 1-10.

Kennett, Steven A. and Wenig, Michael M. (2005), "Alberta's Oil and Gas Boom Fuels Land-use Conflicts – But Should the EUB be Taking the Heat?", Canadian Institute of Resources Law, *Resources*, 91.

Kienetz, Alvin (1983), "The Rise and Decline of Hybrid (Metis) Societies on the Frontier of Western Canada and Southern Africa", in *The Canadian Journal of Native Studies*, III(1), pp. 3-21.

Lehr, John (1982), "The Western Interior: The Transformation of a Hinterland Region", in L.D. McCann (ed.), *Heartland and Hinterland: A Geography of Canada*, Scarborough, Ontario, Prentice-Hall, Canada, pp. 250-293.

Leo, Christopher and Pyl, Mike (2007), "Multi-level Governance: Getting the Job Done and Respecting Community Difference – Three Winnipeg Cases", in *Canadian Political Science Review*, 1(2), pp. 1-26.

Leshchyshen, Bob (Bohdan) (2009), "Ukrainian Credit Union's Penetration of Ukrainian Population in Canada Based on Statistics Canada 2006 Census of Urban Centres", *Coordinator* (Council of Ukrainian Credit Union in Canada).

Lightbody, James (1978), "The Reform of a Metropolitan Government: The Case of Winnipeg, 1971", in *Canadian Public Policy*, 4(4), pp. 489-504.

Lewis, Frank and MacKinnon, Mary (1987), "Government Loan Guarantees and the Failure of the Canadian Northern Railway", in *The Journal of Economic History*, 47(1), pp. 175-196.

Lorch, Brian J. (2004), "Evolving Retail Landscape of Winnipeg: A Geographical Perspective", *Research and Working Paper #43*, Winnipeg: Institute of Urban Studies.

Manitoba Hydro (1997), *History of Electric Power in Manitoba*, Winnipeg: Manitoba Hydro.

Marr, William and Percy, Michael (1978), "The Government and the Rate of Canadian Prairie Settlement", in *The Canadian Journal of Economics*, 11(4), pp. 757-767.

Martin, Thibault and Hoffman, Steven M. (eds.) (2008), *Power Struggles: Hydro Developments and First Nations in Manitoba and Québec*, Winnipeg: University of Manitoba.

McCann, L.D. and Gunn, A. (eds.) (1987), *Heartland and Hinterland: A Geography of Canada*, Second Edition, Scarborough: Prentice-Hall.

Minerals and Metals Sector, Natural Resources Canada (2008), "Nickel", in Evelyn Godin (ed.), *Canadian Minerals Yearbook 2008*, Ottawa: Government of Canada, pp. 32.1-32.32.

Moodie, D. Wayne (1987), "The Trading Post Settlement of the Canadian Northwest, 1774-1821", in *Journal of Historical Geography*, 13(4), pp. 360-374.

Morton, W.L. (1951), "The Significance of Site in the Settlement of the American and Canadian Wests", in *Agricultural History*, 25(3), pp. 97-104.

Morton, W.L. (1957), *Manitoba: A History*, Toronto: University of Toronto Press.

Morton, W.L. (ed.) (1965), *Manitoba: The Birth of a Province*, Altona, Manitoba, D.W. Friesen.

Nicholson, Beverley A. (1987), *Human Ecology and Prehistory of the Forest/Grassland Zone of Western Manitoba*, Ph.D. dissertation, Simon Fraser University.

Norrie, K.H. (1975), "The Rate of Settlement of the Canadian Prairies, 1870-1911", in *The Journal of Economic History*, 35(2), pp. 410-427.

Olar, M., Romain, R., Bergeron, N., and Klein, K. (2004), *Ethanol Industry in Canada*, SR.04.08, Laval, CREA (Centre de Recherche en Economie Agroalimentaire.

Owen, Michelle and Watters, Colleen (2005), Housing for Assisted Living in Inner-City Winnipeg: A Social Analysis of Housing Options for People with Disabilities, Winnipeg: Winnipeg Inner-City Research Alliance.

Panagapko, Doug (2008), "Zinc", in Evelyn Godin (ed.), *Canadian Minerals Yearbook 2008*, Ottawa: Government of Canada, pp. 56.1-56.24.

Pannekoek, Fritz (1981), "The Historiography of the Red River Settlement 1830-1868", in *Prairie Forum*, 6, pp. 75-85.

Persaud, A. Jai and Kumar, Uma (2001), "An Eclectic Approach in Energy Forecasting: A case of Natural Resources Canada's (NRCan's) Oil and Gas", in *Energy Policy*, 29(4), pp. 303-313.

Peters, Evelyn J. (1992), "Self-Government for Aboriginal People in Urban Areas: A Literature Review and Suggestions for Research", in *The Canadian Journal of Native Studies*, XII(1), pp. 51-74.

Peters, Evelyn J. (2001), "Developing Federal Policy for First Nations People in Urban Areas: 1945-1975", in *The Canadian Journal of Native Studies*, XXI(1), pp. 57-96.

Peters, Evelyn J. and Robillard, Vince (2009), "'Everything You Want is There': The Place of the Reserve in First Nations' Homeless Mobility", in *Urban Geography*, 30(6), pp. 652-680.

Pettipas, L.F. and Buchner, A.P. (1983), "Paleo-Indian Prehistory of the Glacial Lake Agassiz Region in Southern Manitoba, 11,500 to 6,500 BP", in J.T. Teller and L. Clayton (eds.), *Glacial Lake Agassiz*, Toronto: University of Toronto, pp. 421-451.

Pettipas, L.F. (1984), "The Paleo-Indians of Manitoba", in J.T. Teller (ed.), *Natural Heritage of Manitoba*, Winnipeg: Manitoba Museum of Man and Nature, pp. 161-173.

Pettipas, L.F. (1998), *Aboriginal Migrations: A History of Movements in Southern Manitoba*, Winnipeg: Manitoba Museum of Man and Nature.

Putnam, Donald F. and Putnam, Robert G. (1979), *Canada: A Regional Analysis*, Second Edition, Don Mills: J.M. Dent and Sons.

Ramsey, D., Annis, B., and Everitt, J. (2002), "Definitions and Boundaries of Community: The Case of Rural Health Focus Group Analysis in Southwestern Manitoba", in *Prairie Perspectives: Geographical Essays*, 5, pp. 187-201.

Ray, Arthur J. (1974), *Indians in the Fur Trade: Their Role as Trappers, Hunters, and Middlemen in the Lands Southwest of Hudson Bay 1660-1870*, Toronto: University of Toronto.

Ray, Arthur J. (1978), "History and Archaeology of the Northern Fur Trade", in *American Antiquity*, 43(1), pp. 26-34.

Richtik, James M., "The Policy Framework for Settling the Canadian West 1870-1880", in *Agricultural History*, 49(4), pp. 613-628.

Robinson, J. Lewis (1983), *Concepts and Themes in the Regional Geography of Canada* (Revised), Vancouver: Talon Books.

Rosenberg, D.M., Bodaly, R.A., and Usher, P.J. (1995), "Environmental and Social Impacts of Large Scale Hydro-electric Development: Who is Listening?", in *Global Environmental Change*, 5(2), pp. 127-148.

Ross, David S.G., "History of the Electrical Industry in Manitoba", *Manitoba Historical Society Transcactions*, Series 3, 1963-64, http://www.mhs.mb.ca/docs/transactions/3/electricalindustry.shtml.

Saku, James C. (1999), "Aboriginal Census Data in Canada: A Research Note", *The Canadian Journal of Native Studies*, XIX(2), pp. 365-379.

Schindler, David W. (2002), "The Eastern Slopes of the Canadian Rockies: Must We Follow the American Blueprint?", in Jill S. Baron (ed.), *Rocky Mountain Futures: An Ecological Perspective*, Washington, DC: Island Press, pp. 285-300.

Scott, Daniel, Jones, Brenda, and Konopek, Jasmina (2008), "Exploring Potential Visitor Response to Climate-Induced Enviornmental Changes in Canada's Rocky Mountain National Parks", in *Tourism Review International*, 12(1), pp. 43-56.

Selwood, John (1996), "The Summer Cottage at Lake Winnipeg", in J. Welsted, J. Everitt, and C. Stadel (eds.), *The Geography of Manitoba: Its Land and Its People*, Winnipeg: University of Manitoba Press, pp. 296-298.

Sharp, Paul F. (1952), "The Northern Great Plains: A Study in Canadian-American Regionalism", in *The Misssissippi Valley Historical Review*, 39(1), pp. 61-76.

Shore, Fred J. (2001), "The Emergence of the Metis Nation in Manitoba", in L.J. Barkwell, L. Dorton, and D.R. Prefontaine (eds.), *Metis Legacy: A Metis Historiographical and Annoted Bibliography*, Winnipeg: Pemmican Publications.

Silver, Jim (2006), *Gentrification in West Broadway? Contested Space in a Winnipeg Inner City Neighborhood*, Winnipeg, Manitoba: Canadian Centre for Policy Alternatives.

Small, Ernest (1999), "New Crops for Canadian Agriculture", in J. Janick (ed.), *Perspective on New Crops and New Uses*, Alexandria, Virginia, ASHS Press, pp. 15-52.

Smith, S.L., Burgess, M.M, Riseborough, and Nixon, F.M. (2005), "Recent Trends from Canadian Permafrost Thermal Monitoring Network Sites", in *Permafrost and Periglacial Processes*, 16, pp. 19-30.

Stabler, Jack C. (1973), "Factors Affecting the Development of a New Region: The Canadian Great Plains, 1870-1897", in *The Annals of Regional Science*, 7(1), pp. 75-87.

Stanford, Quentin H. (ed.) (2003), *Canadian Oxford World Atlas*, Don Mills: Oxford University Press.

Statistics Canada, Population and Dwelling Counts, for Canada, Provinces and Territories, 2006 and 2001 Censuses-100% Data http://www12.statcan.gc.ca/census-recensement/2006/dp-pd/hlt/97-550/Index.cfm?TPL=P1C&Page=RETR&LANG=Eng&T=101.

Statistics Canada, Population and Dwelling Counts, for Census Metroplitan Areas, 2006 and 2001 Censuses-100% Data http://www12.statcan.gc.ca/

census-recensement/2006/dp-pd/hlt/97-550/Index.cfm?TPL=P1C&Page=RETR&LANG=Eng&T=205&RPP=50.

Statistics Canada (2009), *Canadian Agriculture at a Glance*, 96-325-XWE, Ottawa: Government of Canada.

Stone, Kevin (2008), "Coal", in Evelyn Godin (ed.), *Canadian Minerals Yearbook 2008*, Ottawa: Government of Canada, pp. 14.1-14.15.

Stone, Kevin (2008), "Potash", in Evelyn Godin (ed.), *Canadian Minerals Yearbook 2008*, Ottawa: Government of Canada, pp. 36.1-36.12.

Stone, Kevin (2008), "Sulphur", in Evelyn Godin (ed.), *Canadian Minerals Yearbook 2008*, Ottawa: Government of Canada, pp. 47.1-47.6.

Studness, Charles M. (1964), "Economic Opportunity and the Westward Migration of Canadians during the Late Nineteenth Century", in *The Canadian Journal of Economics and Political Science*, 30(4), pp. 570-584.

Taylor, Amy, Bramley, Matthew, and Winfield, Mark (2005), *Government Spending on Canada's Oil and Gas Industry: Undermining Canada's Kyoto Commitment*, Calgary: The Pembina Institute.

Taylor, Scott and Row, Kris (2005), *Home Run: The History of the Winnipeg Goldeyes and CanWest Global Park*, Winnipeg: Studio Publications Inc.

Timoney, Kevin and Lee, Peter (2001), "Environmental Management in Resource-rich Alberta: First World Jurisdiction, Third World Analogue?", in *Journal of Environmental Mangement*, 63(4), pp. 387-405.

Tiwari, R.C. (2008), "Poverty in Winnipeg: A study in Urban-Social Geography", in George Pomeroy and Gerald Webster (eds.), *Global Perspectives on Urbanization*, Lanham, Maryland: University Press of America, pp. 213-234.

Tyrchniewicz, Ed, Carter, Nick, and Whitaker, Nick (2000), *Sustainable Livestock Development in Manitoba: Finding Common Ground*, Winnipeg: Government of Manitoba.

Veeman, Terrence S. and Gray, Richard (2009), "Agricultural Production and Productivity in Canada", *Choices*, 24(4), http://www.choicesmagazine.org/magazine/article.php?article=92.

Wagner, Wayne (2008), "Gold", in Evelyn Godin (ed.), *Canadian Minerals Yearbook 2008*, Ottawa: Government of Canada, pp. 18.1-18.18.

Walker, Ryan C. (2005), "Social Cohesion? A Critical Review of The Urban Aboriginal Strategy and Its Application to Address Homelessness in Winnipeg", in *The Canadian Journal of Native Studies*, XXV(2), pp. 395-416.

Weida, William J., *The CAFO and Depopulation of Rural Agricultural Areas: Implications for Rural Economies in Canada and the US*, International Conference on The Chicken-Its Biological, Social, Cultural and Industrial History, May 17-19, 2002, Yale University, http://www.sraproject.org/wpcontent/themes/default/images/thecafoanddepopulationofruralagriculturalareas.pdf.

Wishart, David J. (2006), "Natural Areas, Regions, and Two Centuries of Environmental Change on the Great Plains", in *Great Plains Quarterly*, 26(3), pp. 147-165.

CHAPTER 11

The North

Balancing Tradition and Change

Gita J. LAIDLER
Carleton University

Andrey N. PETROV
University of Northern Iowa

Introduction: 'The true North, strong and free'

The Canadian anthem 'Oh, Canada' refers to Canada as the "true North strong and free." This reference is not accidental: for many the North is fundamental for Canadian self-identification. Some argue that the idea of Canada is unthinkable apart from the idea of the North. It goes beyond physical facts and economic measures, it is the idea deeply embedded in society's representations, articulations, emotions, ideals and motivations. Some would also say that the North is a distinctive feature of a 'Canadian culture' that distinguishes it from an 'American mass culture.' Describing these views, Rob Shields defined the North as "a resource and economic hinterland, which is simultaneously incorporated in [Canadian] social spatialization as a mythic heartland" (Shield, 1991, 163). In this Canadian 'spatial mythology' the North is associated with truth, purity, freedom and power.

There is no doubt that in its cultural role the North is formative for the Canadian society. It is a region of fragile ecosystems, some of which are yet to be disturbed by industrial activities. It is a homeland, a unique place that continues to support and nurture Aboriginal cultures and traditions. However, it is also important as an economic frontier that concentrates vast resources of minerals, fuels, fresh water, fish and wildlife. Politically, it is an enormous, strategically important territory that extends over lands and waters almost to the North Pole, a defining element of Canadian territory and sovereignty in the Arctic.

At the same time, the North is a very dynamic space. In the recent decades, we have also seen unprecedented transformation of northern political institutions, with Aboriginal peoples gaining more powers in their homelands. We have observed economic shifts, as the Canadian North has suddenly become one of the world's leading diamond producers. Northern biomes are experiencing dramatic environmental changes due to a rapidly changing climate. We are witnessing how receding ice cover and increasing demand for fossil fuels is reconfiguring the geopolitical situation in the Arctic, making it one of the most contested and desired regions of the world. As you will see in this chapter, the Canadian North is seeking to find a balance between tradition and change. You will be introduced to general characteristics of Canada's northern regions, including an overview of the physical environment, history, governance and politics, northern peoples and cultures, and economics. This provides important contextual background to better understand some of the pressing challenges and opportunities facing northern Canadians today in times of change (environmental, economic, social, and political). With the importance of the North to Canada, we hope that this chapter will help you not only to learn the fascinating story of Canada's northern regions, but also to better your understanding of Canada as a country.

Where Is the North?

This is not a trivial question. It depends on who you ask and what criteria you use to base your answer on. If you base your response on bioclimatic conditions you may perhaps, define the North as a combination of two biomes: Arctic and Sub-arctic. The former would be bounded by the 10 °C July temperature isotherm, the latter, laying to the south, would reach the limit of the southern boundary of the boreal forest. Another way to trace the physical boundary of the North is to take the southern limits of the permafrost (permanently frozen ground). In any of these delineations, northern areas are seen as cold, harsh, remote lands with the presence of continuous or discontinuous permafrost.

Another way to define the North would be to focus on the economy. The northern economy is a mix of the modern wage economy and the more traditional subsistence economy. Transportation accessibility and population composition may also be used for this purpose. Louis Edmond Hamelin (1979) suggested the use of both physical and socio-economic characteristics to develop a 'nordicity index.' The index uses multiple parameters to identify the relative nordicity (northerness) of a given area. Several Canadian government agencies also use other multi-criteria methodologies to develop their own definitions of 'northernness' and the North in order to compute northern allowances to their workers (e.g., Isolated Posts and Government Housing Directive) or to process statistical data (e.g. Statistics Canada).

An entirely different approach is to define the North based on perceptions either by northerners themselves, or by other Canadians. It has been said that the North is not a region, but a 'state of mind.' For some, the North can be identified as a romanticized frontier that challenges the imagination of explorers and southerners in general. On the other hand, it is a homeland to Aboriginal Peoples, perceived in ways unique to their respective regional cultures.

None of these delineations can serve as a solid definition, but this diversity should be considered when we try to grasp the essence of the Canadian nordicity. In this chapter we use a political definition of the Canadian North, which is a matter of practicality. We will generally limit our discussion to three Canadian Territories: Yukon, Northwest Territories (NWT for short) and Nunavut (Figure 1). The territories form a special part of the Canadian federation as they have much less political authority compared to provinces, and the federal government exercises constitutional control over them. The Territories (as they are often referred to) are located north of the 60th parallel that forms their southern limit. Being very large in size, the Territories include both continental and island portions (the Canadian Arctic Archipelago). These vast lands constitute 40% of Canadian territory; however, they concentrate only 0.3% of Canada's total population. Most of these areas belong to the Arctic biome. However, as you read this chapter, it is important to remember that 'the Canadian North' extends beyond the three territories to include northern parts of Newfoundland and Labrador, Québec, Ontario, Manitoba, Saskatchewan, Alberta, and British Columbia, in other words the Provincial North. Therefore, many themes described in this chapter will be relevant to the entire North. While focusing on the Territorial North, we try to look at the broader picture of "the North," wherever possible.

The Physical Environment

When you think of the Canadian north, you probably picture a cold place, covered in snow and ice. While it can certainly be very cold, temperatures are not uniform across this vast expanse that is considered "the North", which in fact covers both Arctic and Sub-arctic biomes. While winters are long and dark, the short summers are brilliantly bright, blue, green, and multi-coloured as the ice melts and plants burst to life. To understand the unique the physical environment of the Canadian North (Figure 2) first you need to learn a little about the polar climate, and what characterizes the Arctic and Sub-arctic biomes. This will highlight some of the underlying factors as to why the north is, on average, the coldest region in Canada. Then you will learn about some of the prominent types of northern landforms, and the important role that ice – within the ground or on top of it – plays in defining the northern landscape. Following this is an explanation of seasonal freeze/thaw cycles and how these affect northern waters and oceans. All of these characteristics of climate, landscape, seasonal freeze-thaw, and water availability determine the kinds of vegetation and wildlife that can thrive in this unique environment. This provides important context to understand northern lands and waters as a homeland for Aboriginal peoples and other northerners, and thus how changes in such environments affect not only the natural environment but also the people who live there.

Polar Climate

In the classification of world climates, the polar climate is the coldest, and includes four sub-types (i.e. arctic, sub-arctic, mountain, and ice cap), all of which are found in northern Canada. The Arctic climate is distinguished by an average mean temperature for the warmest month of the year (i.e. July) as being less than 10 °C, meaning this region experiences long cold winters (can be an average of -30 °C or colder in January in many parts of the Arctic) and brief, cool summers. Also because the cold polar air masses cannot effectively absorb moisture from the cold bodies of water, the Arctic tends to be a dry region, with typically less than 300 mm of precipitation falling in a year, and most of that falling as snow. In the Sub-arctic, the summers are distinct and warm, although still short. The continental effect means that winter temperatures can be more extreme in certain inland areas than the Arctic, but the same applies to the warmth in the summer. In contrast, the maritime effect of open ocean along the Pacific or Atlantic coasts can moderate the winter temperatures, although typically the eastern arctic and sub-arctic regions are colder because of the cold ocean currents flowing south from the Arctic Ocean. In general, the position of the treeline reflects the division between Arctic (to the north) and Sub-arctic (to the south) climates, with the mountain (Cordillera) climate of northern British Columbia, Yukon, and the Northwest Territories bordering both Arctic and Sub-arctic climates to the West. The ice cap climate is associated with glaciers on Baffin, Devon, and Ellesmere islands, and so it is isolated within the Arctic climate to the East.

But why is it so cold in these northern regions, and why are there such extreme seasonal variations in temperature? To put it simply, there is a lack of solar energy being received in higher latitudes because of the tilt of the axis of the Earth and how it orbits around the sun. The Earth's axis is titled at 66.5°, and therefore throughout the year as the Earth moves around the sun the higher latitudes receive more daylight than lower latitudes in the summer (because the pole is tilted towards the sun). In the winter, these same latitudes will consequently receive less daylight (because the pole is tilted away from the sun). The Arctic Circle is thus not just an arbitrary line on the globe, its latitude of 66°33' N marks the location at which the sun remains above the horizon for one full day in the summer (on June 21, the summer solstice) – earning the Arctic the familiar label of the "land of the midnight sun." So, that means that in the winter months, the Arctic Circle is also the latitude at which the sun remains below the horizon for one full day in the winter (on December 21, the winter solstice). This means that the further north you go beyond the Arctic Circle, residents there experience more days of the year where the sun does not set as well as more days of the year where the sun does not rise above the horizon. Then, the further south you go, residents below the Arctic Circle will still have long summer days but will not enjoy the midnight sun, although this also means that they will still get to see the sun in the winter. To be clear though, for the towns that are located above the Arctic Circle, when the sun does not rise in the winter it does not necessarily translate into complete darkness. There are often several hours of dusk and twilight where beautiful hues of pinks, blues, and purples emerge as the light from the sun below the horizon is diffused by reflection off particles in the atmosphere, and off the snow-covered ground. Therefore, it is only when you are further north than 72°33' N latitude where polar night (periods of complete darkness) does occur. All this to say that because of the lack of sunlight for parts of the year, and the obliqueness of the sun's rays hitting the northern latitudes in the summer, the northern reaches of Canada do not get much of a chance to warm up. This phenomena is most pronounced in the spring, where although the sun is climbing higher in the sky each day, the snow and ice covering the ground (and the water) reflect most of the sun's energy back to the atmosphere without absorbing it (a process called the albedo effect), which means it takes longer for the air to warm close to the ground. Once the melting does begin, however, it progresses quickly because the ground and water absorb significantly more solar energy than snow and ice (darker, wetter surfaces tend to absorb energy more effectively than lighter, drier surfaces). However, with the global circulation system of atmospheric winds and ocean currents, there is some additional heat transfer from lower latitudes to higher latitudes, which helps moderate some of the build-up of cold polar air masses. This global energy transfer pattern is distributed unevenly though, with the higher latitudes of western North America and Europe receiving more air and ocean warmth than the eastern coast of North America (due to the counterclockwise rotation of the earth).

Northern Terrain

The terrain of the Canadian north as we know it today, from the Rocky Mountains in the West, to the vast expanses of tundra or plains in the central regions, to the mountains, glaciers, and fjords of the East were shaped by past geomorphic and glacial processes, the remnants of which are still visible today. Over the past 25,000 years (the latter part of the Wisconsin Ice Age) glaciation (the formation, movement, and recession of glaciers) was the prime factor in the development of current geomorphic regions and landforms. As the climate cooled, snow accumulated and compacted forming glaciers. These then evolved into massive ice sheets (notably the Laurentide and Cordillera) that were thousands of meters thick, and covered most of North America in the Late Pleistocene period. The importance of this glacial advance it that it created severe erosion, which led to the creation of new landforms (e.g. fjords, glacial fluting) as the massive thickness of ice carved through land and rock. Around 15,000 years ago the climate began warming again, causing these huge ice sheets to melt. So as the glaciers melted and receded, the material trapped in the ice was deposited on the ground, leaving behind a number of types of sorted till features (e.g. eskers, outwash plains, ground moraines, drumlins, or raised beaches). With the melting of the ice sheets, enormous quantities of melt water were also released, which flowed into existing river systems or formed glacial lakes, especially where north-flowing rivers were still dammed by the remaining ice sheet. This influence of historic ice sheets and glacial processes have resulted in the geomorphic regions we find in northern Canada today, including:

- the Canadian Shield (stretching from NWT to Labrador) – the largest geomorphic region in the Canadian North makes it the geologic core of the northern landscape (2.5 billion year old Precambrian rocks), with much of the exposed Shield areas being rough, rolling uplands, which increase eastward to Baffin Island where the strongest uplift has occurred, and where the majority of Canadian glaciers and ice caps are found
- the interior Plains (found between the Cordillera and the Canadian Shield) – as the name suggests, this region is made up of flat or gently rolling landscapes, with its sedimentary rocks housing vast oil and gas deposits (500-100 million year old Cretaceous-age rocks), and the mighty Mackenzie River and its tributaries having been created in the last glacial advance
- the Cordillera (along the West coast in much of British Columbia, Yukon, and a smaller portion of southern Alberta and the NWT) –

this region is complex and mountainous, including majestic mountains, plateaus, valleys, plains, alpine glaciers, and ice fields

- the Hudson Bay Lowland (around the SW portion of Hudson Bay) – is a low, flat coastal plain that has recently emerged from marine sediment deposition and reworked glacial till, it is also an area that is still experiencing isostatic rebound (the land is slowly rising after weight of the ice age ice sheets has lifted).
- the Arctic Lands (northernmost region, mainly in the collection of islands that make up the Canadian Archipelago) – this region has the coldest and driest climate, and it is a complex area where geologic events and geomorphic processes have created a combination of lowlands, hilly terrain, and mountainous regions where the landscape is dominated by periglacial features (widespread scattered bedrock and patterned ground), and where the eastern islands of Ellesmere, Devon, and Baffin are home to most of Canada's glaciers (the origin of most of the icebergs that flow southward with the Labrador Current).

Permafrost

On seasonal time scales, permafrost is most influential in terms of shaping northern lands, as well as determining vegetation distribution, surface drainage, and a variety of landscape features across the Canadian North. As the name would suggest, permafrost is ground that remains at or below the freezing point for at least 2 years (i.e. permanently frozen). However, the depth of this frozen ground can extend for several hundred meters in the High Arctic, or it can be less than 10 meters deep in more southerly locations in the Sub-arctic. As with climate and precipitation, permafrost distribution trends across northern Canada are also uneven, but typically there is a correlation to the mean annual air temperature, resulting in four main types:

- continuous permafrost (over 80% permanently frozen ground, mean annual temperature around -7 °C);
- discontinuous permafrost (30-80% permanently frozen, mean temperature -5 to -7 °C);
- sporadic permafrost (less than 30% permanently frozen, a transition zone where the mean temperature is around 0 to -5 °C); and,
- alpine permafrost (permanently frozen ground in mountainous regions, in other words at high elevations not necessarily high latitudes, found in British Columbia and Alberta).

The mean annual air temperature isotherm at the freezing point is essentially the southern limit of where permafrost (sporadic as it is) may be found. While these are the general rules of permafrost distribution, local permafrost variations also occur in conjunction with: latitude, altitude, surface capability to absorb heat, vegetation cover, snowpack conditions, topography (i.e. elevation and slope/aspect), and drainage patterns.

Although the description of permafrost would suggest that the entire ground surface and subsurface is permanently frozen, there is always a surface layer that thaws each summer after the long summer days arrive and the snow has melted. This "active layer" can also range in depth from several centimetres (in the Arctic) to several metres (in the Subarctic), and it is within this layer that plant roots are confined. Therefore, while ground temperature in the active layer would be above freezing in the short summer, the overall mean temperature is below freezing, and thus still considered part of the permafrost. Also, unfrozen ground – called talik – can be found within the permafrost. The occurrence of talik is generally related to the influence of warmer temperatures from above, so its location and depth is tied to the current or former presence of insulation from of various sizes of water bodies. The presence of permafrost in northern regions also contributes to the formation of a unique aspect of northern landscapes: thermokarst.

Thermokarst landforms are produced by the thawing of the permafrost itself, when the ground loses its solidity and the overlying soil surface collapses. This can result in thaw slumps (e.g. slumping along coastal cliffs, lake shores, river banks when undercut by waves or currents) due to enhanced melting, or from human activities that disturb the ground surface that insulates the permafrost below. Thaw lakes can also form due to surface erosion from permafrost melting, creating an array of elliptical lakes/ponds that dot the expanses of northern tundra (such as on Boothia Peninsula, see Figure 2). Therefore, contrary to the metaphoric notion of the "frozen tundra", arctic and Sub-arctic landscapes are actually in a continual process of change and evolution, with each season of freezing and thawing.

Northern Waters

In the Canadian north, water and heat fluxes are closely linked, since snow, ice, and frozen ground essentially determine the storage and circulation of water. Specifically, snow is an important form of water storage, and plays an important role in the northern hydrologic cycle as it holds 6-9 months of precipitation accumulation and then releases it quickly over several weeks during the spring melt period. Therefore, the hydrologic phases of northern Canada are broken into two distinct portions of the hydrologic cycle. The inactive phase dominates the seasonal round, when water flow is well within the capacity of streams and rivers (or is non-existent, in a frozen state), during most of the summer, fall and winter. The active phase is short but dramatic (in speed, sound and magnitude – ice break-up is very loud and tumultuous process!), and is a characteristic indication of the arrival of spring. This phase is sparked by the increasingly long days of sunlight in the spring. Once the snow pack starts to melt and fragment, winter temperatures are quickly broken and the ensuing warmth melts lake, river, and sea ice quickly. The waterways burst open with the rushing release of water causing streams and rivers to quickly reach their peak seasonal runoff. With this massive influx of water, seasonal flooding occurs as ice pans get jammed between river banks and cause the water to back up. The water from the thawing snowpack and frozen lakes thus runs off into melt streams, tributaries, and then larger rivers, which ultimately end up contributing to the river flow in two of the largest drainage basins in Canada, the Arctic and Hudson Bay drainage basins. The largest rivers in these basins are the Mackenzie, Churchill, and LaGrande Rivers, which empty vast quantities of fresh water into the oceans (creating important estuarial habitats).

Closely following the active phase of the hydrologic cycle, the ground warms and plants respond rapidly by appearing and flowering shortly after, transforming the landscape from white, to brown, to spectacularly green and colourful, over the course of a week or two. Nevertheless, things begin to cool down again within a month or so (usually by the end of August), and water levels return to manageable flows as they return to the inactive phase, followed later by ice formation once again in the fall.

Northern Seas

The marine environment is a critical component of northern life, for humans and wildlife. The cold seas of the Canadian north extend from the Beaufort Sea in the northwest to the Labrador Sea in the east, and include the nearly land-locked water body of Hudson Bay. Although the northern seas are important for coastal life, they have limited impact on the hydrologic cycle due to their low evaporation rates as a result of their cold temperatures during the open water season. In fact, since they are ice-covered for the majority of the year they act more like cold land bodies in terms of their climatic characteristics. In the Arctic Ocean, there is also permanent ice cover known as polar pack ice (i.e. consolidated ice cover which has not melted in the summer for at least two consecutive years). Both the seasonal and permanent ice cover on the ocean are important contributors to climate regulation (with their high surface albedo they maintain the cool polar temperatures). And, the annual formation of sea ice itself drives thermohaline circulation (a key control of global ocean circulation patterns), due to the pattern of convection created during freezing – as the surface water sinks (with higher density at cooler temperatures) it is replaced by warmer water from below, until the surface layer of the ocean is cooled to the freezing point (which is below zero for salt water, in other words the higher the salt content the lower the freezing point). The freeze-up process can begin as early as mid-September in the Beaufort Sea, as late as December/January in Baffin Bay or the Labrador Sea. Consequently, break-up arrives earlier in the eastern Arctic than in the Western Arctic, and within the Archipelago the Arctic Ocean can remain ice-locked all summer. This sea ice environment is relatively stable where it is attached to the land (i.e. landfast), but is also very dynamic, being influenced by winds, tides, and currents (depending on bathymetry, distance from shore, or time of year). Especially in areas of strong currents, or where there is an upwelling of warm water, polynyas can occur (i.e. areas of open water surrounded by sea ice), which are known to be productive "oases" in Arctic waters as access to air and nutrients is abundant in these localized openings which draws numerous animal, bird, fish, and even microscopic species.

Northern Ecozones

Taken together, the climatic, solar radiation, geologic, permafrost, hydrologic, and marine characteristics combine to form several uniquely northern Canadian marine and terrestrial ecozones (i.e. broad ecological zones representing unique ecosystems as characterized by their climate, vegetation, geology, landforms, human settlement and use, wildlife, etc.) within the Arctic and Sub-arctic biomes (large, continental ecosystem units with distinctive climate and soil conditions) (Figure 3). Some of the main characteristics about Canada's northern biomes have already been described above, but as a means of summary Table 1 helps to reiterate those ecosystem components.

The cold lands, waters, and seas of northern Canada are not the icy wastelands that tend to live in peoples' imaginations. The previous sections highlight how climatic and surface conditions create dramatic seasonal shifts in environmental conditions, which have served to create unique ecozones. These have resulted in distinct, regionally variable terrestrial, vegetation, freshwater, and marine conditions thus create particular habitats that support a variety of plant, animal, and bird species.

Table 1. Ecosystem Characteristics for Arctic and Sub-arctic biomes (adapted from Bone (2009)

Charac-teristic	Arctic Biome	Sub-arctic Biome
geographic extent	Nunavut, NWT and northern parts of Yukon, Manitoba, Ontario, Québec, and Labrador	largest natural region in North America, extending from the Rocky Mountains to Labrador) NW-SE trending zone (due to continental effect leads to warmer interior lands and the cold waters of Hudson Bay and Labrador Strait preventing tree growth along E coast)
tempera-ture	coldest in Canada (warmest month has mean temperature of 10 °C)	warmest month has mean temperature of > 10 °C
vegetation	cool summers and permafrost control the type of natural vegeta-tion and soil development tundra is the primary vegetation zone (normal tree growth and soil formation not possible due to temperatures) few plant species survive in such harsh conditions, but those that do are highly adaptable to large seasonal fluctuations in tempera-ture, sunlight levels, and water availability	vegetation demarcates general sub-arctic boundary – i.e. the treeline – the transitional zone between boreal forest and tundra predominantly boreal forest, but range of variations in size and density of forest cover
soil	thin and immature cryosolic soils (formed where there is continuous permafrost, and a shallow active layer)	podzolic (thin, acidic soils formed in cool, wet growing conditions) and gleysolic (formed in waterlogged land such as marches and bogs) soils, associated with discontinuous and sporadic permafrost

Although biological productivity and diversity in northern regions is low compared to warmer climates (with low temperatures, limited precipitation, and short growing seasons being the main inhibiting factors), there is still an array of species that thrive in such an environment, because of their physiological adaptations to the seasonal extremes experienced in high latitudes. More importantly to note though, is that these northern lands, waters, and seas are not only home to uniquely northern plants and animals, but these relatively remote regions of Canada have been the homelands of Aboriginal peoples for thousands of years – since "time immemorial" as many like to say. What seems to be a harsh and unforgiving environment to many of those who have never visited, or lived in, northern regions, is a homeland for northern Aboriginal peoples, which have been sustaining their lives and cultures for many generations.

History of the Canadian North

Pre-contact, Exploration and Fur Trade

Descendants of the Paleo-Indians and Paleo-Eskimos – who are estimated to have crossed the Beringia Land Bridge from Asia into Alaska and the Yukon Territory approximately 20,000 and 9,000 years ago, respectively – are modern day First Nations peoples and Inuit who remain the majority population of Canada's North. The Canadian Arctic has seen several waves of settlement, but the record is very scarce. The last and the most documented wave occurred around 1000 AD, when the Thule people (the ancestors of modern Inuit) replaced the Dorset culture. Inuit and Athapaskan (Dene) tribes occupied the North at the time of the contact with Europeans. Both were primarily hunters: Inuit settled in the coastal areas and used qajaks (typical English spelling is 'kayaks') to hunt seals, whales and other marine mammals; Dene fished and hunted caribou in northern boreal forests. Both groups maintained seasonal nomadism (movement from one place to another) as they followed the migratory patterns of animals.

While it is believed that the first Europeans to establish contact with Indigenous people in the North were Vikings (around 1000 AD), the first well documented contact occurred in 1576, when Martin Frobisher's English expedition encountered Inuit on Baffin Island. Similarly to Christopher Columbus, Frobisher and other explorers were not looking to discover North America, but searching for a new route to the Orient across the Arctic Ocean (the elusive Northwest Passage). Major discoveries in the 17th century were made by Henry Hudson (including the strait and the bay bearing his name). At that time, the contacts between Inuit and European expeditions in the High Arctic were limited to occasional trade. The primary encounters were with British, Dutch and German whalers who began to hunt in the Baffin Bay and Davis Strait in the late 16th century. The search for the Northwest Passage continued with many British expeditions.

In the early 19th century John Ross (1817) and William Parry (1819) led expeditions that advanced farther north than anyone before, but encountered extremely difficult navigation conditions. The British Admiralty soon realized that the frozen Northwest Passage would be of little commercial value. The last expedition of the 'heroic' age was the one led by John Franklin in 1845. When Franklin's crew disappeared, a massive rescue effort was initiated both by land and by sea. Several expeditions were sent between 1848 and 1857 to look for Franklin's party, and confirmed their death. However, this multiyear rescue operation resulted in new discoveries and advancement of knowledge about Arctic waters and islands.

Ross and Parry traveling north and west along the Northwest Passage discovered new whaling grounds. Pressured by depleting whale population in the usual hunting areas, European whalers advanced north, and, thus, further from their home bases. As a result, some whaling expeditions instead of returning home at the wake of winter, 'wintered over' in the Arctic. Permanent stations appeared along the shores of Baffin Island, Hudson Bay, and northern Québec. Here, whalers engaged in active trade with Inuit: Inuit supplied game, sewed clothes, and assisted in navigation, in return they received rifles, knives and other European goods useful in hunting and domestic life. Both commercial whaling and trade were sharply aborted in the early 20th century, when whale prices dropped.

The fur trade flourished in the Canadian boreal and Sub-arctic zones for centuries. Fur traders, such as David Thompson and Alexander Mackenzie, made a significant contribution in exploring inner lands of the Canadian North. However, only in the beginning of the 20th century did the Hudson's Bay Company establish trading posts farther north to gain access to the supply of the Arctic fox pelts. Similarly to other areas, the company built its relationships with Inuit based on barter trade (fox pelts were traded for goods). The fur trade was the dominant activity in the Canadian Arctic until the 1950s, and many Aboriginal households were dependent on it in terms of their hunting gear and home supplies. Overexploitation decreased the number of fur animals and game around trading posts, thus jeopardizing not only trade, but also subsistence of the Aboriginal peoples. As the demand was also in decline, the fur trade diminished.

'Resource Rushes' and Colonization

While pursuits of whale and furs could be considered first "resource rushes" in the Canadian Arctic, they are dwarfed by the magnitude of the infamous Klondike Gold Rush. In 1896 gold was discovered near the Klondike River in Yukon. In the following few years thousands of prospectors rushed to the area to pan for placer gold. A little outpost town of Dawson City, where most prospectors converged, became a city of 40,000 in one year. However, the placer gold was gone in two years and prospectors moved on to Alaska (down the Yukon River). The population of Dawson City plummeted to 8,000 in 1899. The Klondike Gold Rush, albeit short-lived, paved the way to prospecting for gold and other metals across the North. It also prompted the Canadian government to establish formal control over the Arctic frontiers by bringing in the Royal Canadian Mounted Police forces.

Since the middle of the 19th century Anglican and Catholic missionaries started to appear in various parts of the Canadian Arctic. They alongside with the Canadian officials (including Police) imposed European religion, values and laws over Aboriginal societies. As in many other colonies, the British believed in their 'civilizing' mission that was to deliver 'modernization' to 'backward' indigenous cultures. One of the instruments of such 'modernization' was assimilation. An organized assimilation in the Canadian North is associated with residential schools, where missionaries placed Aboriginal children and taught them in English or French. Taken away from parents, children often lost links with their native communities and were no longer able to maintain indigenous way of life. While formal education was provided, this practice was destructive to indigenous cultures and abusive to Aboriginal peoples. Most residential schools were eventually closed in the 1960s.

Other important developments in the Canadian North were associated with military affairs. In the mid-20th century the Arctic suddenly became a battlefield: first, in World War II with the Japanese threat to Alaska, and then in the Cold War with the nuclear threat from the Soviet Union. During World War II the most dramatic changes took place in Yukon with the construction of the Alaska Highway. However, the Arctic became a true battleground shortly World War II ended: the route between USSR and USA across the Arctic is the shortest, and, thus, was the likely path of a Soviet bomber attack. In response, US and Canada built the Distant Early Warning (DEW) Line, a system of 63 radar stations along the 69th parallel. The construction of the system employed 25,000 workers. Many stations were manned by Canadian and US military personnel. Although no longer strategically important (as bombers have been replaced by intercontinental missiles), the system with a reduced number of stations still exists largely to symbolize Canadian sovereignty in the Arctic.

Modern History: Mega-projects, Land Claims and Aboriginal Rights

The 1950s were marked by the new era of resource exploration in the North. An increasing demand for minerals, forest resources and energy from the USA, and around the world, fuelled the new development boom. The "Northern Vision" policy coined by John Diefenbaker called for expanding the use of natural riches of the Arctic by attracting private capital and using public funds. Under the "Roads to Resources" program, the Canadian government constructed miles of roads that connected resource bases in the North with markets in the south. The most ambitious infrastructure project in the Arctic was the Dempster Highway: started in 1959 and finished in 1979, it linked Dawson City in Yukon to Inuvik in NWT.

The government also began to assist corporations (foreign and domestic) in large resource-extraction projects ('mega-projects') by providing incentives and assuming some costs and risks of these endeavors. Mega-projects that included major mining and oil extraction operations, hydropower and forestry projects, have changed both natural and economic landscapes of the North by bringing new development, new people, new settlements and considerable environmental destruction. Mega-projects alongside the expansion of social assistance prompted northern populations to move to towns as well as smaller settlements. Many newcomers (workers from the south) arrived to work on mega-projects. Entirely new towns appeared to house newly arrived northerners, so-called 'instant communities' or 'resource towns.' Attracted by public housing and other welfare benefits, and coerced or forced by government relocation programs, Aboriginal Peoples also concentrated in settlements breaking away from centuries of nomadic life style. In contrast to company towns, these settlements rarely had an economic base or sufficient employment opportunities, so that the integration of Aboriginal residents into the wage economy was difficult. At the same time, hunting now clustered around communities led to the depletion of local wildlife resources. As a result, many Aboriginal communities became heavily reliant on social payments and government transfers to make ends meet. The expanded scope and geographical reach of resource extraction, as well as the rise of Aboriginal Organizations in the 1970s, raised questions and concerns for the social and environmental impacts of developments highlighted above. The Canadian Constitution (1982) recognizes Aboriginal rights associated with original land use, occupancy and self-determination, although it does not legally define them. In the 19th and 20th centuries, Aboriginal peoples and British (and later Canadian) governments signed numerous treaties. Under these treaties, First Nations tribes surrendered their rights to lands to the Crown (the government) in exchange for payments and permission to use unoccupied lands. However, in accordance with the Indian Act a band dissatisfied with the terms of existing treaty can file a specific land claim. In contrast, those bands who did not sign historical treaties, and Inuit (who are not included within Indian Act policies), can negotiate a comprehensive land claim agreement (CLCA) with the government. These agreements are comprehensive as they deal with various issues, including land use rights, political autonomy, compensation payments, subsurface rights, cultural preservation, etc. The idea behind CLCAs is that Aboriginal peoples agree to extinguish their territorial rights that are based on Aboriginal rights in exchange for title to some of the lands and one-time monetary compensation. Most CLCAs, provide additional benefits related to land use, environmental management, employment, and Aboriginal governance.

The recognition of Aboriginal title came after the Supreme Court of Canada decision in 1973 that ended decades of neglect by the government and mining companies. The negotiation of Aboriginal land claims in the Territorial North started in the 1970s with Yukon First Nations, Dene/Métis (split into five separate land claims) and Inuit (Inuvialuit and Nunavut claims). Within the three Territories, after decades of negotiations five agreements with the federal government have been reached (Table 2, Figure 4).

Table 2. Final Comprehensive Land Claim Agreements in the Territorial North

Final Agreement	Date	Territory
Inuvialuit Final Agreement	1984	Northwest
Gwich'in Final Agreement	1992	Northwest
Tungavik Federation of Nunavut Final Agreement	1993	Nunavut
Sahtu/Metis Final Agreement	1993	Northwest
Yukon Indians' Umbrella Final Agreement	1993	Yukon
Tlicho Land Claims and Self-government Agreement	2003	Northwest

The first CLCA in the Territorial North was the Inuvialuit Final Agreement (1984) in Western Arctic. The agreement provided C$45 Million in financial compensation, land use rights, self-governance, measures to protect wildlife, environment and traditional life style. In exchange the Inuvialuit claim has been extinguished. As a result of the agreement the Inuvialuit were able to become more economically integrated into the Canadian society and to spur local business development through the Inuvialuit Development Corporation. Other CLCAs (or modern treaties as they are sometimes referred to) went through similar negotiation processes as the Inuvialuit. The settlement of the CLCA with the Inuit of the eastern Arctic (signed in 1993) resulted in the creation of the newest Canadian territory – Nunavut. The Tlicho Agreement (signed in 2003) is the first agreement in NWT that combined land claim and self-government provisions, giving unique opportunities for Aboriginal self-determination.

Governance and Politics

Territorial Governance

When Canada was formed in 1867 its Arctic lands were still in British possession. The acquisition of the Territorial North started with the purchase of Rupert's Land and the North-Western Territory from the Hudson's bay Company in 1870 and was not finalized until British transferred the Arctic Archipelago to Canada in 1880. Initially, these new lands – then known as the North-West Territories – were administered directly from the Nation's Capital, Ottawa. However, growing population on the wake of the Klondike Gold Rush prompted the creation of the Yukon Territory: in 1898 Yukon became a separate territory. Still, the government was led by a commissioner appointed in Ottawa. The southern boundary of the Northwest Territories took shape when Saskatchewan and Alberta (in 1905) and Manitoba (in 1912) were assigned lands south of the 60th parallel. In the same year the name of the territory was changed to Northwest Territories (NWT). In 1999, in fulfillment of the Nunavut Land Claims Agreement, the eastern part of NWT was separated into a new territory – Nunavut (Figure 1).

NWT was under the direct control of Ottawa as it was administered by the Commissioner and the council appointed and seated in the capital. In 1967 the seat of the Territorial government was moved to Yellowknife (and remains there since then). Gradually, the council (the Legislative Assembly) has become an elected body comprised of the representatives of the people of the Territory. The Cabinet (the Executive Council) is now headed by the Premier and includes ministers, all elected by the Legislative Assembly. The Commissioner no longer has administrative functions, but serves a symbolic role, similar to provincial lieutenant-governors.

The recent history of NWT illustrates a progressing devolution, a process of acquiring more self-governance and autonomy from the federal center. Devolution is associated with a growing strength of the territory (in terms of population, economy and political clout) and with an increasing willingness of the federal government to delegate its administrative functions to territorial authorities. Notably, the territorial powers (in contrast to provincial) do not stem from the Canadian constitution: rather, they are assigned (delegated) to the territorial government by Ottawa. In other words, the territories are inferior in their constitutional status to Canadian provinces. The territorial government in NWT (as well as in Yukon and Nunavut) has authority over education, taxation, community and social services, transportation, and wildlife, among others. However, the territories do not have access to tax revenues from their natural resources, which are controlled by Ottawa. Lacking a major revenue stream, the territorial governments are dependent on transfer payments from the federal government, which fund 60-90% of territorial budgets.

The political system in Yukon is similar to those in provinces and is party-based. In contrast, NWT and Nunavut do not have political parties and operate under a unique system of consensus government. This non-partisan system is inspired by Aboriginal governance traditions based on consensus. Members of the Legislative Assembly (MLAs) are independents elected from districts by a simple plurality voting. The MLAs elect the Premier and the Cabinet members (Executive Council), none of whom has a party affiliation. The consensus principle insists that all legislators caucus together, with no separation into the opposition and a ruling party.

Political Change, Fate Control and Aboriginal Self-determination

The creation of Nunavut and settlement of comprehensive land claims are the signs of a political sea of change in the North. Devolution and establishment of Aboriginal self-government are transforming the political landscape of the region. Overall, one can say that northerners, especially Aboriginal Peoples, witnessed dramatic gains in 'fate control,' i.e. their ability to determine their own destiny. Fate control is a multifaceted concept that includes personal and collective empowerment of people by enabling them to make and implement their own decisions. This includes a decision-making authority in political, economic, and civic affairs and a capability (legal, financial, etc.) to implement these decisions. There is no doubt that in the last forty years northerners gained in all aspects of fate control: acquiring more political and governance powers (both in the North and in the Canadian federation), receiving more control over natural resources, lands and resource exploitation, improving access to financial resources and economic development tools, and defending the right to preserve their cultures and traditions.

As mentioned earlier, comprehensive land claim agreements (CLCA) are now settled in most of the territories: in Nunavut, Yukon and in Inuvialuit, Sahtu, and Gwich'in regions of NWT (Figure 4). There are a few outstanding claims in central and northern NWT (Deh Cho, Akaitcho and South Slave) and southern Yukon With CLCAs Aboriginal peoples were able to get control over land use planning and development, wildlife management, property taxation, and natural resource management, among others. They also gained access to payments from the federal government that are used to ensure economic prosperity and development of the Aboriginal communities. As result, various forms of self-government appropriate for their cultures have been established.

A growing fate control is also associated with devolution of federal authority to territorial governments started a half-century ago. However, this process in not complete: there are demands, especially in Yukon, to grant the Territories a provincial status. Albeit this is unlikely to happen in the near future, this is a serious issue that may transform Canadian federalism. Another emerging political issue in the North is the relationship between public (serving all northerners) and Aboriginal governments. In NWT and Yukon they are clearly separated, but Nunavut represents a more complicated case where territorial and Aboriginal governance structures are highly intertwined with the public government serving an 85% Inuit population.

A prominent process in the North is an increasing self-determination of Aboriginal people. The principle of self-determination is fundamental for understanding rights of Aboriginal people to govern themselves, according to their beliefs and traditions. In the North there are two distinct types of self-governance arrangements: implementation of self-governance through public government (e.g., Nunavut) and Aboriginal self-government (e.g., Yukon). Whereas public government serves all people of its jurisdiction, regardless of their ethnicity, in areas with strong Aboriginal majorities such government usually incorporates many of the traditional governance practices. In areas with permanent Aboriginal minorities the Aboriginal self-governments usually exist in parallel with the public government (that is elected by all residents) to exclusively serve Aboriginal people. These structures have been established to ensure a greater degree of autonomy for Aboriginal minorities. For example, the self-government agreements involving the Yukon First Nations include law-making authority in managing internal affairs and lands and the power to make decisions in accordance with indigenous values and institutions. A third form of self-governance comes with settling comprehensive land claim agreements. CLCAs often create special structures responsible for fulfilling the terms of the agreement. For example, in Nunavut, the lands and financial assets received as a result of CLCA are monitored by an organization elected by Inuit, Nunavut Tunngavik Inc. Representing the vast majority of population, this organization exerts a significant political power in Nunavut.

Nunavut

Nunavut has been carved out of NWT to fulfill the land claim agreement between the Inuit of the Eastern Arctic and the federal government. As such, Nunavut is the newest territory in Canada created on April 1, 1999. Nunavut means 'our land' in Inuktitut, and Inuit constitute 85% of its population. The new territory is the largest administrative unit in Canada. The consensus-based Nunavut government and the Inuit land claim organization (Nunavut Tunngavik Inc.) work together to ensure cultural vitality and economic prosperity of Inuit in the territory, as well as to ensure that territorial governance agrees with principles of Inuit living. According to the Nunavut Land Claims Agreement and the Nunavut Act, the territorial government exercises control over land use planning and development, wildlife management, property taxation, and natural resource management, among others. The land claims agreement provided Nunavut with a one-time payment of $1.2 billion (managed by Nunavut Tunngavik Inc.) and a share of government royalties from Nunavut natural resources. The creation of Nunavut is a fascinating example of devolution and innovative self-governance in the Arctic. However, Nunavut also faces a number of challenges in order to reduce its dependence on federal transfer payments and meet its goals of 80% Inuit representation in the civil service, to match the numbers of Inuit population in the territory.

Northern Peoples and Cultures

Aboriginal Cultures

As has been alluded throughout the descriptions of the physical environment and the historical context, the lands, waters, and seas have been used and occupied by northern Aboriginal peoples for centuries. Through this long term engagement and close interactions with northern lands and wildlife, northern Aboriginal peoples have acquired detailed and sophisticated knowledge of their homelands in order to sustain themselves, create shelter, provide heat and light, make clothing, travel, and meet the challenges of daily life in a semi-nomadic society (historically, and traditionally).

For the purposes of clarity and consistency, terminology used in this chapter reflects the standards used by Indian and Northern Affairs Canada (INAC) (www.ainc-inac.gc.ca), the Canadian federal government department responsible for Aboriginal and northern affairs in Canada:

- Aboriginal peoples – The descendants of the original inhabitants of North America. The Canadian Constitution recognizes three groups of Aboriginal people – Indians (First Nations), Métis, and Inuit. These are three separate peoples with unique heritages, languages, cultural practices, and spiritual beliefs.
- Indians (First Nations) – The term "Indian" is still used in Canada within government policies and legislation, in relation to the designa-

tion of: Status Indians, non-Status Indians, and Treaty Indians (all related to the Indian Act and the negotiation of Treaties and Reserve lands). However, the term First Nations came into common usage in the 1970s to replace the word "Indian", which is generally considered offensive. Although the term First Nations is widely used, it has no legal definition. Within Canada, the expression 'First Nations peoples' refers to people either with or without "Status". In addition, some Aboriginal groups have adopted the term "First Nation" to replace the word "band" in the name of their community. Despite the legal implication of the term Indian, we use the term First Nation throughout to refer to this group of Aboriginal People, as more respectful and accepted terminology.

- Métis – People of mixed First Nation and European ancestry who identify themselves as Métis, as distinct from First Nations people, Inuit, or non-Aboriginal people. The Métis have a unique culture that draws on their diverse ancestral origins (e.g. Scottish, French, Cree, and Ojibway).
- Inuit – An Aboriginal people in northern Canada, who live in Nunavut, Northwest Territories (Inuvialuit Settlement Region), Northern Québec (Nunavik), and Northern Labrador (Nunatsiavut). The word means "people" (the singular form of Inuit is Inuk) in the Inuit language (Inuktitut).

Traditionally, the regions of the Canadian North were populated solely by Aboriginal peoples. Generally speaking, Inuit occupied the lands (and waters/ice) beyond the treeline in the arctic tundra, while First Nations peoples occupied the boreal zone of the Sub-arctic. Both groups used the vast expanses of land and water as their sustenance, with semi-nomadic lifestyles reflecting seasonal patterns of wildlife availability and land/ice use. Because of the low carrying capacity of the northern environment, people tended to travel and live in small, spread out groups based on family relations and kinship ties, for the majority of the year, with special celebrations or seasonal activities facilitating larger congregations. As Inuit describe in relation to their traditional lifestyles, they rarely spent more than a few years at any given seasonal camp because it was important to "let the land cool down" (i.e. to let the land and/or wildlife recover from their influences). Developing an understanding of the diversity and richness of Aboriginal cultures in northern Canada (not to mention the rest of Canada) could easily fill a chapter – or a book – unto itself. It is thus impossible to avoid generalizing in the context of this short section, and we can only provide a very general overview of the three broad groups of Aboriginal peoples who are the original inhabitants of the Canadian Territories: Inuit, Athapaskan (Dene First Nations), and Métis.

Inuit have occupied the arctic regions from Alaska to Greenland (including Canada), and along the Chukchi Peninsula of Russia, for approximately 5,000 years. They were primarily marine peoples, with the exception of only a few smaller inland-dwelling groups. Therefore, the sea provided the most important resources for Inuit, who relied on ringed and bearded seals as primary sustenance, as well as walrus, narwhal and beluga, along with caribou and arctic char inland. As such, their lives were highly tied to seasonal cycles of environmental change and wildlife migration, in particular the seasonal formation of sea ice and snow cover to enable effective travel by dog team and qamutik (sled). Cultural traditions, beliefs, and knowledge were passed on through oral history, and social networks were developed typically within extended family groups as well as through practices such as food sharing and observance of strict taboos. Their traditional language Inuktitut is still widely spoken today, and is the strongest Aboriginal language in Canada in terms of use and proficiency in both adults and children. Various regional cultural groups developed regionally specific practices and dialects of Inuktitut, which to a very generalized extent are now reflected in the various Inuit groups across Canada – all of which have also successfully negotiated comprehensive land claims with the Canadian Government. To learn more, visit the website of the national Inuit organization Inuit Tapiriit Kanatami (www.itk.ca).

Athapaskan peoples have occupied the western Sub-arctic regions of Alaska and Canada for approximately 15,000 years. With many regional cultural and linguistic groups defined within the broad grouping of Athapaskan, their traditional territories spanned present-day Yukon and the NWT, but also likely expanded into northern BC and Nunavut. Like the Inuit, traditionally a hunting culture, their seasonal rounds were tied to extreme seasonal variations and the movements of caribou herds, within or at the edge of, the boreal forest (although some did also travel into the tundra to hunt barren-land caribou). Cultural values and beliefs were also passed on orally through elders of the society, in the form of stories and legends. In the Yukon there are eight language groups generally within two major language families: Athapaskan and inland Tlingit. In the NWT, the Dene (as they prefer to be called) speak variations of Chipewyan or Slavey, among other more localized languages, also part of the Athapaskan language family. To learn more, visit the Yukon (www.cyfn.ca) and NWT (www.denenation.com) organizations representing Athapaskan peoples.

Métis peoples are usually discussed in the context of Atlantic or French Canada; however, there was also a distinct group of Métis who developed from engagements with Dene and Cree First Nations peoples and English and Scottish Hudson's Bay Company employees as they expanded northwards with the fur trade. There was also northward expansion of Métis peoples from the Canadian Prairies. They have worked together with other northern First Nations peoples to settle joint land claims agreements, particularly in the NWT (Table 1). To learn more, visit the Métis National Council (www.metisnation.ca), and also search for individual provincial or territorial Métis organizations.

While today's life in northern regions is centralized in communities with homes, infrastructure, and services, not too unlike what we enjoy in southern Canada, it is important to recognize, and learn from Aboriginal peoples who continue to maintain a strong connection to the land and waters of northern Canada alongside global technological, economic, political, and social connections. Therefore, the North is still very much a homeland, and Aboriginal peoples continue to remind us that we are not separate from the environment, but part of it.

Ethnic Diversity

While the North still has a strong Aboriginal population, with nearly 50% Aboriginal population in the Territories, migration of southern Canadians and immigration of others from foreign nations, began to grow mainly through the 1950s-1980s. So, today's northern population is increasingly multicultural. For Aboriginal peoples, the Canadian North has been a homeland for generations. For outsiders, the draw to the north tended to coincide with three influential historical factors: i) economic opportunities; ii) sovereignty and government services/control; and iii) exploration and adventure.

More recently, important northern attractions for non-Aboriginal people include the high-paying wages of northern employment, cost-of-living allowances, subsidized housing, and prospects for rapid advancement. However, despite these substantial attractions, the influx of people to the north often rapidly turns over due to challenges with lifestyle transitions, job uncertainty, distance to family, pursuit of education opportunities, or achieving their financial goals. Due to the high proportion of people who only remain in the North for 1-5 years after arriving, there is a tendency for migrants to be labeled as "temporary residents". Such trends are especially typical for larger regional centres, which form the hubs of government and economic activities (and thus employment opportunities). However, the effects of this high turnover are felt hardest in smaller, more remote communities, where there are more direct interpersonal effects and where impacts on local services are greater. Therefore, people from all over the world can be found living in the Canadian North, but Aboriginal peoples continue to make up a high percentage of the northern population, and with an increasingly young population (e.g. 39% of Inuit are under the age of 15), this reality is unlikely to change despite fluctuations in in-migration.

Demography and Settlement

Although Canada's northern regions still have small and relatively sparse populations in comparison to more southerly locations, northern population has been slowly and steadily increasing since the early 1900s, both in numbers and in diversity. Just above 107,800 people lived in the Territories in 2008: 43,300 in Northwest Territories, 33,100 in Yukon and 31,400 in Nunavut. Given the large area of the Canadian North, the region is very sparsely populated with population density about 0.02 persons per square kilometre. Three territories constitute only 0.3% of total Canadian population. This population is also unevenly distributed. Almost half of northerners live in three capital cities: Whitehorse, capital of Yukon (20,416) Yellowknife, capital of NWT (18,700) (Figure 5), and Iqaluit, capital of Nunavut (6,184), with close to 60% living in the ten largest urban settlements (Table 3). The rest reside in much smaller communities (from a few thousand to just a few dozen people) scattered across the North (Figure 1). Very few Aboriginal peoples still maintain a nomadic way of life – most now live in settlements.

A very important characteristic of northern population is the presence of two distinct population groups (Aboriginal and non-Aboriginal), with relatively equal numbers if considered across the North. Aboriginal population dominates in Nunavut (85%) and constitutes a half of NWT total population. However, non-Aboriginal population dominates in Yukon (75% vs. 25% Aboriginal). Indeed, non-Aboriginal northerners are the descendants of settlers who arrived to the North in the past decades, or are in-migrants themselves. They predominantly concentrate in large towns, especially territorial capitals, where they find wage employment. In contrast, Aboriginal people tend to live in smaller communities. The population of the North demonstrates several distinct demographic characteristics. First, it has young age structure and male predominance – and, thus, is very different from the Canadian population structure. The youthfulness of northern population is explained by high birth rates and influx of migrants, who tend to be younger male workers. Secondly, Northern population has very high natural increase (a difference between births and deaths) as it has higher than average birth rates and relatively low death rates.

Table 3. Largest urban settlements by population size

Urban center/ settlement	Territory	Population in 2006 (Census)	Change, 2001-2006; %
Whitehorse	Yukon	20,416	+7.4
Yellowknife	Northwest	18,700	+13.1
Iqaluit	Nunavut	6,184	+18.1
Hay River	Northwest	3,648	+3.9
Inuvik	Northwest	3,484	+20.4*
Fort Smith	Northwest	2,364	+8.2*
Rankin Inlet	Nunavut	2,358	+8.3
Arviat	Nunavut	2,060	+8.5
Rae-Edzo	Northwest	1,894	+22.0*
Cambridge Bay	Nunavut	1,477	+12.8
Percent (%) of territorial pop.		58.1	

Source: Canadian Census, 2006. Note: * data must be interpreted with caution due to undercounts in 2001.

However, the life expectancy, especially among Aboriginal northerners, is not very high. The infant mortality – a well-established measure of ‘demographic’ well-being – in the North exceeds Canada’s average, indicating that the Territories lag behind other regions of the country in terms of well-being. Thirdly, the North exhibits intensive population turnover: a lot of people come (in-migrate) to the North, but many people leave (out-migrate). The turnover creates population as well as labor supply instability.

The population of the North grew dramatically in the last 100 years (Table 4). This increase was the result of the interplay of fertility, mortality and migration. There are four main characteristics of the population dynamic in the Canadian North:

- The increase in total population. According to the available historical data, population of the North doubled over the last 100 years. Population growth came from two sources: natural increase and migration. An intensive in-migration of southerners to the Canadian North started at the end of the 19th century. After World War II, population growth was provided not only by mass migration, but also by relatively high natural growth among Aboriginal population and newcomers. The population growth was especially intensive in the 1940-1980s.

Table 4. Population Change in the Canadian Territories, 1871-2006

Year	1871	1881	1891	1901	1911	1921	1931
Pop.	48,000	56,000	100,000	47,000	15,000	12,000	13,500
Year	1951	1961	1981	1991	1996	2001	2006
Pop.	25,000	37,600	69,000	85,000	96,000	93,000	107,800

Source: Bone, 1992, 2003; Census, 2006.

- The "boom-and-bust" character of the population dynamic. Since the population dynamic in the North is dependent on migration, the region experienced fluctuations caused by the economic character of migration. Usually, large "jumps" in the number of inhabitants were connected to resource booms, at the end of which population declined and settlements closed (Table 4). This makes northern population very unstable. In sum, while in the aggregate the population of the region almost always increased, some areas experienced intensive growth but others saw rapid or gradual depopulation.
- The decline and subsequent increase of Aboriginal population. It is easy to notice that in the early 19th century northern population declined (Table 4). Why? After contact with the Europeans the number of Aboriginal people quickly fell as they were affected by diseases brought by the Europeans. The Aboriginal population was in decline until the 1950s, when the trend reversed. Two main factors provided this change: movement to settlements and improvements in health care. Since then, the number of Aboriginal people has been growing rapidly due to their high natural increase (for instance, in the mid-1980s the rate of natural growth among Aboriginal population was nearly four times greater than the national average).
- Regional differentiation in population trends is associated with varying population composition and migration patterns. Most Aboriginal communities experience high growth rates because of the natural increase, whereas non-Aboriginal settlements are strongly affected by migration flows. Among the latter group, some mining towns appeared almost overnight as a company commenced its operation (so called 'instant towns'), and quickly winded down after the mining was halted.

Northern Economies

Northern Economy: Mixed and Dual

The pre-contact economy in the Canadian North was based on subsistence hunting, fishing, and gathering with a considerable amount of barter exchange within and among families. In other words, in a traditional Aboriginal economy a household was (and is) not only a unit of consumption, but a unit of production (a "micro-enterprise"). With the arrival of the Europeans, Aboriginal groups were gradually involved in what is typically known as "modern", "capitalist," "cash" or "wage" economy (as opposed to Aboriginal, traditional, subsistence economy). In contrast to a subsistence economy focused on providing living for a family, in the "modern" sector goods are produced not for family consumption, but for sale. Households also sell labor (on full or part-time basis) to receive wages. The subsistence and "modern" sectors (modes of production) are not isolated from each other: cash from sales is used for a household's benefit and goods produced for subsistence may also be sold. In addition, a household may receive income from government transfers (pensions, subsidies, etc.). On the other hand, it also may obtain resources from extended networks of households through sharing and exchange. The two "modes of production" coexist in a system termed a mixed economy. The mixed economy has long existed in the North: it started when fur traders began to hire Aboriginal northerners to bring fur pelts in exchange for European goods.

Albeit related, traditional and cash economies are not completely integrated. In the earlier history of the North, when the fur trade dominated the capitalist sector, the link between the two modes of production was quite explicit as traditional and commercial hunting went hand-in-hand. Later, when the fur trade lost its dominance to large-scale resource extraction, this relationship weakened because subsistence activities ceased to have a direct connection with the industrial economy. Households participating in a traditional economy have only few ways to engage in the resource-based "modern" sector, and are often restricted to seasonal and/or unskilled employment. The decoupling effect (i.e. a disjunction between the subsistence and capitalist modes of production) leads to a dual economy in the North, in which coexisting traditional and "modern" sectors have narrow linkages. The decoupling effect has severe consequences for northern communities: it prevents northern residents, especially Aboriginal, from benefiting from industrial development, such as mining and oil extraction. Many communities remain uninvolved in the 'modern' sector as companies prefer to bring labor and supplies from outside the North, because local residents do not have necessary skills and local supplies are too costly (or do not exist). As we will see, the elimination of the decoupling effect is a primary purpose of economic development efforts in the North today.

A mixed, subsistence-based economy is still the way of life for many Aboriginal northerners, especially in smaller communities: over 70% of adult Inuit participate in subsistence activities (such as hunting, harvesting and fishing), and traditional food accounts for 75% of the diet in

Inuit households. Subsistence household production aided by a seasonal wage employment is a very typical arrangement for Aboriginal families. In contrast, non-Aboriginal residents typically work in a capitalist (wage) sector.

A Diamond in the Rough: the Resource Economy

Furs became the first major commodity from the Canadian North to be sold at the European markets. By the mid-19th century the Hudson's Bay Company and its competitors created an extensive network of fur trading posts across the entire North. Fur and whale in the Arctic coastal areas were the first staples extensively shipped from the Canadian North to the south and then to Europe. Staples refer to unmanufactured, raw materials, which are primary commodities for most northern regions. It could be furs, mineral ores, oil, lumber and other resources that are extracted and shipped elsewhere, unprocessed. The transition to mineral staples was marked by the Klondike Gold Rush, when the North became the global supplier of other staples: gold, iron, nickel, copper, lead, zinc, diamonds, oil and natural gas became main northern commodities of the 20th century.

Economist Harold Innis who studied fur trade and other resource-based activities in the North concluded that staple-producing regions inevitably become dependent on their more developed counterparts that provide supply for the raw commodities. Similarly to Canada being dependent on Europe's (and then US) demand for furs, lumber, and minerals, the North is dependent on southern regions of Canada and global markets. Staple regions become, therefore, hinterlands (peripheries) of more advanced heartlands (cores), which concentrate population and industry (such as southern Ontario). The heartlands exchange staples from hinterlands for manufactured goods they produce. A staple-reliant hinterland economy experiences uneven terms of trade (increasing amounts of raw materials are required to be sold to buy the same amount of manufactured goods) and eventually faces a crisis, when demand from the heartland is exhausted or a staple is depleted. As a result, the development gap between hinterlands and heartlands increases. To narrow this gap, hinterlands need to break away from resource-reliance and invest in their own industrial and service sectors, foster alternatives sources of income and build community economic capacities.

Northern Resources

The Canadian North concentrates considerable deposits of fossil fuels and minerals. A critical fact about these resources is that the most impressive reserves are yet to be used. In other words, the North is the world's last resource frontier with vast unused (and, perhaps, undiscovered) reserves of various raw materials. Mineral riches of the North became well-known after the Klondike Gold Rush over a century ago. Since then, prospectors and mining companies found considerable deposits of gold, lead, zinc, silver and other metals (Figure 6). In the 1930s major gold operations commenced near Yellowknife, where Con and Giant Mines operated for more than 60 years, until they were closed in 2003 and 2004. The Lupin Gold Mine located on Contwoyto Lake in Nunavut operated between 1982 and 2006. While most gold mines in the North are currently closed, the growth in gold prices inspired several new projects that would either recover last available reserves in old mines or commence new mining operations.

The North concentrates considerable deposits of lead and zinc both on the continent and on the islands. Until recently, major mines included Faro (1969-1998), Nanisivik (1976-2002), and Polaris (1981-2002), all large mines in the Arctic Archipelago. At one point Faro was the largest open-pit lead-zinc mine in the world. However, they succumbed to an economic downturn (see below). In the recent years, mining companies have been working on several new projects, for example Izok Lake (zinc, copper, silver and gold) and High Lake (copper and zinc). Large deposits of iron ore found on Baffin Island (Mary River) are yet to be developed.

The North contains very large resources of oil and natural gas (Figure 6). Major deposits exist in the Beaufort Sea and in the Sverdrup Basin, and there are indications of more deposits in other areas. However, these deposits are far away from the markets. The Beaufort Sea is more accessible and, perhaps, its oil and natural gas are first in line for further exploration and use. Still, high transportation and operation costs prevent the vast petroleum reserves of the Arctic becoming available to global consumers. It will take years to build the required infrastructure, whether on- or off-shore. At the same time, receding sea ice can help in making Arctic oil basins more accessible.

Diamonds were discovered in the North relatively recently. First diamonds in NWT were discovered in 1991. By 2007 three major mines: Ekati, Diavik and Snap Lake were in production delivering 12% of the world diamond production (by value). The deposits are expected to last for 20 years. With large volume of production, three mines became the backbone of the NWT economy. A smaller diamond mine – Jericho – opened in 2006 in Nunavut. The mining companies invested billions of dollars not only in mining operations, but in local infrastructure, local businesses and communities.

Resource Boom-and-Bust Cycles

The instability of staple prices (be it oil, metals or lumber) causes dramatic changes in the hinterland economy. Jumps in staple prices cause a pattern of erratic economic fluctuations known as boom-and-bust cycles. When prices (demand) are high, the staple economy is flourishing, when they fall the staple economy is declining. A growing demand for mineral resources from the US and global markets prompted the Canadian government to focus on resource development in the North in the 1950s and 1960s. This boom generated most prominent mega-projects that are discussed in the next section. Major hikes in oil prices in the 1970s and 2000s prompted oil companies to seek oil in the Arctic (e.g. in the Mackenzie Delta). Now, with evidence of receding ice, there are emerging opportunities for offshore drilling. High demand and value of diamonds allowed mining companies to invest billions of dollars in diamond mines in the Northwest Territories. Boom-and-bust cycles can also be caused by depletion in the staple resource. A prominent example of this situation is the Klondike Gold Rush with its drastic growth in gold mining in population in 1898-1899 and no less dramatic decline in the early 1900s.

Since resource extraction costs in the North are typically higher than in other regions because of remoteness and difficult conditions, even a moderate decline in staple price can make production unprofitable. Idling of a mine is expensive, and most companies prefer to abandon an operation altogether. With the mine closed, surrounding mining towns quickly wind down, and the local wage economy collapses. The Canadian North presents a lot of illustrations of this type of resource bust. For example, in the late 1990s such a cyclical wave almost wiped out mineral production in Yukon, where between 1996 and 2001 it declined from $407.5 million to $41.1 million. At the same time, the mining town of Faro (the largest open pit lead-zinc mine in the world at the time) shrunk from over 2,000 to 400 residents. The Yukon Territory lost about 12% of its population. This story is not unique, many mining operations in the Arctic survive only a few years (e.g. Hope Bay silver mine 1973-1975, Cullaton lake gold mine 1981-1985).

Mega-projects

In the 1950s Canada's Prime Minister John Diefenbaker declared the new "Northern Vision" policy that aimed to explore and develop the natural riches of the Canadian North. Learning from experiences of the Gold Rush and other earlier development endeavors, which collapsed as quickly and chaotically as they emerged, the Canadian government took a lead in investing in northern infrastructure and development. Since the 1960s the North has seen a rise of so called 'mega-projects.' Mega-projects are large resource-extraction projects (mostly in mining, oil extraction, hydro-electricity production, and forestry) controlled by multinational corporations and typically subsidized by the state. Given remoteness and severe environments of the North, mega-projects usually required investments exceeding $1 billion and many years to complete. The multinational corporations involved with a project typically created a consortia to share risks or negotiated 'deals' with the government that provided certain concessions, additional investments, and assumed a part of risks and costs of a mega-project. The most famous mega-projects in the North (although outside the Territories) are the James Bay Hydro-electric project in northern Québec, Churchill-Nelson River project, uranium mines in northern Saskatchewan, and Alberta tar sands. In the Territorial North, the largest projects have been the Norman Wells project (1985-present) and diamonds projects (1996-present all in the Northwest Territories. Other mineral projects included lead-zinc mines in Pine Point, Faro, Polaris, and Nanisivik and gold mines near Yellow-knife and the Lupin mine (Figure 6).

Mega-projects were designed to bring northern resources to global markets and serve as main economic 'engines' in northern development as they deliver investments and jobs. In fact, each project creates a short-term economic boom and greatly benefits communities that are located next to it. However, in the construction stage most of the invested money is spent outside the North (e.g., for buying construction materials). In the operation stage many of the jobs are taken by outsiders who come to reside in the North temporarily. Most of the revenues from resource extraction are appropriated by the company and even wages paid on site largely benefit an outsider labor force that pays taxes in the south. In addition, most of the processing of raw materials is done outside the North, thus, failing to create more jobs for northerners. For instance, in Yukon before mine closures in the late 1990s only 50 cents of every dollar earned by the mining sector stayed in the local economy (in a form of purchases, earnings, payments, etc.), while 50 cents leaked outside the Territory. This capital leakage is the principle shortcoming of mega-projects. It is an illustration of the aforementioned "decoupling effect" that separates "modern" capitalist economy from local economies of the North. Another problem is that mega-projects are based on non-renewable resources, which are prone to volatile price fluctuations and have a limited life span. Due to high operation costs, northern mining and other resource activities require high resource prices on the global market and/or a very high quality of deposits. A mega-project quickly collapses (or suspends operations) following the decrease in global prices for a certain resource. For these reasons, mega-projects conducted by multinational corporations and supported by the government were unable to deliver economic sustainability to northern communities, and have sparked concerns around the environmental and social impacts of such developments.

Environmental Impacts of Industrial Development

Arctic ecosystems are remarkably vulnerable to disturbance and damage, and, at the same, time, take a very long time to recover. The fragility of Arctic environment is related to its cold temperature regime, presence of permafrost and long compensatory recovery periods. The first effects of economic activity on northern environment were already noticeable in the 19th century when the fur trade and whale hunt depleted the stocks of fur-bearing animals and bowhead whales. In the 20th century, the local environmental impacts were brought by industrial and military projects, especially mining.

There are several ways in which industrial development damages northern ecosystems. Construction of highways and pipelines not only remove ground vegetation, but also introduce obstacles to wildlife migration routes (e.g. caribou). They also affect permafrost and may cause a major disaster if a pipeline disintegrates with frost heave. On the other hand, aerial disturbance and pollution occurs with mining and oil drilling. Large mining operations are extremely destructive to natural vegetation, soil and wildlife. They are also sources of air, water and soil pollution. The Diavik Diamond Mine extracts two million tons of ore annually and has a footprint of 10 sq. kilometres. Fortunately, diamond operations do not use toxic agents to process the ore. Gold mining requires the use of such agents however, and as a result gold mines generate significant pollution. Their waste deposits then slowly release toxins into the environment, which can often present a pollution 'time bomb.'

For a long time environmental degradation was considered an 'acceptable price' for resource development. The companies were not held accountable for damaging the environment during or after their mining operations. The volatility of staple markets prompted many mining activities to be suddenly ceased, leaving behind piles of waste, which threatened biological life and infiltrated into water. For example, when the Giant Mine near Yellowknife shut down in 1997, it left behind 270 thousand tons of toxic arsenic trioxide. In another case, the contamination with PCB (Polychlorinated biphenyls) has been detected in 18 out of 21 abandoned DEW Line military sites, which now require a massive and expensive clean-up. In addition, mega-projects were run with limited regard for the desires and concerns of local populations, especially Aboriginal northerners. The lack of attention and care from governments and companies resulted in significant damage to northern ecosystems and land/marine based human activities.

Only in the 1970s an inquiry into the Mackenzie Valley Pipeline project led by Justice Thomas R. Berger prompted the Canadian government to toughen the rules. The Berger Inquiry (1974-1976) helped to raise public awareness about the environmental effects of industrial mega-projects in the North. The Berger Inquiry also clearly pointed to the connection between environmental sustainability and vitality of Aboriginal economy and culture. It laid the way for future inquiries and assessments, by supporting the rights of Aboriginal Peoples to demand guarantees of environmental and cultural preservation and request compensation from the companies. The Inquiry has become a land-mark in processes for environmental assessment in terms of considering social as well as environmental impacts, and undertaking extensive community consultations. Now such projects require assessment of environmental impacts to be done prior to their approval. For instance, new diamond mining projects in NWT proceeded only after reaching environmental agreements between the mining companies and federal and territorial governments. Aboriginal peoples who have signed land claim agreements exercise control over environmental issues through co-management boards, and are thus also required to be consulted in environmental impact assessments. Many advocate a wider use of traditional knowledge of Aboriginal people in environmental assessment and monitoring.

The Current Status and Future of Northern Wage Economies

As economies in other parts of Canada, northern economy consists of four sectors: primary (fishing, trapping, forestry, mining), secondary (manufacturing, construction), tertiary (services) and quaternary (information sector). Given the prominence of the resource-based (primary) activities, one would expect them to constitute the largest share in northern economy and employment, but this is not the case. Most northerners are employed in the services sector that includes the public sector, such as government jobs. This pattern is not uncommon for hinterlands of post-industrial countries like Canada, where remote areas enjoy a similar level of private and public services as do other regions. The attempt to provide a full range of services stems from the policy of the Canadian state to ensure welfare of its entire people. With growing well-being of the northern population, more and more services are required to satisfy local demand. This includes not only public, but also private services such as banking, insurance and real estate.

What is the role of the primary sector then? The primary (staple) sector forms an economic base for the wage economy of the North. The base sector is an activity that feeds other sectors of economy by providing demand for supplies from other local industries. The base activity is usually the most productive, export-oriented activity, a stem of an 'economic tree.' Because the base sector is connected to other industries, an investment in a base sector generates impacts in other sectors. For example, a construction of a new mine may require construction supplies, fuel, vehicles, etc. to be purchased on a local market. As a result, additional employment is also generated in industries outside mining. Because of in-migration to fill new jobs in the mine, a growing population will require more teachers, shop assistants, bus drivers and, thus, will create extended impacts in sectors that have no direct relationship with mining. This effect is known as the multiplier effect. The larger the multiplier, the more benefits that remain in the community. However, as mentioned before, a large portion of revenue from resource extraction in the North is spent in other regions (capital leakage), leaving northern communities without the expected economic benefits. This is one, among others, disadvantage of a staple economy.

For the Canadian Territories, staples, namely oil extraction and mining remain the pillars of regional economy (Figure 7). Mining and oil sectors together constitute about 30% of northern economy (in terms of GDP). Although cyclical nature of the resource production adversely affected mining in the 1980s and 1990s, the income generated by the staple sector has been increasing, especially due to growing diamond excavation. The increase was especially dramatic in the Northwest Territories, where major diamond companies built three large mines: Ekati (1998), Diavik (2003) and Snap Lake (2008). The diamond production offset the unsteady trends in oil and gas extraction and gold mining. Still, most of the diamonds are shipped out of the NWT as rough or unworked and the capital leaks from the Territory to benefit corporations and regions in southern Canada and abroad.

In the last few decades, northern Canadians achieved substantial progress in advancing their economic interests and delivering more benefits and development to local communities. Territorial, local and Aboriginal authorities were able to engage the increasingly socially responsible corporate businesses in negotiations on tangible benefits for northern communities in exchange for popular consent to resource-extraction activities. The struggles of the past around mega-projects, where local, especially Aboriginal stakeholders were sidelined and felt a necessity to resist, paved the way to a new partnership between corporations, communities and governments. Big business realized that development in the north, along with the improvement of infrastructure and community well-being, is a good investment as it provides a stable and fruitful environment for doing business. Although these relationships are still problematic, there has been a relatively positive experience with Impacts and Benefits Agreements (IBA). IBAs are the contracts between a resource company and a local community that identify benefits and other deliverables (such as guaranteed jobs and payments) that a community receives from a company. With an IBA, a company secures the community's consent for its development activity. As a result of IBAs, more northerners, especially Aboriginal Peoples, receive professional training and enter into the high-wage resource sector workforce, as well as having a say in how development takes place.

Public administration is the second largest sector in the northern economy. Many northerners work for territorial, local, municipal and federal governments. As a result, transfers from the federal government account for over 80% of territorial revenues (Nunavut is up to 90%), i.e. the federal government is funding the territorial governments. This creates a high degree of dependency of territorial budgets on federal support, a problem exacerbated by the dependency of many northerners on their government wages. However, it may not be entirely correct to say that the Territories are subsidized by the federation, since the Canadian government is benefiting from royalties and taxes associated with resource extraction in the North.

In the recent decades, the Canadian North witnessed an increase in the services sector that provides services for people and businesses. The FIRE (finances, insurance and real estate) sector is the leading services sector associated with both the resource sector and the local consumer economy. Despite a steady growth, this and other high-wage service sectors are dependent on the stability of the staple economy and well-being of northern residents. When resource extraction and household consumption decline, services quickly lose revenues and jobs.

Manufacturing in the North is very limited. One reason is the cost of parts and supplies required for the manufacturing process: they have to be delivered from far away and must be able to operate in the arctic climate. Another problem is the lack of qualified labor force: manufacturing needs highly skilled workers, but this group of professionals in the North is very small. This situation illustrates the staple thesis: a resource-dependent hinterland is a source of raw materials that are being shipped unprocessed to the core (south). In contrast to the mining sector, manufacturing usually has high multiplier effect on other industries and services. That is why northern governments work hard to attract more manufacturing activity to the region. Recently, several diamond processing plants opened in Yellowknife. They cut and polish stones from the NWT diamond mines which keep more of the revenue in the territory (the value of processed, jewellery quality diamond may double that of the rough stone).

Living on the Edge: Traditional Land/Marine-based and Household Economies

Traditionally, the survival of northern peoples depended on their successful hunting of small or large game (land or marine mammals, birds, and fish), harvesting of wood, plants and berries, and the conversion of these materials into food, clothing, shelter, heat, light, and transportation. So, up until the 1960s, the majority of the northern economy was still relatively land(ocean)-based, with mobile, multi-family hunting groups working together to fulfill basic daily needs. Some of these products became valuable trade commodities with European explorers or entrepreneurs (e.g. trade in fox and other desirable furs with the Hudson Bay Company facilitated trade expansion and community settlement across northern regions). However, much of the engagement in hunting and harvesting activities prevails today (along with earning wage income or receiving transfer payments), as seen in the number of subsistence and commercial enterprises rooted in regional and cultural traditions – despite no longer directly needing them for survival (i.e. there are stores to buy food and clothing, and housing and related services are very much like anywhere else in North America – just on a smaller scale). This contributes to local livelihoods, as well as passing on hunting techniques, safe travel and navigation practices, and methods for preparing skins or meat, all of which are important elements of northern Aboriginal cultures – although there are many variations in the respective techniques, practices, and methods, depending on the region and cultural group in consideration. However, beyond the transmission of cultural knowledge, these endeavours also make important contributions to household economies in a number of ways:

– they provide opportunities for full time hunters or fishers to make a living, either commercially or through provisions for the family;
– they provide valuable raw materials for many Aboriginal crafts;
– they provide essential "replacement value" (i.e. healthy food, rich in nutrients and protein, that would otherwise need to be purchased at the store at high prices and often lower nutritional value);
– they support the ethic of food sharing and communal feasting common to many Aboriginal cultures; and
– they provide local opportunities for economic diversification, to minimize impacts of market fluctuations.

Extended families tend to pool and share food, cash, and labour as required, combining market and non-market activities in a mutually supportive manner. Therefore, the household is considered the primary unit of production, distribution, and consumption, and harvesting is still among the most important economic activities in most northern communities. In terms of the economic context though, the cash value of these activities is considerably lower than the social (i.e. ethic and culture of sharing), replacement (i.e. protein replacement for store-bought food), and cultural (i.e. passing on traditional practices, environmental knowledge, and healthy, active lifestyles) values of engaging in such activities.

Although specific enterprises and products will vary across northern Canada, being tailored to local wildlife, resources, and cultural practices, some examples of northern economic ventures rooted in tradition include:

- country food – referring to wildlife hunted or plants harvested that would have been part of a traditional diet, is still an important part of modern northern diets and there have also been efforts to develop country food products and stores to sell goods and to offset fluctuations in demand for furs or pelts
- commercial fishing – recently (i.e. the last few decades) the development of commercial fisheries in northern waters has received a lot of attention from economic development programs, with a few mid-size fisheries able to maintain consistent operations and productivity.
- fur, leather, by-products, and crafts – are important and interlinked sectors that seek to continue or replace markets lost because of bans and protests, using a variety of wildlife species with the use of skins, fur, feathers, bones, antlers, etc. for the purposes of creating clothing, jewellery, toys, footwear, or other handicrafts that reflect traditional skills as well as modern demand and styles.
- sport hunting and fishing – many visitors to the north are attracted by the opportunity to hunt and fish in unique environments, and seek unique species – although there is also an increasing demand for "non-consumptive" or more experiential wildlife experiences. Nevertheless, a limited number of opportunities (based on available quotas) are provided by local Hunters and Trappers Associations to hunt for large game such as polar bears, caribou, moose, and muskox, under the supervision of local or commercial outfitting guides as a way to supplement local incomes, improve cultural awareness, and encourage the ongoing pursuit of learning and applying traditional hunting techniques (e.g. sport polar bear hunters can only travel by traditional means of a dog team, and not by the more common means of transportation – the snowmobile).

Many of the activities relating to country food harvesting and sharing, as well as fishing and developing products from fur or other animal parts, are also still undertaken for strictly subsistence purposes, with food or material goods being distributed within family or community networks. Nevertheless, there have also been varying degrees of success experienced in transforming these more traditional pursuits into viable economic opportunities.

Challenges and Opportunities for the Land/Marine-based Aboriginal Economy

When undertaking new entrepreneurial ventures, especially those trying to combine traditional pursuits with demand in a modern economy, there are inevitably a number of challenges and opportunities that arise. A number of these are outlined generally in Table 5. Despite the many challenges in engaging in land/marine-based economies, those individuals and companies that have been relatively successful have usually incorporated some combination of the following key factors:

- learning from the old and trying something new – blending elements of tradition and innovation to help maintain cultural continuity as well as encourage interest from external markets
- keeping it small – having local leadership or community-based enterprises, with target markets within the community or the region, seem to have the most success (i.e. small- to moderately-sized enterprises)
- tailoring to market access – distance to market will not cease to be a challenge, so those enterprises who have been able to take this into account, and consider this in the development or marketing of their products, have had greater success
- emphasis on outcomes beyond wealth – although the accumulation of wealth is the conventional driver for economic development, those community- or region-based enterprises that develop products/experiences to also meet cultural, educational, and health needs, along with a creation of profit, tend to be more successful in the long run.

Cultural Economy: Arts and Crafts

Aboriginal arts and crafts are not only important elements of Native culture, but also commodities that can deliver economic profit (this economic sector is known as 'cultural economy'). In fact, artisans and handcrafters can provide incomes to support an entire community. The commercial production of arts and crafts from bone, ivory, soapstone, and hides was important since the 1950s. According to the Survey of Living Conditions in the Arctic, 18% of Aboriginal residents of the Canadian Arctic manufactured crafts for sale. Almost one-third of all Aboriginal people reported receiving some income from selling pieces of traditional art. Involvement in commercial handcrafting and artisanship was the highest in Nunavut (22%), especially in some communities, like Cape Dorset. In another survey, 30% of Inuit living in Nunavut reported deriving a part-time income from their sculpture, carving and print-making. Interestingly, the region encompassing Baffin Island (including Iqaluit and Cape Dorset) has been the most creative rural area in Canada with 230 artists per 6,700 residents.

Table 5. Summary of Challenges and Opportunities for the Aboriginal Economy (adapted from Myers (2000))

Opportunities	Challenges
(i) learning experiences – by undertaking economic ventures rooted in tradition, it provides valuable opportunities to continue to pass on (and contribute to the evolution of) traditional oral knowledge around environmental conditions, wildlife health and migration patterns, hunting tools and techniques, safe travel and navigation practices, and so on, to younger generations to foster cultural learning, and a greater sense of pride, identity, and confidence (ii) cultural continuity – through the ongoing practice of traditional harvesting methods, as well as through the consumption of country foods and engagement in related activities (e.g. sharing food, community feasts, skin preparation, creation of handicrafts) cultural continuity can be fostered (iii) promotion of sustainable practices – by pursuing hunting and harvesting activities rooted in tradition there is simultaneously a promotion of renewable resource development and valuable incentives to undertake sustainable environmental practices to ensure that the lands, waters, oceans, and wildlife are not degraded or depleted	(i) land/marine based enterprises are subject to market fluctuations – embargoes on certain animal species or products can prevent the exportation of some products beyond Canada, and have previously led to devastating collapses in the fur economy (e.g. fox and seal) (ii) costs of hunting are high – due to high dependency on gas-powered vehicles (e.g. snowmobiles, boats, ATVs), high cost of acquiring equipment and vehicles in remote communities, usually meaning that full-time hunters need to have additional financial support, either from a wage-earner in the family or from transfer payments such as hunter or family support programs (iii) meeting federal regulations can be challenging – there are strict federal regulations around the storage, processing, and sale of country foods (including commercial fisheries), and given the lack of infrastructure and distance to market this can be restrictive in terms of expanding to including southern Canadian and international markets (iv) distances to market tend to be far – a great deal of the costs involved with running a northern business relate to the costs of acquiring sufficient equipment and tools to undertake the business, as well as to ship products to target markets, which can severely restrict expansion beyond local or regional enterprises (v) shifting roles for family members – in traditional societies gendered divisions of labour were common, mainly out of necessity, where men typically traveled and hunted to provide for the family, and women took care of camps and children, as well as food and skin preparation – in the current context these traditional roles are in a continual state of flux, with a general trend emerging in some families where women's roles have evolved to be primary wage earners (i.e. holding down full time jobs in the community), in order to support the hunting lifestyle and procurement of country foods that some of the men still undertake

Commercial arts and crafts are a substantial and growing sector of northern economy. It is estimated that these activities contribute $30 million in earnings. Most of the purchases are made by tourists. Whereas they bring millions in sales to Aboriginal communities, the challenge for Native artists in the North is low wages. Crafters and artists receive much smaller wages than their counterparts in other regions and than other workers in their own areas. The average earnings of an artist in Nunavut are just above $20,000 compared with an average wage of $38,000 for all workers in the Territory. In fact, this is below the official poverty line. As a result, arts and crafts are predominantly part-time activities for women: 40-80% of artisan products in Nunavut are done by Inuit women, many of whom are over 60 years old.

Engaging younger people and men is an important strategy for furthering cultural economy in the North. Improving access to markets and stimulating the demand for native arts are necessary to increase the attractiveness of artisan occupations.

Tourism, Nature, and Heritage

Tourism in the Canadian north often centres on the spectacular parks and expansive protected areas that are found across the territories and provinces. This is a small but growing sector of the northern economy, and one that is considered a viable option in the promotion of both cultural and environmental sustainability. Many of these tourism destinations relate to adventure or back-country hiking, qajaking, canoeing, skiing, rock climbing, and so on, as well as opportunities to learn about and participate in traditional cultural activities (e.g. dog-sledding, fishing, spring camps, various forms of hunting, visiting sacred or cultural sites, among other things). Tourism operations mainly involve seasonal (spring/summer) employment, from individual outfitters to running permanent lodges. The high costs of travelling to these regions can be prohibitive to receiving large numbers of visitors, but it is believed to have considerable potential for growth, with increasing transportation options (through environmental and infrastructure changes) and improved marketing (e.g. using the internet to advertise). In addition, northern cruise tourism is becoming more popular, both with increased interest and decreased sea ice cover, so there is a sort of double-edged sword to environmental change on the economic front.

Community Development: Reconciling Traditional Economy and Capitalism

Northern, especially Aboriginal, communities often lack a sustainable economic base. They typically have a small population, a limited production mix, and a remote location. All in all, northern communities represent marginal markets dependent on public subsidy or a nearby resource activity. In the latter case, although some community members may benefit from employment associated with a mine, the community itself usually experiences only minor gains due to the capital leakage discussed earlier. Since it is hard to attract investment and new development, northern communities have to rely on their internal entrepreneurial resources and government support to ensure economic vitality. Community development as a holistic concept presumes that a community controls its economic destiny, that wages and benefits remain in the community and that a community builds up its business infrastructure and local expertise.

Community development is a combination of 'top-down' and 'bottom-up' processes. 'Top-down' economic development efforts are usually related to government programs that assist remote communities in building infrastructure, setting up professional training, and providing credit for local businesses. A common problem with 'top-down' efforts is their patchiness and lack of strategic focus. These programs provide funding for selected purposes, but are unable to bring long-term benefits without community support and engagement. 'Bottom-up' processes of fostering development, which come from within the communities are required to achieve economic vibrancy in northern settlements.

Co-operatives are an example of bottom-up efforts to foster local economic growth. The northern co-operative movement started in the 1950s and was related to traditional activities, such as harvesting, crafts and fishing. Today co-operatives exist in various economic sectors from fur harvesting to handicrafts, and from fisheries to retail stores. Co-operatives are also involved in operating post offices, freight hauling, airline agencies, hotels, and coffee shops, residential and commercial real estate ventures, and, more recently, in cable television services. Co-operatives are run by local residents and often operate without profit and receive assistance from governments or Aboriginal organizations in the form of capital costs and infrastructure.

Another successful approach to community development is realized through comprehensive land claims agreements (CLCA) and impact and benefit agreements (IBA). CLCAs often have stipulations for investment in local development, for example, through creation of regional development corporations that manage compensatory payments to the benefit of Aboriginal communities. An IBA is a mechanism of ensuring that northern communities share some of the positive economic impacts stemming from new resource mega-projects. Communities and corporations sign these agreements, where they prescribe community benefits from proposed activities. This may include provisions for employment quotas, local purchases, professional training, use of local contractors, etc., all of which channel additional investment to the community and limit capital leakage. Currently, seventeen IBAs are in effect in NWT, six in Yukon and four in Nunavut.

Despite some improvement in living conditions in northern communities, most of smaller settlements still face substantial challenges. The unemployment rates in the territories vary between 10 and 16% (2006) with some Aboriginal communities having unemployment in excess of 20%, compared to 5-8% in territorial capitals (Canada's average in 2006 was 6.6%). Many northerners, especially Aboriginal Peoples, have part-time or seasonal employment or do not participate in the wage economy at all. For decades, northern Aboriginal communities have been unable to narrow wage and educational gaps with the rest of population. Aboriginal northerners have fewer options to receive formal education and professional training, and thus tend to have limited employment opportunities and lower wages. This gap is one of the reasons why many Aboriginal people and communities cannot benefit from mining mega-projects that require skilled and experienced labor. Professional training (funded by Aboriginal organizations, government or private companies) is one of the ways to alleviate the education gap. At the same time, it is important to realize that schooling in Native communities is also geared towards preserving and cultivating traditional knowledge and culture.

Conclusion: Times of Change

In northern regions, change is nothing new. Seasonal variations and annual fluctuations of environmental conditions, wildlife, weather, and social relations have been a part of life since earliest human expansion into the Canadian North. However, the changes these days seem to be more on the forefront of public consciousness, in the North but also around the country and around the world. It's hard to pick up a newspaper or turn on the radio now without hearing something about climate change, pollution, or new economic developments. Especially in terms of climate change, think of how often you've seen references to melting glaciers, ice caps, or sea ice, or threats to the survival of polar bears as a species. The Arctic, and northern regions more generally, have thus come increasingly into play with regards to global climate change, and the impacts humans (especially in developed countries) are having on the environment. What is mentioned less often however – despite the incredible efforts and increasing influence of Inuit leaders and activists such as Siila Watt-Cloutier (former Inuit Circumpolar Council president) and Mary Simon (Inuit Tapiriit Kanatami president) – is the far-reaching impact these changes are having on the lives and livelihoods of northerners, who have for generations depended on and thrived in an environment dominated by cold, snow, and ice, and the plants and animals these conditions support.

Environmental Change

Environmental change can occur due to any number of natural or human-induced causes, but the ones most prominently affecting northern environments today relate to climate and economic change. Why is the Arctic so important in relation to global climate change? Why is it that melting snow and ice are featured so prominently in accounts of dramatic impacts of a globally warming climate? Well, there are a few important reasons:

- higher latitudes are more affected, and respond to warming faster, than lower latitudes, and because of this higher latitudes are an indicator of global change trends (i.e. what the future may hold for lower latitudes);
- because of the impacts of warming in higher latitudes, they contribute to global feedback processes (i.e. enhanced warming around the world); and,
- the northern environment, economy, and culture are highly intertwined, therefore environmental impacts due to warming directly affect many other aspects of northern life.

But then why do high latitudes play such an important role in global climate change trends? Again, there are a number of important contributing factors based on the unique northern environment:

- a warmer climate tends to melt snow and ice, which exposes more of the land and ocean (darker surfaces) which in turn absorb more solar energy (i.e. reduced albedo);
- this increase of extra trapped energy goes directly into warming, rather than evaporation;
- in higher latitudes atmospheric warming can occur faster because the atmospheric layer that causes surface warming is shallower and thus warms easier;
- as sea ice retreats, solar heat is absorbed more easily and transferred more rapidly to the atmosphere;
- changes in atmospheric and oceanic circulation can also increase warming.
- For all of the above reasons, a warming polar region (Arctic or Antarctic) can contribute to enhanced warming around the world – called a positive feedback. However, this same influence can also occur in relation to cooling (i.e. if temperatures cooled, and more snow and ice formed, then it would contribute to global cooling).

The Arctic Climate Impact Assessment (ACIA) undertaken in 2004 was a landmark study evaluating the global circumpolar arctic region to evaluate the effects of climate change on northern regions around the world. Their key findings for the circumpolar Arctic include the following:

- the Arctic climate is now warming rapidly and much larger changes are projected;
- Arctic warming and its consequences have worldwide implications;
- Arctic vegetation zones are very likely to shift, causing wide-ranging impacts;
- animal species diversity, ranges, and distribution will change;
- many coastal communities and facilities face increasing exposure to storms;
- reduced sea ice extent is very likely to increase marine transport and access to resources;
- thawing ground will disrupt transportation, buildings, and other infrastructure;
- Aboriginal communities are facing major economic and cultural impacts;
- elevated ultraviolet radiation levels will affect people, plants, and animals; and,
- multiple influences interact to cause impacts to people and ecosystems.

Many of these conclusions are also true for the Canadian north (although they may manifest differently in different regions across Canada and across the circumpolar North). Some of the Canadian-specific implications of a changing climate are identified in the chapter on "Northern Canada" as part of a federal government commissioned report on From Impacts to Adaptation: Canada in a Changing Climate).

Environmental impacts of focus involve the foundations of the arctic and sub-arctic ecozones, such as: permafrost, sea ice, freshwater rivers and lakes, snow cover, glaciers and ice sheets, and sea level rise and coastal stability. Each of these in turn has impacts on economic development and adaptation, as well as community health and well-being. Table 6 highlights a few examples of how environmental changes manifest, and how they affect other sectors.

Climate Change and Economy

How might climate change affect economic perspectives of the North? Most economists believe that the impact of a warmer climate will vary across economic sectors and regions. The offshore oil and gas sector is expected to benefit from less extensive and thinner ice. However, onshore operations will likely be more costly, as it will become more difficult to build on thawing permafrost (leading to a lack of solid ground for construction). The same applies to mining operations, they will likely to become more expensive. These activities will also be negatively affected by storm surges and erosion.

The infrastructure in the North may experience more floods, avalanches, and mudslides. The winter road season will be shortened, and the quality of roads will generally decline as permafrost becomes less stable. However, seas and rivers are expected to be ice-free and navigable for longer periods of time.

Even tourism will be impacted by major disturbances in the landscapes, considering that most of Arctic tourism is nature-based. Other nature-dependent activities, such as fishing and hunting, will also be deeply affected as northern ecosystems change. Indeed, for the same reason, climate change will have dramatic consequences for the subsistence economy of Aboriginal peoples.

However, it is not only the climate change that will transform northern economy. The North is being actively involved into the global economic system. As a main resource frontier of the world, it has already been, and will continue to be, profoundly touched by globalization. Opening access to Arctic resources, growing demand for fuel and minerals, as well as fresh water, fish and lumber, will inevitably bring more development and investment to the region. The question is whether this is going to deliver relatively short-term impacts on economic development, transportation, and infrastructure, while bringing more far-reaching risks to arctic environments and northern livelihoods. For example, an increase in marine transportation may lead to cargo and fuel spills, wildlife disturbance by underwater and airborne noise, and destabilization of ice. These disturbances may affect fishing and hunting in ways that jeopardize the viability of a traditional marine-based economy.

Table 6. Examples of Northern Environments Impacted by Climate Change, and Its broader Implications

Feature	Impacts of a changing climate	Broader implications
permafrost	– recent warming of permafrost temperatures in many regions of northern Canada – increased depth of active layer in summer months – regional and temporal variations (i.e. more noticeable in western Arctic) – projections suggest increases in these trends	– thawing of the permafrost has potential to release large pools of carbon (enhancing climate warming trends) – thawing permafrost can reduce landscape stability and compromise overlying infrastructure (which is typically built into the permafrost to provide a strong foundation) – affects hydrology – potential increased infiltration and groundwater storage/flow, decreased runoff – increased ground temperature and deeper active layer could promote enhanced vegetation growth and the potential encroachment of new species – shifts in diversity and accessibility of vegetation impact several foraging mammals including a variety of caribou species as well as muskox
glaciers and ice sheets	– general trends have been observed regarding the retreat of glacier fronts and decreased glacier volume (mass balance) since about 1920, although there are large regional variations in such trends, with seeming less dramatic trends in the Canadian High Arctic as in more southerly latitudes or alpine regions	– melting glaciers and ice sheets contribute to global sea level rise with input of water into ocean systems – melting Canadian glaciers though, will have little contribution to sea level rise in comparison to larger ice sheets melting (e.g. Greenland and Antarctica) – changing glacier melt trends affect the magnitude and timing of river flows and drainage patterns

sea ice	– annual average area of sea ice in Northern Hemisphere has decreased by approximately 3% per decade for the last 25 years – annual maximum coverage has shrunk around 2% per decade, while annual minimum area has declined more rapidly (5.5% per decade) – dramatic decreases area occupied by multi-year ice – predictions suggest this trend will only continue, and some anticipate ice-free arctic summers by 2050 – record lows in ice coverage in 2005 and 2007 – trends in later freeze-up timing in the fall, earlier break-up timing in the spring, thinning sea ice, and longer open water seasons in the summer	– reduced albedo contributes to positive global feedback, serving to enhance warming – reduced sea ice cover enables more potential for ship travel (which in turn means enhanced interest, and potential feasibility, for economic development) – concerns for increased ship travel (potential for accidents, contamination, wildlife disturbance, increased oil and gas exploration, etc.) – positive aspects of increased ship travel (increased community resupply, raw material export, scientific and tourist cruises, etc.) – reduced extent and quality of habitat for marine mammals such as a variety of seals, polar bears, and walrus – as well as migratory and sea birds (with related implications for northern residents who depend on these species) – reduced extent, stability, and useable timeframe of sea ice as a travel and hunting platform for northern residents – enhanced travel dangers, both for ships and on-ice travel – changes in water temperature can affect a variety of marine fish species
sea level rise and coastal stability	– enhanced climate warming affects global sea level through thermal expansion – global mean sea level rise is predicted between 0.2 and 0.5 m by 2100 – again with many regional variations	– sea level rise can contribute to increased flooding and/or coastal erosion, which threatens low-lying coastal towns or cities – the western Canadian Arctic is at much higher risk than the eastern regions – when combined with decreased ice extent (i.e. accompanying enhanced wave action) and warming ground temperatures (e.g. deeper summer thaw of permafrost) can also speed up coastal erosion and undermine stability – more risks of floods associated with storm surges

Source: Adapted from Furgal and Prowse, 2008.

It is clear that given the potential vulnerability of northern natural and social systems, precautionary measures need to be taken into account to understand potential impacts of new development on permafrost conditions, wildlife, water, surface vegetation and traditional land- and marine-based activities. Not surprisingly, consideration of climate change and related impacts are now being required for inclusion into environmental impact assessments.

Social/Cultural Change and Related Impacts

Despite the delayed influence of European settlement and colonization in the Canadian North, the pace of social and cultural change since the mid-1800s has been incredibly rapid. Transitions (voluntary or coerced) from traditional, semi-nomadic, subsistence lifestyles to community settlements, wage-based (and mixed) economies, with global connections through radio, television, and now Internet, were dramatic, rapid, and often times (intentionally or not) detrimental to local cultures and social networks. Northern Aboriginal peoples are highly adaptive, but they have done so over the long term and largely under their own powers and capabilities. The cultural differences between Aboriginal groups and the influx of European explorers, traders, and missionaries, and later Canadian government officials and service practitioners, led to a rapid unbalance in power relations, lack of authority in decision-making, and the promotion of assimilationist agendas.

Many complex and interrelated factors were (and still are) at play to support such an evolution in northern relations (i.e. concern for citizen health and well-being (e.g. cases of family starvation), lack of understanding of traditional values and practices, drive to educate Aboriginal peoples to incorporate them into mainstream society, sporadically intense resource development drives, the Cold War and military interests, the goal to provide infrastructure and services similar to the rest of Canada, and various levels of willingness to negotiate land claims and self-government agreements – or rectify past treaties).

We must continue to emphasize that such changes manifested in a variety of regionally- and culturally-specific ways; however, broadly speaking the rapid pace of social/cultural change (implicated with economic and political ideals) contributed to the disenfranchisement of many Aboriginal groups, and contributed to many of the socio-economic challenges faced in northern communities today. Nevertheless, the resilience of northern Aboriginal peoples persisted, and they were able to turn around some of the most challenging of experiences (e.g. forced re-location, residential schools, involvement in the market economy etc.) to fight for their inherent rights (through long term land use and occupancy), negotiate some of the most revolutionary land claims in Canada (and indeed perhaps in the world), and regain control over local and regional governance. Many efforts are now ongoing to re-envision educational programs that are more appropriate to geographic and cultural contexts (and languages), to bridge the knowledge transmission and cultural gaps between elders who grew up with igloos and youth who grew up with iPods, and to foster pride in current cultural practices which meld traditional practices with modern circumstances. And yet, these are not easy measures to implement, or issues to overcome, and efforts continue to be undermined by increasing pressure for non-renewable resource development, environmental change and uncertainty related to climatic changes, and changing federal political priorities.

New Geopolitics and Emerging Borders in the Arctic

Discoveries of vast mineral reserves in the Arctic, especially in the bottom of the Arctic Ocean, are changing the geopolitical role of the region. The economic and geopolitical importance of the Arctic is also growing as Arctic seas become more and more ice-free and, thus, potentially available for mass navigation in the future. The marine route from Asia to Europe across the Arctic (through the Northwest Passage) is several times shorter than using traditional waterways. In addition, receding ice opens up access to oil, natural gas and other mineral reserves in the Arctic Ocean. Therefore, control of Arctic seas is a key component of the modern geopolitical game. The Arctic Ocean is bounded by six countries (Russia, United States, Norway, Iceland, Greenland/Denmark and Canada), who all lay claims to particular parts of the Arctic waters. However, there are very few established maritime boundaries. Delineation of the maritime zones is a matter of dispute.

Despite its long presence in the Arctic, Canada is facing challenges to its Arctic claims. In 2001 Russia formulated a formal claim over the underwater Lomosonov Ridge beyond the Siberian shelf. In a symbolic move, the Russians also planted a Russian flag at the bottom of the ocean under the North Pole to the bolster their presence in the region. Nevertheless, Canada has formulated its claims to the Arctic lands and waters based on historical title, customary law and the United Nations Convention on the Law of the Sea (UNCLOS). First of all, Canada bases its sovereignty over the Arctic Archipelago on both cession (a transfer of the islands from the UK) and effective occupation. This claim is largely unchallenged, except for Hans Island, a tiny piece of land between Canada's Ellesmere Island and Greenland (a territory of Denmark). Both Canada and Denmark consider the island their territory, but, given the small size of the island, this dispute is minor.

A more complicated case is Canada's control over the Arctic waters. Canada claims a 12-mile territorial sea seaward of the straight baseline surrounding the Arctic Archipelago. Canada considers this area to be within its territorial waters. Further, Canada claims a contiguous zone of 12 miles beyond the territorial sea. Canada has also established a 200 nautical mile exclusive economic zone around the Archipelago, where it declares rights to exploration and use of natural resources. Lastly, Canada asserts its rights in the continental shelf. Under the Oceans Act, the length of the continental shelf extends to 200 miles from the baselines. However, Canada retains its right to claim an extended continental shelf (as permitted under UNCLOS). On the continental shelf, Canada will be able to exercise control over living and non-living natural resources, including subsoil. Canada has also established legislation to extend its environmental protection to the Arctic seas enclosed between 60th and 141st parallels.

Canada's claims to Arctic waters are being challenged at least in two aspects. First, several countries, most vocally the USA, dispute the Canadian sovereignty over the Northwest Passage. The Passage is a maritime path connecting the Davis Strait to the Bering Strait. While it is predominantly ice-covered, it is forecasted to be available for summer navigation in the foreseeable future. Canada considers the Northwest Passage as being within its internal waters and under its sovereignty, meaning that Canada has authority to establish controls and conditions to protect its safety, security, environmental, and citizen interests. The opponents, who refuse to accept Canadian control over the Passage, contend that the Passage is rather an international strait defined as a water corridor linking two open seas that can be freely used for navigation by all countries. The weak point of this argument is that the Passage has hardly been used for international traffic, and therefore may not qualify under the definition of an international straight. The United States and Canada are also in a dispute around the delineation in the oil-rich Beaufort Sea. While Canada believes that the maritime boundary should follow the 141st meridian, the US argues that it should follow a line at equal distance to the closest land point of each state.

New Days for the North?

The environmental, historical, and social/cultural descriptions described in this chapter provide important context to understand the changing nature of the Canadian North, and the interrelations between the environment, society, and external influences. Northern Aboriginal peoples have strong and deep-rooted cultural traditions that evolved in the context of their northern homelands. However, the emphasis for northern Aboriginal peoples and cultures should be placed on their innovation and adaptability, to enable their survival through many kinds of change (environmental, technological, cultural, political, economic, etc.). Many traditional practices have been adapted over time, through experience and passing on knowledge over generations, although the core of the activity may remain very similar to the original practice or intent. Furthermore, with the incorporation or development of new technologies and ideas also comes the creation of new traditions. For example, there is still a strong hunting culture within northern Aboriginal peoples for cultural, subsistence, and commercial purposes, but the modes of transportation and use of weapons have changed significantly with the introduction of the snowmobile and the firearm, respectively.

So, what does the future hold for the North? As a part of a rapidly changing Arctic, the Canadian North is facing formidable challenges and unique opportunities, and there is not necessarily any contradiction or incompatibility between tradition and change. Change in nothing new for the North, in fact, change is a part of tradition. Northern people and ecosystems will find ways to adapt to this change, just as arctic social systems have demonstrated their resilience by surviving through centuries of assimilation and colonialism. Traditional knowledge and institutions, informed by modern technologies and experiences, will serve as the foundation for adaptation. This adaptation, however, has its limits. In this light, there is a need for mitigating the consequences of environmental and social change, which may compromise the very basis of northern ecosystems and societies. Institutionalizing sustainable environmental practices, integrating traditional knowledge with modern science in decision-making, and monitoring of climate and human development conditions are some of the urgent mitigation strategies that must be implemented in the Canadian North.

Northern communities of today reflect the (sometimes too rapid) changes that have impacted northern regions over the past 60 years, and yet aspects of traditional life and culture are ever-present in most households through beliefs, social engagement, lifestyles, and livelihoods. The increasing pace and uncertainty around environmental change, combined with social changes, poses new challenges for northern communities. It is that achievement of balance between strong tradition, and adaptability to change, that will help northern communities find effective solutions to current and future challenges.

Review Questions

1. What are some of the defining physical characteristics of the Arctic environment?

2. What are some of the characteristics used to distinguish the Arctic and Sub-arctic biomes?
3. What are the three main historical periods identified in this chapter? How does understanding this history help you understand other sections in the chapter? Give examples.
4. Describe the gradual process of devolution and increasing fate control as it pertains to the Canadian northern Territories.
5. What are the three main Aboriginal cultural groups that have made the Canadian North their homeland? What languages do they speak? Use the websites provided to locate one northern community within each of these broad cultural groups.
6. Using materials from this chapter and additional sources, describe the history and culture of Canadian Inuit.
7. What are Comprehensive Land Claim Agreements? What agreements are currently in place? Why are they important?
8. What is a mixed economy? Explain and provide examples.
9. Explain the notion of staple economy and discuss advantages and disadvantages of resource mega-projects. What role (if any), in your opinion, does globalization play in economic development of Canada's northern regions?
10. Identify and discuss some of the major implications of climate change for Arctic peoples and environments.

Recommended Further Reading

Main Sources

ACIA (2004), *Arctic Climate Impact Assessment*, Cambridge: Cambridge University Press.

Arctic Human Development Report (2004), Stefansson Arctic Institute, Akureyri, Iceland.

Arctic Social Indicators Report (2009), Stefansson Arctic Institute, Akureyri, Iceland.

Bone, R.M. (2010), *The Canadian North*, 3rd ed., Toronto: Oxford University Press.

Damas, D. (2002), *Arctic Migrants/Arctic Villagers: the transformation of Inuit settlement in the central Arctic*, Montréal/Kingston: McGill-Queen's University Press.

French, H.M., and Slaymaker, O. (ed.) (1993), "Canada's Cold Environments", Canadian Association of Geographers Series in *Canadian Geography*, Montréal: McGill-Queen's University Press.

Furgal, C., and Prowse, T.D. (2008), "Northern Canada", in *From Impacts to Adaptation: Canada in a Changing Climate*, Lemmen, D.S., Warren, F.J., Lacroix, J., and Bush, E. (ed.), Ottawa: Government of Canada, 57-118.

Hamelin, L-E. (1979), *Canadian Nordicity: It's Your North, Too*, Montréal: Harvest House.

Pielou, E.C. (1994), *A Naturalist's Guide to the Arctic*, Chicago: The University of Chicago Press.

Wonders, W. (ed.) (2003), *Canada's Changing North*, Montréal and Kingston: McGill-Queen's University Press.

Other Sources and References

Abele, K. and Coates, K. (2001), Northern Visions. New Perspectives on the North in Canadian History, Broadview Press.

Abele, F., Courchene, T.J., Seidle, F.L., and St-Hilaire, F. (eds.) (2009), *Northern Exposure: People, Powers and Prospects in Canada's North*, Montréal: The Institute for Research on Public Policy.

Anderson, R. and Bone, R. (eds.) (2003), Natural Resources and Aboriginal People in Canada. Readings, Cases, and Commentary, Captus Press.

Bennett, J., and Rowley, S. (2004), *Uqalurait: An Oral History of Nunavut*, Montréal: McGill-Queen's University Press.

Berkes, F. (2008), *Sacred Ecology*, New York: Routledge (Taylor and Francis).

Berkes, F., Huebert, R., Fast, H., Manseau, M., and Diduck, A. (eds.) (2005), *Breaking Ice: Renewable Resource and Ocean Management in the Canadian North*, Calgary: University of Calgary Press and Arctic Institute of North America.

Burt, P. (2000), *Barrenland Beauties: Showy Plants of the Canadian Arctic*, Yellowknife: Outcrop Ltd. – The Northern Publishers.

Duffy, Q.R. (1988), *The road to Nunavut: the progress of the eastern Arctic Inuit since the Second World War*, Montréal and Kingston: McGill-Queen's University Press.

Grace, S. (2003), *Canada and the Idea of North*, McGill-Queen's University Press.

Glomsrod, S. and Aslaken, I. (eds.) (2006), *The Economy of the North*, Statistics Norway: Oslo-Kongsvinger.

Dufrence, R. (2007), *Canada's Legal Claims Over Arctic Territory and Waters*, Ottawa: Parliamentary Information and Research Service.

Hamilton, J.D. (1994), *Arctic Revolution: Social Change in the Northwest Territories, 1935-1994*, Dundurn Press.

Innis, H. (1999), *The Fur Trade in Canada: An Introduction to Canadian Economic History*, Revised and reprinted, Toronto: University of Toronto Press.

Krupnik, I. and Jolly, D. (eds.) (2002), *The Earth is Faster Now: Indigenous observations of Arctic environmental change*, Arctic Research Consortium of the United States, Fairbanks, Alaska.

Laidler, G.J., Dialla, A., and Joamie, E. (2008), "Human geographies of sea ice: Freeze/thaw processes around Pangnirtung, Nunavut, Canada", *Polar Record*, 44: 335-361.

Laidler, G.J., Ford, J.D., Gough, W.A., Ikummaq, T., Gagnon, A.S., Kowal, S., Qrunnut, K., and Irngaut, C. (2005), "Travelling and hunting in a changing Arctic: Assessing inuit vulnerability to sea ice change in Igloolik, Nunavut", *Climatic Change*, 94: 363-297.

Loukacheva, N. (2007), *The Arctic Promise: Legal and Political Autonomy of Greenland and Nunavut*, Toronto: University of Toronto Press.

MacDonald, J. (1998), *The Arctic Sky: Inuit astronomy, star lore, and legend*, Toronto and Iqaluit: the Royal Ontario Museum and the Nunavut Research Institute.

McMillan, A.D. (1995), *Native Peoples and Cultures of Canada*, Vancouver: Douglas and MacIntyre Ltd.

Myers, H. (2000), "Options for appropriate development in Nunavut communities", *Études/Inuit/Studies*, 24(1): 25-40.

Nickels, S., Furgal, C., Buell, M., and Moquin H. (2006), *Unikkaaqatigiit – Putting the Human Face on Climate Change: Perspectives from Inuit in Canada*, Ottawa: Inuit Tapiriit Kanatami, Nasivvik Centre for Inuit Health and Changing Environments at Universite Laval, and the Ajunnginiq Centre at the National Aboriginal Health Organization.

Oakes, J. and Riewe, R. (eds.) (2006), *Climate Change: Linking Traditional and Scientific Knowledge*, Winnipeg: Aboriginal Issues Press, University of Manitoba.

Petrov, A. (2008), "A talent in the cold? Creative class and the future of the Canadian North", *Arctic*, 61(2), 162-176.

Rea, K.J. (1976), *The political economy of northern development*, Science council of Canada background study, 36, Ottawa: Science Council of Canada.

Rich, E.E. (1961), *Hudson's Bay Company, 1670-1870*, New York: Macmillan.

Shields, R. (1991), *Places on the Margin: Alternative Geographies of Modernity*, London-New York: Routledge.

Solomon, P. (ed.) (2004), *The Dynamics of "Real Federalism" Law, Economic Development, and Indigenous Communities in Russia and Canada*, Toronto: CERES, University of Toronto.

Statistics Canada (2003), *Aboriginal Peoples Survey 2001 – Initial findings: Well-being of the non-reserve Aboriginal Population*, Ottawa: Statistics Canada, Housing, Family, and Social Statistics Division.

Usher, P.J. (2003), "Environment, race and nation reconsidered: reflections on Aboriginal land claims in Canada", *The Canadian Geographer*, 47(4): 365-382.

Wenzel, G.W. (1991), *Animal Rights, Human Rights: ecology, economy, and ideology in the Canadian Arctic*, London: Bellhaven Press.

White, G. and Cameron, K. (1995), *Northern Governments in Transition: Political and Constitutional Development in the Yukon, Nunavut and the*

Western Northwest Territories, Montréal: Canada, Institute for Research on Public Policy.

White, G. (2008), "'Not the Almighty:' Evaluating Aboriginal Influence in Northern Land-Claim Boards", *Arctic*, 61(suppl. 1): 71-85.

Young, J. (ed.) (2007), *Federalism, Power and the North: Governmental Reforms in Russia and Canada*, Toronto: CERES, University of Toronto.

Appendix

Figure 1. The Canadian Territorial North and populated centres

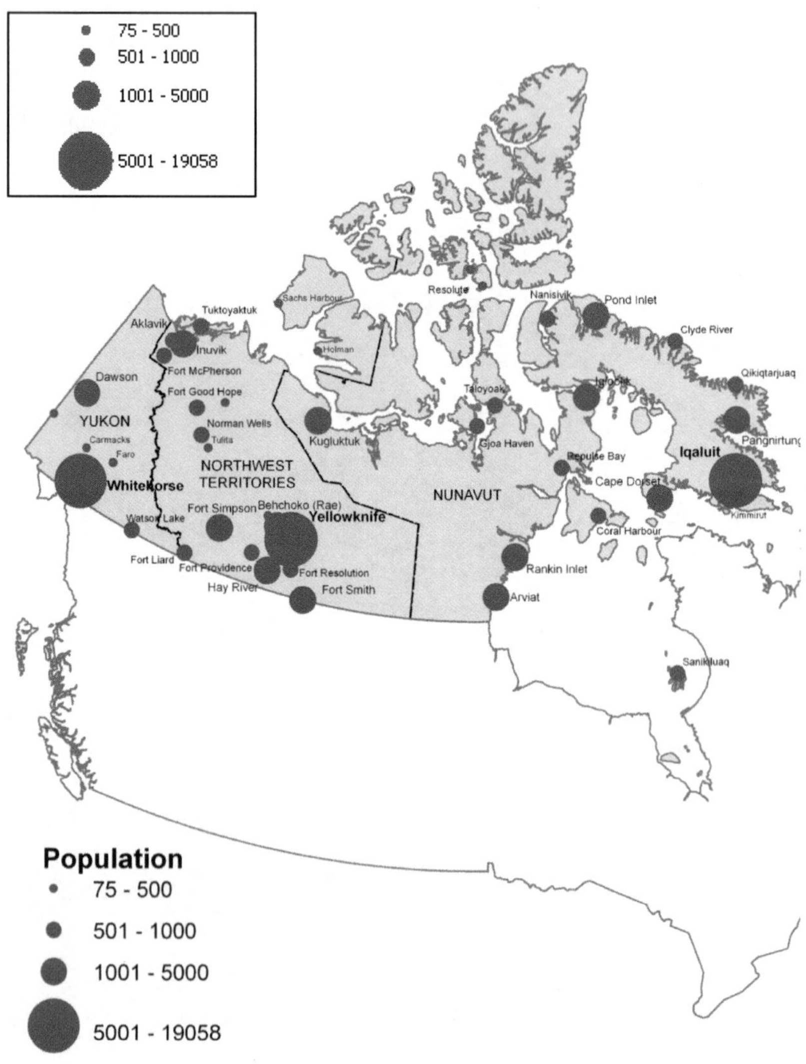

Source: Statistics Canada, 2007. Portrait of the Canadian Population, 2006 Census. Catalog Number 97-550-XWE2006001.

Figure 2a. Arctic scene. Winter in Cape Dorset, Nunavut (2005)

Photo by G.J. Laidler.

Figure 2b. Arctic scene: Summer on Boothia Peninsula, Nunavut (2001)

Source: Photo by G.J. Laidler.

Figure 3. Arctic and Sub-arctic biomes

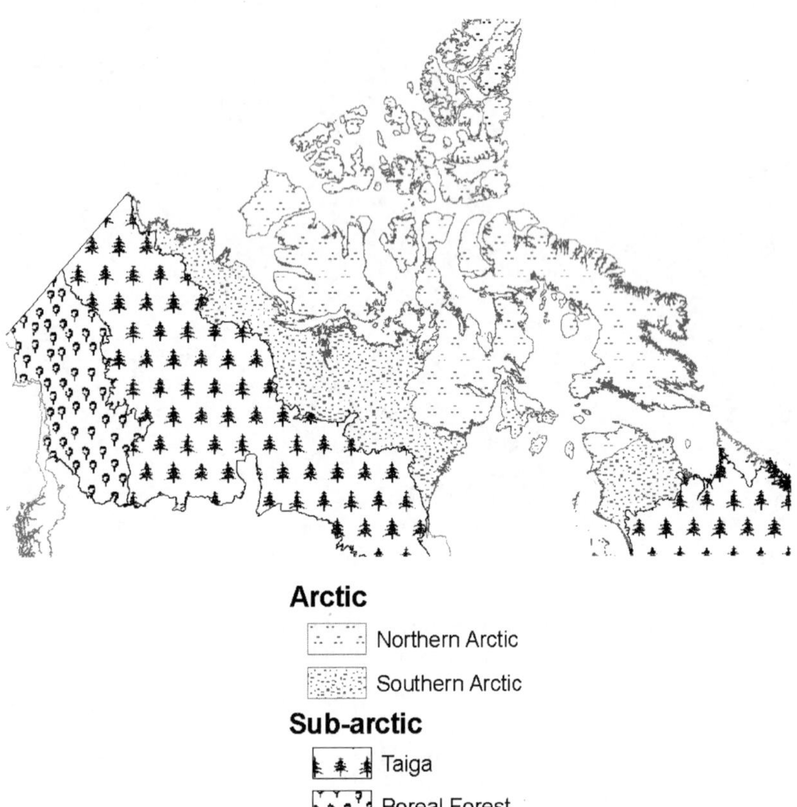

Source: Natural Resources Canada, 2003. The Atlas of Canada. Ottawa, Ontario, Canada.

Figure 4. Comprehensive Land Claim Agreements (CLCA) in the Territorial North

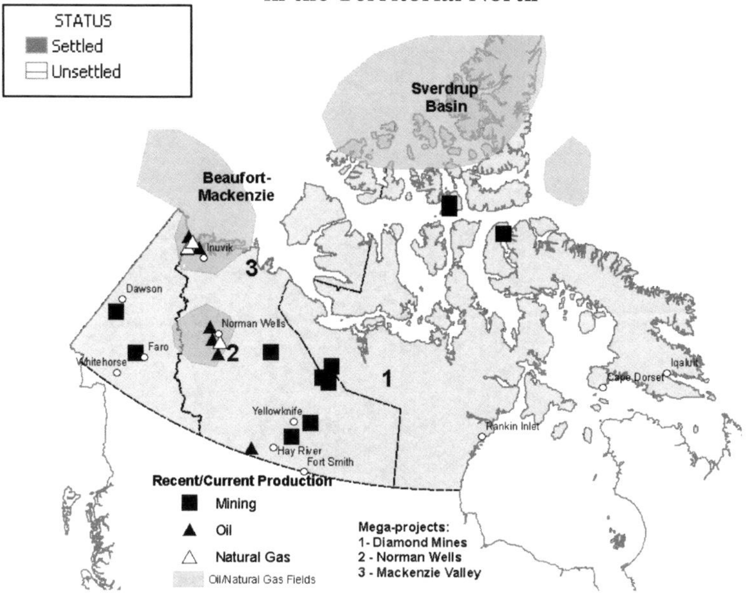

Source: Indian and Northern Affairs Canada.

Figure 5. Yellowknife, the Capital of NWT (2007)

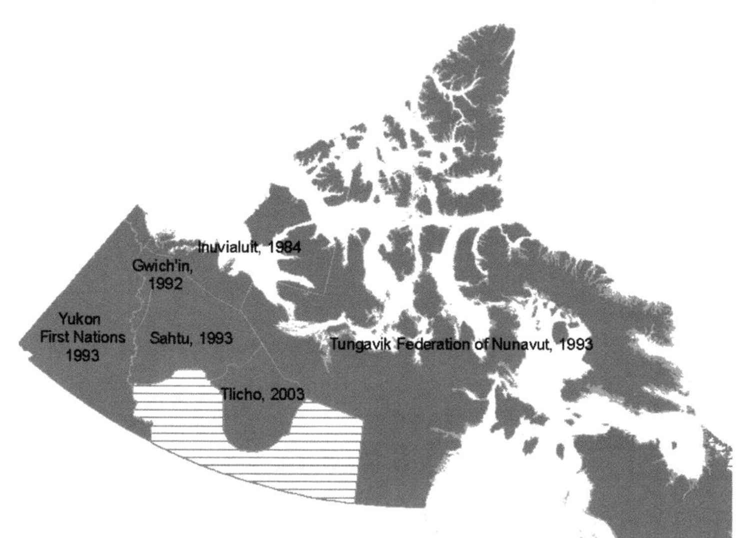

Source: Photo by A. Petrov.

Figure 6. Mineral Resource Based Economy of the North

Source: (multiple) A. Petrov.

Figure 7. Value added by industry in the Canadian Territories – 2002

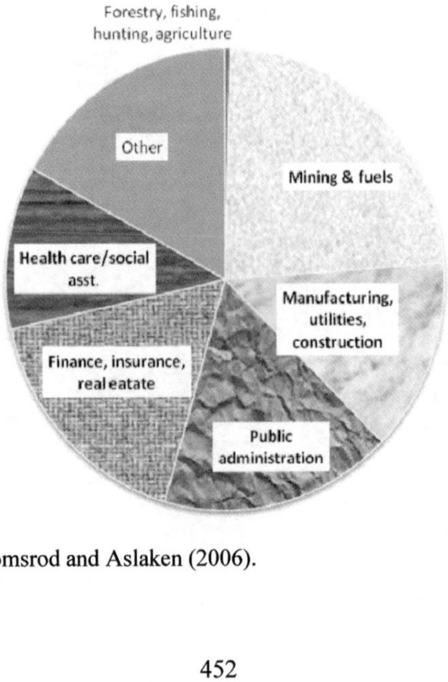

Source: Based on Glomsrod and Aslaken (2006).

CHAPTER 12

British Columbia

Geographies of a Province on the Edge

David A. ROSSITER

Western Washington University

Introduction: Society, Space, Environment, Region

For many Canadians, British Columbia (BC) sits at the edge of their country. While obviously true in a physical sense (the province abuts the Pacific Ocean), the notion that BC marks a series of national limits goes well beyond its location on the map (Figure 1). As the well-worn nicknames 'Lotus-land' and 'Left-coast' might indicate, the social, cultural, and political-economic characteristics of the province have also served to induce a sense of BC as distant from the centre of some mainstream Canada (were it possible to define such a thing). Further, with its reputation as a region of pristine and rugged natural beauty, much of the province has been represented as a space of wilderness; wilderness as the other to an increasingly urban world (Braun, 2002; Rossiter, 2004). And finally, the concentration of that urban world in BC's southwest corner stands as Canada's main gateway to the cultures and economies of the Pacific Rim, thus placing the province in the midst of political-economic and cultural realms quite distinct from traditional trans-Atlantic arrangements. In all of these ways, British Columbia has come to be seen as, in Jean Barman's memorable phrase, 'the west beyond the west' (Barman, 1996).

Despite all of this edginess, however, BC is in many ways central to Canada's social and political-economic life. With a population of 4.3 million people in 2007, the province was home to just over 13% of Canadians (Statistics Canada, 2008a). Of those people, nearly two million lived in metropolitan Vancouver (Figure 2), Canada's third largest urbanized area after Toronto and Montréal. As with these other metropolises, the citizens of Vancouver hail from an extremely diverse set of backgrounds and reflect national immigration and settlement programs. Beyond the densely urbanized southwestern corner of the province, natural resource landscapes have been constructed; landscapes of economic production that are in close relationship with the technologies and financing of metropolitan Vancouver. Scholars have noted this linkage between urban and rural/resources geographies in numerous regions across Canada, labeling it a heartland-hinterland relationship. In this sense, BC fits in the grand tradition of staples (raw natural resource) development that was outlined by the economic historian Harold Innis in the first half of the 20th century (Innis, 1995). In combination, the service-based industries of the cities and the resource-based industries in rural areas of the province produced a gross domestic product of over $163 billion in 2007, ranking fourth amongst all the provinces, following only Ontario, Québec, and Alberta. Overall, BC made up 12.5% of Canada's total GDP of $1.3 trillion in 2007 (Statistics Canada, 2008). Taken together, then, the demographic and economic characteristics of the province place it near the centre of national experience, not at the margins.

Figure 1. BC within Canada

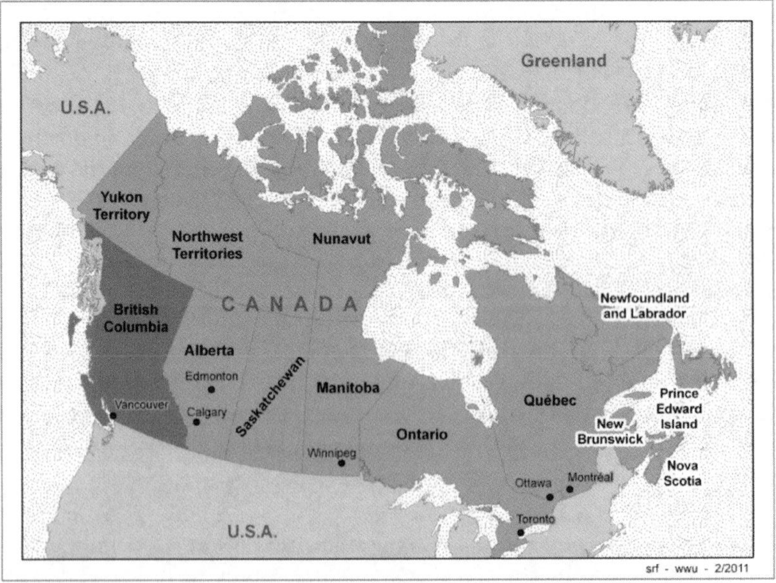

Source: Stefan Freelan, Western Washington University, 2/2011.

In this chapter, my aim is to unpack the ways in which the interplay between the factors that give BC its edge-like qualities and the characteristics that place the province near the centre of national life have served to produce a unique regional geography. In doing so, it is important to be mindful of three critical points regarding the study of regions. First, and most fundamentally, an ontological question is raised in the identification of any region: why *this* region with *these* boundaries? Regions are not a priori objects; they do not pre-exist their identification by humans. Rather, they are spatial categorizations that are constructed on the basis of selected geographical characteristics (both natural and social); these include geology, climate, vegetation, economy, politics, and culture. In the case of BC, then, the dominant characteristic used to define the region is political. The boundaries of the region are those of the Province, and they are lines that have emerged out of a colonial history with roots in the imperial activities of the late eighteenth century (Gough, 2008). With significant constitutional control over natural resources, social policy, and economic activity, Canada's provinces form powerful jurisdictions. As a result, the production of BC as a political unit has in many ways forged its social and economic character as well. In tracing the production of the political unit that is modern BC, however, it would be wrong to focus solely on political geographies and the social and economic spaces that have emerged within them; these spaces have been produced in a reflexive relationship with the environmental characteristics of the region. Thus, we must also pay attention to the interplay of political, social, and economic space with the province's environmental characteristics if we are to make sense of the emergence of BC as an important Canadian region. Indeed, the regional approach allows us to trace the ways in which regions are in fact socio-natural products that emerge as communities structure relationships with their environments.

Secondly, depending upon defining characteristics, regions can be identified across multiple scales. As a consequence, smaller regions can be nested within larger ones. In the case of British Columbia, such divisions are ubiquitous. The urbanized southwest corner of the province (colloquially referred to as the Lower Mainland) is often considered a region apart from the rest (much more rural) of the province. Or, drawing on environmental rather than socio-economic characteristics, British Columbians frequently distinguish between the Coast and Interior regions of their province. And, within several ministries of the provincial government, the administration of services and activities, ranging from health care to forestry management, occurs through regional divisions based upon social and/or environmental characteristics. Thus, as the regional geography of BC is explored in the pages that follow, it will become apparent that, just as Canada is in many ways a country of regions, so is BC a province of regions.

Figure 2. BC major centres

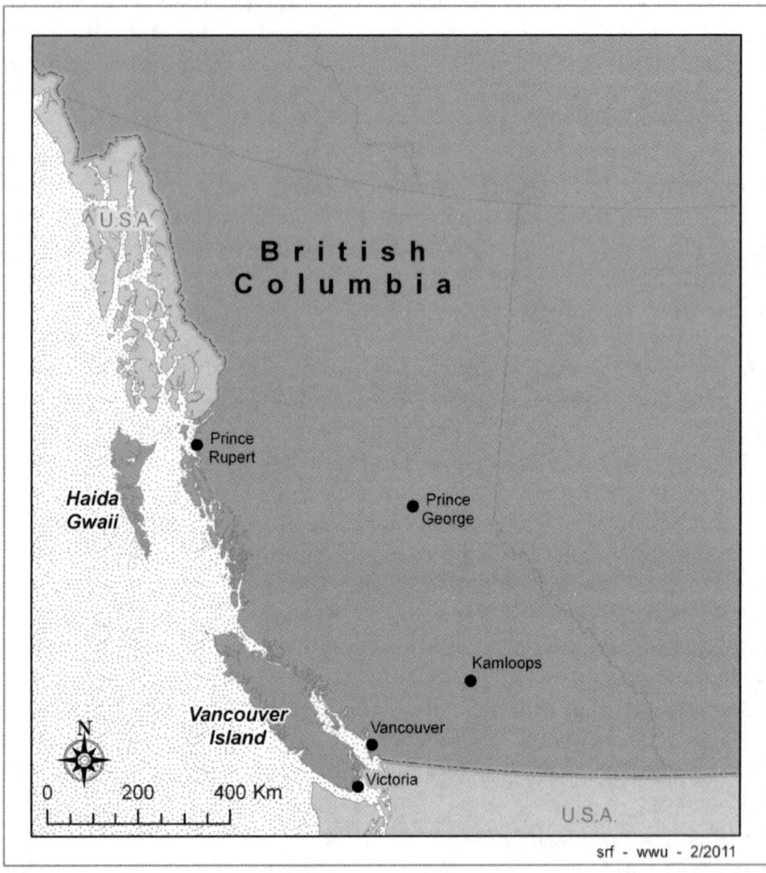

Source: Stefan Freelan, Western Washington University, 2/2011.

Finally, no matter what boundaries are drawn around which geographical characteristics, no self-contained region can be identified on the face of the earth; everywhere there are flows and connections between the spatial units that we identify as regions. As any biologist will tell you, the ecological systems of the lowlands of the Fraser River (the location of BC's Lower Mainland) do not observe the border between BC and Washington State in the United States of America; all manner of organic and inorganic substances cross the international border on an ongoing basis. Likewise, BC's economy and society, while regulated and administered by the Province, is deeply integrated into national and global networks: the province's forest, mineral, and agricultural products (to name a few) are sold to markets around the Pacific Rim; its peoples hail from and remain connected to dozens of countries and hundreds of cultures; and its landscape aesthetics are appreciated through digital (and other) media in offices and living rooms across the globe. Therefore, any exploration of BC's geographical character will necessarily involve thinking through its connections to spaces and places that are conventionally considered to be well beyond the region.

In tracing the production of BC as a distinct Canadian region, I proceed in roughly same manner as the other authors covering the 'regions' in this text by: 1) providing a broad overview of BC's physical geographies; 2) describing some of the processes and patterns that have made up the province's human geographies; 3) considering the mutually-constitutive characteristics of urban space and the development of the modern provincial economy. My aim is to sketch the major geographical factors that have shaped modern BC. While necessarily selective given the space constraints imposed by a single chapter, the narrative that follows will allow readers to place BC into the larger Canadian and, indeed, international contexts. With their geographical imaginations thus provoked, it is my hope that readers will press on and get to know the province in greater detail through both study from afar and experience on the ground.

Physical Geographies

Bounded by the rough waters of the Pacific Ocean to the west and the rugged peaks of the continental divide to the east, BC's natural landscape and climate have long been central to the forging of both provincial identity and political economy. With a land area of 925,186 km^2 and a total area (including water bodies) of 944,735 km^2, BC makes up approximately 9.5% of Canada's total area (Stanford, 2003: 39). While this share of national territory does not stand out in terms of sheer quantity, for Canadians, and many around the globe, the qualities that mark BC's physical landscape do resonate as remarkable. The province's literal position at the northwestern edge of North America (where over millions of years successive plates of the earth's crust have violently accreted to form rugged mountainous terrain, and where moisture-laden Pacific storms have provided the conditions necessary to support a temperate rainforest) provides geological and climatic explanation for BC's dramatic environmental character. However, this dramatic character also works to reinforce the image of BC as a province on the edge, as a place where modern society bumps up against a wild and undomesticated nature. Whether in terms of the aesthetics attached to modern tourism or the products derived from resource extraction, BC's physical geography provides a base upon which the complexity of human societies have played out. Thus, it is worth spending a few pages outlining the principal characteristics of the province's geology, topography, climate, and vegetation.

a. Geology and Geomorphology

The underlying rock structure of BC (its geology) and the development of landforms from that rock over hundreds of millions of years (its geomorphology) form the basis of the province's much remarked upon physical environment. With rock ranging in age from a relatively young 40 million years old (Cenozoic) to a more weathered 250 million years old (Mesozoic) and type from igneous (intrusive, extrusive, and metamorphic) to sedimentary, the rugged mountains, deep fjords, and high plateaus that mark the province have resulted from an extraordinarily complex set of geological processes. And, it is a complexity that has emerged out of the province's position on the edge. In the geological sense, BC's edginess comes from its position at the western edge of the North American plate of the earth's crust (lithosphere). This edge, over the course of hundreds of millions of years, has been subject to the powerful and violent tectonic forces unleashed by the movement of the plates of the earth's crust. As the westward drifting North American plate has encountered the Pacific plate, the latter has been forced under the former (subduction). As a result, rock has been folded, faulted, and exposed to volcanic activity on a massive scale all across the interface of these two plates. And, as this tectonic activity occurred over millions of years (and, it continues today), fragments of the larger plates (terranes) were forced up and against the North American plate (accretion). These accreted terranes form the basis of BC's mountain ranges from the coast to the interior west of the Rocky Mountains. The Rockies themselves mark the old western edge of the North American plate upon which the terranes accreted (McGillivray, 2000: 21-28). Thus, even in geological terms, BC has been produced through its edge-like qualities; its foundation is at one and the same time connected to an older core and extending away from that core in a process of complex and ongoing change.

In an effort to impose some order on this complex natural history, Brett McGillivray has identified five geological categories that cover BC's territory (McGillivray, 2000: 23-24). In the first category are the rugged Coast Mountains and the more rounded landscapes of the south central interior. Consisting largely of intrusive igneous rock that has been exposed through glacial processes, these formations are approximately 100 million years old. Due to subduction action off of the coast, the Coast Mountains also make up a zone of significant volcanic activity. As a result, extrusive igneous rock formations dot the area, thereby contributing to the geological complexity. The second category encompasses much of the remainder of the interior of the province, as well as the northern end of the Queen Charlotte Islands (Haida Gwaii). These areas consist of sedimentary rocks (built up under water and accreted on to the North American plate as a terrane) overlain with igneous rock resulting from lava flows dating from between approximately 40 and 50 million years ago. Thirdly, the region in the far northeast of the province (the Peace River region) stands as BC's flattest and least dramatic landscape. It is made up of sedimentary rock, is a part of the North American plate, and shares its geological structure and history with the Great Plains. The uplifted sedimentary rocks of the Rocky Mountains form the fourth category. The rocks are also a part of the North American plate and were folded and uplifted to elevations of more than 4,000 metres in a series of events that occurred between 250 and 65 million years ago. Finally, ancient metamorphic rock a billion years old or more makes up the core of the mountainous Kootenay region in the southeast of the province. While each of these categories is a simplification of an infinitely more complex geological reality, they are useful building blocks from which to proceed as we try to make sense of BC's geographical character.

While the geological character of BC is extremely diverse, however, almost all of the province has experienced a common geomorphological event: complete coverage by glacial ice during the Last Ice Age, or late Wisconsin Ice Age (so named after the US state that marked the southern limit of the ice sheets), that was at its maximum 18,000 years ago (Bone, 2008: 36-37). The grinding and scouring of rock as the ice advanced, as well as the re-deposition of these materials as the massive glaciers retreated between 18,000 and 10,000 years ago, stand as common processes in shaping the entire province's landscape. From the rugged alpine landscapes of arêtes and cirques in the Coast Mountains to the long valley bottom lakes of the interior's Okanagan Valley, evidence of glacial activity is ubiquitous across the province. Indeed, remnants of the giant ice sheets remain in the high alpine reaches of the Coast Mountains, the interior ranges, and the Rocky Mountains. In a sense, these mark the historical-geographic edge of the last epoch that saw major transformation of BC's geological character.

b. Weather and Climate

As a largely mountainous region, BC experiences an extraordinary variety of weather. Following the dominant hemispheric air movements and the jet stream, weather systems form over the Pacific Ocean and arrive on the coast of BC, the edge of the continent, more or less laden with both moisture and energy. As these systems reach land they are forced up by the immediately mountainous terrain. The resultant orographic precipitation is heavy along the coast, leaving that part of the province with a maritime climate with temperature moderated by the relatively warm waters of the Pacific Ocean. On the leeside of the Coast Mountains, however, the orographic regime results in a rain shadow effect where precipitation levels drop off dramatically from those experienced on the coast. As systems continue to move eastward with the jet stream, they pick up more moisture from inland water bodies and the orographic effect is repeated across successive mountain ranges. Thus, the Okanagan Valley of the south and central interior stands as one of the driest regions of the province as it rests in the rain shadow of the Coast Mountains. In contrast, the Monashee and Selkirk mountains to the east receive higher levels of precipitation as the air masses that reach this region have been somewhat recharged with moisture after having moved through the Okanagan valley. In addition to affecting the precipitation regime, the province's mountain ranges make for highly variable temperature regimes. As a general rule, as elevations rise, temperatures fall. But, beyond this simple formulation, variability reigns. Air masses are variously trapped and/or channelled by valleys, with the result being the development of numerous micro-climates across relatively small regions. Taken together, the complex weather patterning can once again be thought of as an effect of the province's position on the edge of the continent. The combined effect of a rugged topography and positioning as first landfall for weather systems coming off of the Pacific Ocean, then, has produced a weather profile for BC that reinforces its identity as a unique region.

While the influence of the Pacific Ocean and the province's mountain ranges play a central role in shaping the climate of BC, their influence is moderated by hemispheric and global air mass circulations. Of particular interest are two dominant pressure systems off of the west coast of North America. The Pacific high to the south generates general clear and stable weather, while the Aleutian low to the north generates storms with high winds and heavy precipitation. These two pressure systems shift to the north with the jet stream in the winter months and to the south in the summer months. As a consequence, from October to March, coastal BC is subject to the storms and high precipitation produced by the Aleutian low. Conversely, for the remainder of the year, the Pacific high dominates and the weather, particularly on the south coast, tends to be mild and dry. The significant seasonal variation in precipitation levels is well demonstrated by the weather data collected for the area around the city of Vancouver in the province's southwest corner. Out of a total average annual precipitation of 1,167.4 mm, only 235 mm falls between April and September (Stanford, 2003: 19). As a result, it is commonplace to think of the coast (particularly in its southern reaches) as having only two types of weather: rainy and not rainy. With the maritime influence moderating temperatures, such an attitude is not unreasonable for inhabitants of coastal BC; indeed, Vancouver's annual average temperature range is only 15 °C, with July and August as the warmest months at 18 °C and January as the coldest at 3 °C (Stanford, 2003: 19). However, the interior of the province does experience much wider variations. Away from the coast and the immediate influence of Pacific air masses, the interior regions of the province experience continental climate regimes with both colder and drier winters and warmer summers than the region proximate to the Pacific. Taken as a whole, then, BC exhibits a highly variable climate. And, at root, this diversity results from the meeting of Pacific weather systems with the rugged and soaring edge of the North American continent.

c. Water, Soils, and Vegetation

Water has played an important role in shaping BC's landscape and environmental character. Fed by the copious precipitation arriving with storms off of the Pacific Ocean and stored in the lakes, snowpack, and glaciers of the cordillera, BC's hydrological system has contributed to both the shaping the province's landforms and the provision of the requisite habitat for its flora and fauna. Large rivers such as the Thompson, Fraser, and Skeena (Figure 3) have carved deep valleys through the heart of mountain ranges and high plateaus, the sediments carried by their waters producing wide and flat deltas where they meet the Pacific Ocean. The banks and benches of these river valleys and the lowlands surrounding the deltas contain soils that are high in nutrients, and thus valuable for agricultural production in a landscape that is otherwise quite hostile to agriculture (McGillivray, 2000: 34). The Okanagan Valley and the lowlands surrounding the delta of the Fraser River in the province's southwest corner are particularly fertile examples in this regard and thus, serve as the principal agriculturally productive parts of the province. It is in large part from the flow of water and the concomitant transport of sediments and nutrients that BC's arable lands and habitable areas have been produced.

As the plentiful precipitation that re-charges the system moves through the soil layers on mountain slopes and valley bottoms near the coast, it combines with the relatively mild maritime climate to produce conditions conducive to the growth of a temperate rainforest ecosystem. Growing in podzolic soils, the trees of the coast are of magnificent size and age. Douglas fir, hemlock, and Western Red cedar are the dominant species, and, where stands of old-growth forest still exist (having escaped the blades of industrial forestry) one can find trees of several meters in diameter and many centuries in age.

Figure 3. BC major rivers

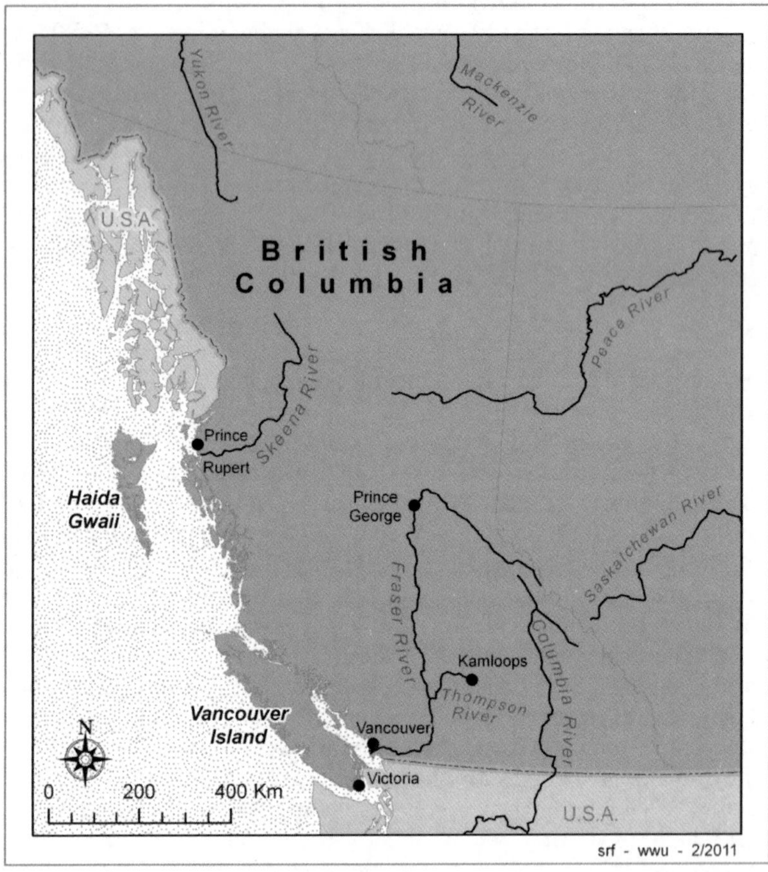

Source: Stefan Freelan, Western Washington University, 2011.

The south and central interior, by contrast, are marked by pockets of grasslands and coniferous tree coverage by species that are smaller than those found on the coast. Here, the lodge pole pine and spruce species dominate, growing in a complex array of soil types. Finally, the northern third of the province is made up of boreal forest; black spruce dominated forest in a northerly climate that results in a dense forest landscape composed of small, hardy conifers (Bone, 2008: 53; McGillivray, 2000: 34-36). In each of these three generalized zones (the coast, the south central interior, and the boreal north), climatic conditions combine with water and soil regimes (which are in turn predicated upon geological character) to produce distinct vegetation regimes. Thus, in the forest cover (or in the grasslands) we can catch a glimpse of the combined effect of geology, geomorphology and climate on variably providing the requisite conditions for life.

Settlement and Resettlement

The province's physical geographies have long been central to human engagement with BC's environment. That engagement, however, has been variable across both time and different cultures. An illustrative example: in the midst of an overland journey from the settlements around the Red River (in present day Manitoba) to the Pacific Ocean in the summer of 1863, the amateur British explorer Viscount Milton and his entourage encountered a thickly forested river valley. The party had crossed the vast prairies of Rupert's Land, traversed the Rocky Mountains through Yellowhead Pass, and descended into a part of the British Colony of British Columbia that, to that point, had hardly been encountered by people of European background. In an account of his travels that was first published in 1865, Milton described the scene that confronted them as they headed south through the valley:

> No one who has not seen a primeval forest, where trees of gigantic size have grown and fallen undisturbed for ages, can form any idea of the collection of timber, or the impenetrable character of such a region... timber of every size, in every stage of growth and decay, in every possible position, entangled in every possible combination (Milton and Cheadle, 2001: 286-287).

Fifty-eight years later, in 1921, the same valley was a subject of the BC Forest Branch's *Annual Report*. Timber cruisers, who were conducting 'Forest Reconnaissance' for the provincial government, noted:

> In the bottom lands a good stand of spruce balsam is found, while on the slopes hemlock and cedar come into the mixture and balsam drops out. The total area of merchantable timber... was found to be 10,110 acres. A total stand of 193,000,000 feet was found on this area, consisting of 40 per cent. spruce, 40 per cent. cedar, 10 per cent. hemlock, and 5 per cent. each of fir and balsam (British Columbia, 1922: 20-21).

Offered about three generations apart, these accounts of the valley seem to reflect significantly different views of the forested landscape. Milton's account places him squarely within the forest, and everywhere he sees chaos and barrier to his passage. His words communicate mixed emotions of fear and wonder. To his mind, this was an unknown land; he was in the midst of a wilderness marked by unending forest. By contrast, the *Annual Report* reflects knowledge of and control over the forest valley. It portrays the landscape from the perspective of a detached observer, surveying from above. It suggests a conquest of wild nature by human reason. Despite these differences, however, the descriptions are a part of a single legacy; they are both voices of European colonialism in the northwest corner of North America that viewed the territory as an "edge of empire" (Harris, 2004) subject to domestication and cultivation. Aboriginal peoples understood the spaces of their homelands in quite different terms.

The valley that these excerpts address was carved by what is presently known as the North Thompson River. Slicing southwards through the heart of central BC, the valley bottom and river have served as both a means of travel and source of subsistence for generations of Native peoples. Beginning in the second half of the nineteenth century, however, a settler society, rooted in the British Empire, began to lay down a network of modern communications and transportation in order to accommodate capitalist exploitation of the valley's natural resources (Harris, 1997: 161-193). It was a process that began in the second half of the 18th century along the coast and eventually played out all across the territory that ultimately became Canada's west-coast province. As a result, Native human geographies were disrupted by those of newcomers, often to the point of rendering them invisible and doing great damage to aboriginal societies. Central to this process were the perceptions and understandings, environmental and otherwise, brought to bear by settlers and how these shaped colonial society and space (Clayton, 1999; Owram, 1992). Milton embodied the European encounter with 'the wilderness' which marked so much of the contact history of the 'New World.' He was a man of the centre of Empire; a man who had a definite sense of the global reach of the British Crown, and yet found himself in a country of which he could not make sense. In clear contrast, the description in the Forest Branch's *Annual Report* points to an intellectual mastery of the landscape that had so terrified and frustrated the band of explorers that passed through little more than half a century prior. In the intervening years, the valley went from an unknown (to Europeans) region on the margins of the British Empire, to one arranged around modern western science and capitalism.

Rather than call colonial processes and the new human geographies that were produced "settlement," as is common practice, historical geographer Cole Harris has coined the term "resettlement" (1997). Thus described, the nature of colonialism is portrayed in a much more accurate light; the replacement of one people's human geographies by those of another. In the case of British Columbia, aboriginal depopulation through exposure to epidemic disease combined with racism and the momentum of European, Canadian, and American capital (at first centered on fur trades, and later on minerals, timber, and fish) to produce, initially, the British colonies of Vancouver Island (established 1849) and British Columbia (1858) and, ultimately, The Province of British Columbia, the seventh province to join the Canadian federation (1871). It would be erroneous, however, to regard this history as having erased Native cultures and spaces in BC. Rather, the province's First Peoples have, in the face of over a century of physical, social, political, and cultural marginalization, retained their identities and, particularly over the last generation, have begun to powerfully re-assert claims to land, resources, and governance. For the remainder of this section, I trace this story of cultural contact and geographical change by describing the outlines of pre-contact Native human geographies, the spaces of contact and colonialism, and the emergence of the human geography of modern BC over the course of the 20th century.

a. First Space

The first humans to use and occupy the lands that are now called BC were the ancestors of the aboriginal peoples currently residing in and around the province. How these people arrived in the region is a point of some debate. Anthropologists and archaeologists have constructed a general case (one which has multiple and contested variants and time-frames) that argues that the first human inhabitants of North America crossed into the continent from Asia by way of a land bridge across the Bering Strait as the glaciers from the late Wisconsin Ice Age retreated, providing ice free corridors south through the eastern portion of the cordillera (and, some believe, along portions of the Pacific coast). By this measure, the ancestors of BC aboriginal peoples can claim approximately 10,000 years of ongoing use and occupation of the territory without facing much dispute (Wynn, 2007). Alternately, aboriginal peoples' creation accounts identify their occupation of their homelands as dating to "time immemorial"; indeed, for Native societies, the people are of their land (Carlson, 2001). Despite the differences within and between these explanatory paradigms, it is clear that human geographies marked the region for millennia before any European knowledge of, or actions in, the area. And, they were complex human geographies, produced by aboriginal societies, economies, and polities with a total population estimated at between 80,000 and 200,000 people on the eve of contact with Europeans (Harris, 1997: 20-30).

The cultural geography of aboriginal peoples in the territory that is now BC is extremely complex, particularly along the coast; an environment that supported the flourishing of multiple aboriginal groups. These groupings, distinguished by language and dialect, have come to be referred to as First Nations. More than cultural groupings, however, First Nations served (and continue to serve) as units of social, economic and political practice and they interacted with one another for mutual benefit. Thus, members of both coastal Nations such as the Stó:lō (Fraser lowlands and south coast), Nuu-chah-culth (west coast of Vancouver

Island), Haida (Queen Charlotte Islands, or Haida Gwaii), and Tsimshian (north coast) and interior Nations such as the Nlha7kápmx (Thompson River Valley) and Shuswap (southern interior) engaged their neighbours in trade and social and political alliances, relationships that served to keep cultural geographies in an ever-dynamic state as marriage, war, and slavery blurred the boundaries between groups.

While inter-nation trade and communication provided opportunities to increase groups' social, political, and economic capital, however, daily life for Native peoples in the Pacific Northwest centered upon the annual round of resource procurement; activities pursued by extended families who were members of the larger Nations. The Stó:lō peoples of the lower Fraser River provide a useful example. From spring through fall, families moved through a series of fishing, hunting, and gathering sites along the river, in its estuary, and up the mountain sides of its watersheds. These sites, rights to them inherited by families though a matrilineal system, supplied subsistence throughout the growing season. However, one particular resource was also stockpiled for the winter: salmon. Caught in late summer and early fall at sites along the Fraser and dried upon racks in the sun, salmon provided a source of protein throughout the cool, dark, and wet winter months (Carlson, 2001). As a consequence, the Stó:lō (and other coastal groups) were able to remain sedentary during the winter, thereby providing an opportunity for families to come together in large village sites and engage in art and social and political life. The art that northwest coast Native peoples are famous for emerged out of this set of circumstances. Thus, with complex geographies that depended upon an intimate engagement with the physical environment, aboriginal peoples in the territory that would become BC had certainly settled it in their own way.

b. Spaces of Contact and Colonialism

Although never closed and insular, in the middle of the 18th century the aboriginal geographies of the Pacific Northwest began to be exposed to outside peoples and places on a scale unknown up to that point. Russian, Spanish, British, and American explorers and fur traders had, by the turn of the 19th century, brought knowledge of the region back to Europe, thereby facilitating imperial designs on its lands and resources. While the Russians were the first to chart the northern portion of the coast and interact with the Native peoples in that area and the Spanish were among the earliest Europeans to contact the lands and people around what is now called Vancouver Island and the Georgia Basin, it was the British who ultimately dominated the Enlightenment-era exploration and mapping of the region. Most famously, Captain James Cook's voyages in the last decades of the 18th century served to render the spaces of the Island intelligible for imperial audiences in London; audiences who were aroused by the potential wealth that might be produced by engagement in trade with the local Native peoples whose supplies of luxurious sea otter furs had been noted by Cook's crew members (Clayton, 1999). In the years immediately following Cook's voyages, one of his officers returned to the region to continue to chart its waters and shoreline (Hayes, 2005). Captain George Vancouver, by filling in the blank spaces of European knowledge around and across from the Island named after him, further enabled British commercial activity in the area and at the same time solidified British claims to the territory (Clayton, 2000). Despite the increasingly dominant position of the British, during the first decades of the 19th century the coast remained as an international fur-trading realm. After the 1821 consolidation of two competing fur-trade companies under the banner of the Hudson's Bay Company (and coincident with the near extinction of the otter along many areas of the coast pushing the trade inland), however, commercial competition was reduced to that between the Russians and the British to the north and the British and Americans to the south. By 1846, the international boundaries marking out Russian territory to the southern tip of the Alaskan panhandle at 54°40' N the north (sold to the US in 1867) and the US territory of Oregon below 49°N (Gough, 2008) had been established. Over the course of a single century, then, aboriginal territorial claims and relationships that had been built up over millennia were destabilized, and the possibility for the construction of radically new human geographies was set firmly in place.

The resettlement of BC by non-Native peoples began in earnest with the land-based fur-trade; a trade that was dominated by the Hudson's Bay Company. With operations based out of Fort Victoria (established in 1841) on the southern tip of Vancouver Island, the company's men ranged across the island and adjacent mainland and traded blankets, firearms, and metal implements with Native groups, in exchange receiving a range of pelts from inland species. Beavers and martens were the new sea otters. As the Company's men pursued this trade, they did so within a very particular geography. While people of European background controlled the small spaces of the traders' forts, beyond these islands territorial control rested largely with Native peoples. Historian Robin Fisher has argued that during this period, as well as during the earlier maritime trade, Native societies fared rather well; wealth was increased and many Native leaders were able to cement political alliances and the health of their leadership through exploitation of the trade (Fisher, 1977). However, Harris has cautioned that the assertion of European control through violent displays at and near forts that induced fear in aboriginal people was a part of the story and must not be overlooked (Harris, 1997: 31-67). Despite these different emphases, both analyses of the trade's impact upon Native peoples point to a human geography dominated by aboriginal populations that have yet to lose control of their resources and territories to peoples of European and other backgrounds.

Formal efforts at colonizing the region with British settlers began in 1849. Seeking to keep the administrative burden on the Crown as low as possible, and assuming that the Hudson's Bay Company knew the country best, the British government granted the fur-trading corporation a ten year charter with exclusive trade rights under the condition that the Company engage in efforts to colonize Vancouver Island. The Chief Factor at Fort Victoria, James Douglas, was made Governor of the colony and charged with settling the area around the Fort and exploring the island for new commercial activities (Barman, 1996: 53-54). Douglas did not have much success, beyond the establishment of a small and insular transplanted British society in the immediate vicinity of Victoria. Beyond this, the Island remained a vast resource frontier that was exploited by Native peoples and fur-traders. And then, in the spring of 1858, the situation changed dramatically. The discovery of gold in the Fraser River on the mainland across the Strait of Georgia from the Island caused a stampede to the region by fortune seekers, veterans of the 1849 rush in California among them, looking to stake a claim. The small town of Victoria was flooded with miners purchasing supplies en route to the mainland, and the generally sleepy nature of the Colony of Vancouver Island was shattered over the course of the summer. Fearing that American miners would run lawlessly across the Fraser Valley, and cognizant of the tenuous nature of British claims to a territory that featured virtual no physical expressions of British possession, the Colonial Office in London granted Crown Colony status to its mainland territory, naming it the Colony of British Columbia and choosing a strategic upland site on the north bank of the Fraser River for the capital city of New Westminster. The following year, the Hudson Bay Company's charter to the Island expired, and, in light of the Company's poor record in colonizing the area, Vancouver Island was also proclaimed a Crown Colony. By 1866, the two colonies had been united into the Colony of British Columbia (with Victoria serving as capital), with that old veteran of the fur trade, James Douglas, serving as Governor for the massive territory (Barman, 1996; 72-98). In less than a decade, the rush to the region unleashed by the discovery of gold on the Fraser served to solidify British colonial aspirations. Small mining settlements, fishing operations, and lumbering operations quickly emerged on the Island and lands accessible from the Fraser River and the southern coast (Rossiter, 2007). While still outnumbered across the entire territory, non-Native peoples' human geographies began to radically alter aboriginal worlds. The full force of dispossession of was on the verge of being released by a combination of the burgeoning settler state and the momentum of private investment in resource production.

As British settlers began to occupy what was previously Native space, they did so largely in the absence of formal treaties transferring title and control from one party to the other, a stipulation of British settlement in North America since the Royal Proclamation of 1763 recognized prior aboriginal title to and occupation of clearly used lands. The only exceptions were the handful of so-called Douglas treaties signed by the Governor around Forts Victoria and Rupert in the 1850s. Outside of these agreements (in effect, purchases) concerning small parcels of land (Harris, 2002: 18-30), no other treaties were signed with Native peoples. Scholars have debated the reasons for this, with a general consensus forming around a combination of racist views of Native rights to what was seen by the British as 'unimproved' land, a pragmatic concern for the cost of treaty negotiation, and a sense that much of the territory was empty and, thus, free for the taking (a sense that was facilitated in part by the emptied landscapes left in the wake of epidemic diseases in the late 18th century and in part by the extensive, as opposed to intensive, land uses that marked aboriginal human geographies) (Harris, 2002: 45-69). Thus, right up to the eve of BC's joining the Canadian federation in 1871 and then beyond, Native peoples (particularly those on the coast and in the south and central portion of the colony) were marginalized within their own territory, set up on tiny reserves laid out by representatives of settler society, and generally considered to be inferior members of a newly emerging society that was based upon British social, political, economic, and cultural systems.

c. Provincial Space

Upon joining Confederation in 1871, BC was a sparsely populated (by non-Native peoples) resource frontier. Political and administrative functions were carried out in the young cities of Victoria and New Westminster, with resource extraction shaping the development of settlements at Nanaimo (coal) on the east coast of Vancouver Island and Burrard Inlet (logging and milling) on the mainland (the future site of the City of Vancouver). With promises of both a transcontinental rail link to central Canada and the assumption of BC's substantial debt by the federal government in Ottawa, politicians in Victoria deemed the deal to be a good one. And, as Prime Minister John A. McDonald's dream of a federation stretching from the Atlantic to the Pacific hung in the balance, concerns in Ottawa regarding the disposition of 'Indian lands' in BC seem to have been set aside. Over the next decade, the young province's settler population would swell to half of the overall population of 50,000 (as gleaned from the results of the 1881 Provincial Census) as people came in search of opportunity in Canada's newest province (Harris, 1997: 138).

Young provincial society in BC was organized around British notions of civilization and society, with Anglo-Saxon 'whiteness' being a consciously applied cultural marker. It was a marker laid down by members of a dominant culture who lived in close proximity to minorities of quite different backgrounds. Of the roughly 25,000 settlers enumerated in 1881, 4,200 were identified as 'Chinese' (Harris, 1997: 138). These non-European immigrants had come to the place that they referred to as 'Gold Mountain' in order to work on the construction of the Canadian Pacific Railroad through the seemingly impassable mountains of the Canadian cordillera. Finding opportunities beyond the railway, Chinese people integrated into BC society in cities and towns by starting businesses to serve the needs of settler society. As a distinctly non-British settler group, Chinese migrants to BC set the stage for a historical-geography of immigration to BC that would have connections to places all across the world.

As peoples hailing from places as diverse as Sweden, Russia, Italy, and India made their way to BC in the last years of the 19th and early decades of the 20th century, they found themselves in a multicultural, yet socially and economically stratified society. White men of British background made up the social, political, and economic elite, and then prestige worked backwards on a sliding scale depending up, generally, darkness of skin color. This race-based stratification of society was made manifest in the spaces of both the province's cities and resource settlements. The experience of Japanese Canadian sawmill workers in the early 20th century is exemplary (Kobayashi and Jackson, 1994). The workforce in BC sawmills in the period around the turn of the twentieth century was marked by two racial groupings: White and Asian (Hak, 2000: 150). The term 'Asian' was a large category deployed by white British Columbian society in identifying a range of peoples from the eastern and southern portions of the continent of Asia. In the case of sawmills, however, while there were workers of both Chinese and Japanese backgrounds, the latter were present in twice the number of the former in 1902[1]. Further, it was a significant and growing presence. In that year, the total number of Japanese Canadians working in BC sawmills was roughly 500. By 1917 that figure had jumped to about 700 (Kobayashi and Jackson, 1994: 41-44). The experience of Japanese Canadian sawmill workers was conditioned by the construction of

1 This discrepancy was in large measure due to the enactment of Federal anti-immigration legislation in 1886 that taxed each Chinese arrival. In the years following the tax would become more onerous, making it increasingly difficult for migrants from China to settle in Canada (Kobayashi and Jackson, 1994: 45).

'Asians' through official and popular discourse. First and foremost, people of Chinese and Japanese origin were seen to constitute an 'Oriental threat' or 'Yellow peril' to mainstream British Columbian society. This was felt largely along the lines of the possibilities, or lack thereof, for assimilation into 'mainstream' BC society. Kobayashi and Jackson cite a typical statement from a submission to the Royal Commission on Chinese and Japanese Immigration of 1902 by a sawmill worker from Vancouver to demonstrate this: "My principal objection to them is that they do not assimilate, cannot assimilate, with our race, and that our country should be for men [*sic*] of our own race, instead of being overrun by an alien race." It was felt that Asian migrants had no allegiance to mainstream society and that their presence was in conflict with the interests of white settlers.

Further, Asians, particularly Japanese, were seen as constituting a direct economic challenge to white British Columbians. Consider the words a Provincial Immigration Officer from Vancouver Island in 1902: "I consider Japanese cleanly in habit, industrious and intelligent. I believe them to be more dangerous competitors in the business of the country than the Chinese" (Kobayashi and Jackson, 1994: 46). Showing durability beyond official discourse, however, these positions were also clearly articulated in more popular portrayals. Early 20th century settler George Godwin's autobiographical novel *The Eternal Forest* serves as a useful example. During a passage describing the economy of the lower Fraser Valley in the first decade of the twentieth century, the character of an old settler moans: "Traveller up this morning from Sapperton, bad times, people buying Chinese eggs all along. Chinese peddlers selling the garden truck they grow on Lulu Island, cutting our throats" (Godwin, 1994: 43). And, in the case of the Japanese, the narrative points to fear of more than mere economic competition. Commenting on the sale of local land to two Japanese men, another character claims, "(i)t's the Japs' purpose to get this Province by peaceful penetration… Yes, the Japanese (are) out to get the whole Pacific slope" (Godwin, 1994: 92-93). Racial stereotypes such as these, sometimes differentiating between Chinese and Japanese and sometimes not, formed the foundations for Japanese Canadians' work experience in BC sawmills, a space where they were in economic competition with other members of settler society. It was a competition, however, where the rules did not apply evenly to all of the players.

Kobayashi and Jackson (1994: 54-56) identify two related consequences of such racialization with regard to Japanese Canadian participation in sawmill labour. First, the outright hostility towards 'Asians' within society at large allowed for, in the kindest terms, a blind acceptance of a racist wage scale on the part of the majority of the BC public. In 1902, unskilled white labourers earned between $1.50 and $1.75 per day at The Royal City Planing Mills at New Westminster, while Japanese workers performing the same work earned $1. This differential was repeated in mills that were popping up along the south coast and in the southern interior. The existence of such unequal wage rates had major effects on the workplace. Overall wages were suppressed and racial tension was exacerbated by the perception among white workers that the 'Asians' were undercutting their ability to earn a living wage. Second, the tensions and hostilities created and reproduced by the racialization of Japanese Canadians as 'cheap labour' resulted in strategies of resistance by Japanese people to the ascription of a place in settler society: the creation of self-help societies, leaving sawmill employment for independent farm or fishing work, and the establishment of Japanese union organizations. Thus, in response to the prevailing conditions of work, Japanese Canadian sawmill workers in the first decades of the twentieth century constructed their own geography, their own social and economic spaces. Mill work was sought for short-term employment until one could move on to a farm or fishing boat. Social and economic circumstances necessitated the creation of ethnic societies and organizations that would provide workers with some power and security. Through such moves, Japanese Canadian labourers were going about making BC their place, too.

The production of social and economic space through and in response to racism occurred many times over throughout the 20th century. From small cities such as Prince Rupert on the north coast (fishing) and Kamloops in the central Okanagan Valley (ranching and timber) to the young metropolitan societies growing up around Victoria and Vancouver, immigrants from around the world came to BC looking for some way to get ahead. Whether through the chance for a steady wage in order to send money back home to support family members or the opportunity to rebuild lives in the aftermath of the two World Wars, BC was seen as a space of possibility. For some, those possibilities were happily realized. For others, experiences of bigotry, loneliness, and cultural disconnection led to dashed hopes. In this regard, BC shares in a Canadian experience. Indeed, immigration policy is administered by the Government of Canada. However, BC's position at the edge of the country and on the Pacific Rim has shaped the geography of migration to its shores (in both the federal era and prior). And, the province's resource extraction dominated economy, again an effect of BC's edge-like environmental profile, has shaped the spaces immigrants have made for themselves upon arrival in the Pacific Northwest of North America. As such, BC's historical geographies of settlement, from Aboriginal to European to Chinese, represent the ongoing interaction of multiple cultures with a unique environment; an interaction that has served to produce a unique human geography within Canada.

Urban Development and the Modern Spatial Economy

While the vast majority of BC's territory might be considered to be either rural or wilderness space, the bulk (approximately two thirds) of the province's citizens reside in the urbanized southwest corner of the province in Victoria on Vancouver Island and in the area surrounding Vancouver that is commonly referred to as the Lower Mainland. Located in the broad and fertile delta formed by the Fraser River as it empties into the Strait of Georgia, the highly developed region surrounding the City of Vancouver is BC's metropolitan hub. The centre of the province's financial life, the region contains polished office towers, trendy and gentrified condominium blocks, and sanitized suburban strip malls, all of which seem to stand in stark contrast to the image of a province built upon the work in the woods and mines. This is an opposition, city versus wilderness, which lies at the core of BC's edge-like geographical character, however. Vancouver, founded upon a sawmill site in 1886 as the Pacific terminus of the Canadian Pacific Railway (McDonald, 1996), grew into the modern metropolis that it is today from its position as the nexus between the resources landscapes of the province and the markets, both Canadian and international, into which the products of those landscapes are sold. In the city are found the financial, insurance, real estate, accounting, and other services required by individuals and corporations engaged in resource extraction across the province; an extraction of resources that are mostly publicly-owned, with access granted to producers through tenure and royalty systems that feed revenues into provincial coffers. This relationship between an economic 'core' and resource 'periphery' has been coined the 'heartland-hinterland' relationship (McCann and Gunn, 1987), and it can be noted across multiple regions of Canada.

It is also a geographical relationship that one can detect at multiple scales. So, while Vancouver has served as the heartland for the timber, fish, and mineral producing hinterlands of the province, all of BC might be thought of as a hinterland to the cities of central Canada whose manufacturing industries bought many of the raw resources produced in the western provinces. Alternately, one can trace heartland-hinterland relationships at the regional level *within* BC. While Vancouver dominates industry and finance and Victoria holds down the political centre of gravity, smaller cities act as regional service centres to local resource industries: Prince Rupert historically anchored the fisheries of the north coast; Prince George started out as the principal fur trade post of the upper Fraser River region and developed into the centre of the northern logging and milling industry; Kamloops services the forestry and ranching industries of the south central interior; and Nelson, once the focus of the mining operations of the province's southeast corner, now serves as the headquarters for outdoor tourism in the area. While these regional cities' fortunes have waxed and waned along with both the health and technological development (Rajala, 1998) of the various resource industries active in BC, the fundamental geographical pattern that they relate remains as a key to understanding the province's geography: the wealth of the urban system is dependant upon the production of hinterland regions. In a sense, then, such dependence might point to the resource producing regions as the true 'heartlands.' Such a rearranging of terms would certainly undo the assumptions of an implied geographical hierarchy that lies in the background of an otherwise useful way of conceptualizing the spatial economy of BC and, indeed, Canada.

Beyond, and, indeed, because of, Vancouver's position as the financial centre of BC, it has served as the province's (and, in many ways, Canada's) 'gateway' to Pacific Rim societies and economies. And, while social, cultural, and economic ties have linked BC to the markets and peoples of Asia for over a century, recent experiences of accelerated and intensified international flows of people, capital, and information have created a metropolitan region of extraordinary cultural diversity and vibrancy (McGillivray, 2000: 209-224). These are also experiences, however, that have been accompanied by the challenge of creating a harmonious and cohesive society amongst peoples of a variety of quite different backgrounds. In this respect, life in metropolitan Vancouver is not unlike that in Calgary, Toronto, or Montréal. Again, however, the city's position at the edge of the Pacific Rim has greatly shaped its experience of the phenomena ubiquitously known as 'globalization'; it has influenced the origins of the people who have come to the province, the industries in which they were engaged, and the host culture into which they arrived. Thus, what began as the service centre of BC's resource industries and the nexus of Canada's Pacific trade grew, over the course of a century, into a major world city deemed worthy of hosting the 2010 Winter Olympic Games.

Conclusions: Centres, Edges, and Claims to Nature and Space

A vast and complex region of Canada, BC is well conceived of as exhibiting an edginess that distinguishes it from the rest of the country. From the rugged landforms created by the meeting of two plates of the earth's crust to the role modern Vancouver plays as a cultural interface between Canada and the peoples of South and East Asia, the province serves to connect the northern part of North America to a significant portion of globe. At the same time, and thus enabling this sense of connection at the edge, BC has remained tied to the centre of Canadian historical-geographical development; its key location in completing the dream of a bi-coastal nation, the very British roots of its young colonial and provincial societies, its role in providing raw resources for the rapidly industrializing eastern cities of the late-19th century and the rising prairie farming communities of the early-20th century, its significant share of the present-day national population, and its important contribution to the country's gross domestic product all point to a province very much central to the success of the Canadian federation. It has been in the push and pulls between centre and edge, then, that the geographies of Canada's west coast province have been forged.

The tale of BC's historical-geographical development that I have related in these pages hinges upon successful claims to nature and space in the province. These were claims that were variously backed by displays of military, commercial, social, political, and economic power by claimants. Thus, British officials were able to claim Native lands and resources, private industrialists were able to rapidly and devastatingly exploit the province's bountiful environment in the pursuit of short-term profit, and self-identified members of mainstream 'white' society were able to create geographies of racial exclusion around work places and in public spaces. Each of these claims was advanced by their proponents in opposition to the wishes of other people. As such, it is important to always remain cognizant of the politics of nature and space that accompany the development of any region. In the case of BC, claims to territory rooted in the sovereignty of the Crown are currently under formal challenge by dozens of First Nations across the province; challenges based upon the treaty requirements outlined in the Royal Proclamation of 1763 that went largely unheeded in BC. As these cases are heard and settled, the geography of the province is being re-written. In a very real sense, the geographical change unleashed by colonialism has yet to settle to a conclusion. Likewise, the regime of industrial resource extraction that the inheritors of Britain's west coast colony constructed has come under attack on basis of, beginning in the early-20th century, labour relations (Hak, 1989) and, since the 1970s, ecological impact (Rossiter, 2004). Both of these challenges to long standing uses of nature and space seem likely to produce profound re-workings of the province's geographies. Predicting what these might look like is a difficult task; an optimist might envision a future where Native and newcomer claims to space are reconciled with a resultant society focused upon the balancing of economic (in the broadest sense of the word) and ecological health. Given BC's familiarity with living on the edge, perhaps these seemingly distant goals will be within reach sooner rather than later.

Review Questions (with Websites)

- Where is the highest point in BC?
http://www.attractionscanada.com/britishcolumbia/index.html.
- Where is Black Tusk and how was it formed?
http://gsc.nrcan.gc.ca/urbgeo/vanrock/garibaldi_e.php.
- What is the annual level of precipitation for Prince Rupert, BC?
http://bccommunities.ca/princerupert/index.php.
- What significance does the Raven hold for the Haida Nation?
http://www.haidanation.ca/.
- Where is Barkerville, BC, why was it founded, and what is its function today?
http://www.barkerville.ca/.
- What was the *Komagata Maru* incident?
http://www.vancouverhistory.ca/archives_komagatamaru.htm.
- What was the Canadian Government's policy towards Japanese-born Canadians during World War II?
http://history.cbc.ca/history/?MIval=EpisContent.html&lang=E&series_id=1&episode_id=14&chapter_id=3&page_id=3.
- What is the dominant ethnicity in the City of Richmond, BC?
http://www.richmond.ca/discover/demographics/Census2006.htm.
- In what year did the Nisga'a Treaty become law?
http://www.richmond.ca/discover/demographics/Census2006.htm.
- What are the main details of the climate change policy released by the BC Government in 2008?
http://www.livesmartbc.ca/government/index.html.

References

Bone, Robert (2008), *The regional geography of Canada*, 4th ed., Don Mills: Oxford University Press.

Braun, Bruce (2002), *The intemperate rainforest: nature, culture, and power on Canada's west coast*, Minneapolis: University of Minnesota Press.

British Columbia (1922), *Report of the Forest Branch of the Department of Lands, for the year ending December 31st 1921*, Victoria, William H. Cullin, Printer to the King's Most Excellent Majesty.

Carlson, Keith Thor (ed.) (2001), *A Stó:lö – Coast Salish historical atlas*, Vancouver: Douglas and McIntyre.

Clayton, Daniel (1999), *Islands of truth: the imperial fashioning of Vancouver Island*, Vancouver: University of British Columbia Press.

Clayton, Daniel (2000), "On the colonial genealogy of George Vancouver's chart of the north-west coast of North America", *Ecumene*, 7(4), 371-401.

Fisher, Robin (1977), *Contact and conflict: Indian-European relations in British Columbia, 1774-1890*, Vancouver: University of British Columbia Press.

Godwin, George (1994), *The eternal forest*, Robert S. Thomson (ed.), Vancouver: Godwin Books.

Gough, Barry (2007), *Fortune's a river: the collision of empires in northwest America*, Madeira Park: Harbour Publishing.

Hak, Gordon (1989), "The socialist and labourist impulse in small-town British Columbia: Port Alberni and Prince George, 1911-33", *Canadian Historical Review*, LXX, 4, 519-42.

Hak, Gordon (2000), *Turning trees into dollars: the British Columbia coastal lumber industry, 1858-1913*, Toronto: University of Toronto Press.

Harris, Cole (1997), *The resettlement of British Columbia: essays on colonialism and geographical change*, Vancouver: University of British Columbia Press.

Harris, Cole (2001), *Making native space: colonialism, resistance and reserves in British Columbia*, Vancouver: University of British Columbia Press.

Harris, Cole (2004), "How did colonialism dispossess? Comments from an edge of empire", *Annals of the Association of American Geographers*, 94(1), 165-182.

Hayes, Derek (2005), *Historical atlas of Vancouver and the Lower Fraser Valley*, Vancouver: Douglas and McIntyre.

Innis, Harold. A. (1995), "The Importance of Staples Products in Canadian Development", Daniel Drache (ed.), *Staples, Markets, and Cultural Change*, Montréal: McGill-Queen's Press, 3-23.

Kobayashi, Audrey and Peter, Jackson (1994), "Japanese Canadians and the racialization of labour in the British Columbia sawmill industry", *BC Studies*, 103, Fall, 33-58.

McCann, L.D. and Gunn (eds.) (1987), *Heartland and hinterland: a geography of Canada*, 2nd ed., Scarborough: Prentice-Hall.

McDonald, Robert A.J. (1996), *Making Vancouver: class, status, and social boundaries, 1863-1913*, Vancouver: University of British Columbia Press.

Milton, Viscount and W.B., Cheadle (2001), *The North-West passage by land*, 6th ed., Toronto: Prospero Canadian Collection.

McGillivray, Brett (2000), *Geography of British Columbia: people and landscapes in transition*, Vancouver: University of British Columbia Press.

Douglas, Owram (1992), *The promise of Eden: the Canadian expansionist movement and the idea of the west, 1856-1900*, Toronto: University of Toronto Press.

Rajala, Richard (1998), *Clearcutting the Pacific rainforest: production, science, and regulation*, Vancouver: University of British Columbia Press.

Rossiter, David (2004), "The nature of protest: constructing the spaces of British Columbia's rainforests", *Cultural Geographies*, 11(2), 139-164.

Rossiter, David A. (2007), "Lessons in possession: colonial resource geographies in practice on Vancouver Island, 1859-1965", *Journal of Historical Geography*, 33(4), 770-790.

Stanford, Quentin H. (ed.) (2003), *Canadian Oxford world atlas*, Don Mills: Oxford University Press.

Wynn, Graeme (2007), *Canada and Arctic North America: an environmental history*, Santa Barbara: ABC-CLIO.

CONCLUSION

Canada in the 21st Century

Robert BONE

University of Saskatchewan

Canada in the 21st century remains a country of regions. Yet, Canada and its regions have evolved over time – and they continue to evolve – creating their own sense of regional consciousness as well as reconfiguring their place within Canada and the global economy. In this process of nation-building, regions form a critical building block with compromises between Ottawa and its provinces a necessary element in keeping the federation united.

Canada's size, geography and political structure foster regionalism. While the nature of regional consciousness varies across the country, this sense remains strongest in Québec. Yet, as strong as regional commitments appear, just below the surface of Canadian psyches as expressed in these regions lies a latent national pride that occasional emerges. One such appearance took place during the 2010 Winter Olympics. These games, held in Vancouver, ignited a sense of national pride rarely seen and, at least for that instance, Canada ceased to be a country of regions.

Canada, as with any federal state, is not without internal tensions. Regional disagreements often spring from sensitive issues buried in Canada's past or from geography that has favoured some regions over others, thus creating *have* and *have-not* provinces. Yet, even with these regional inequalities, political solutions are found. While not perfect, these solutions – in reality compromises – reveal the nature of Canadian identity. The federal Equalization Program – as an illustration of political negotiations and compromise – attempts to soften these regional differences by providing annual payments to *have-not* provinces from general tax revenues – most of which are generated by the *have* provinces.

The Changing Nature of Canada's Regions

While the foundation for Canadian regionalism lies anchored in its physical geography and historic origins, contemporary events provide fresh insights into the dynamic nature of these regions. Each region, for example, has reacted differently to the reshaping of Canada's economy in the first decade of the 21st century. Three powerful external events dominated in this reshaping process and they impacted Canadian regions differently. These world events were – and more importantly remain in play as Canada enters the second decade of the 21st century:

- The failure of the United States – Canada's major trade partner – to quickly recover from the 2008/09 global recession.
- The seemingly bottomless demand for resources by China and other emerging countries.
- The hollowing out of North American manufacturing belt.

Regional Reactions

Ontario contains the largest economy and population of Canada's regions. More than other Canadian regions, Ontario is closely integrated into the US economy. Considered the 'province of opportunity' for many decades, Ontario's sudden fall to 'Have-not' province status indicated the dramatic and unanticipated impact of these three events on Ontario. But just how did these events affect Ontario? In recent decades, Ontario prospered in a continental economy, driving largely by its automobile industry. More recently, global events have not been kind to Ontario and much of its manufacturing has suffered from off-shore imports with the net result of a shrinking manufacturing sector, smaller labour force, and a higher unemployment rate. Without a doubt, much of Ontario's success was due to its automobile manufacturing sector which was part of a North American automobile production and sales system. Not surprisingly then, Ontario's decline in manufacturing was related to the fierce competition from foreign producers and growing number of manufacturers either outsourcing or relocating to lower cost countries. A final blow came to Ontario with the global recession of 2008/2009 which saw a near collapse of the Big Three automobile manufacturers. Saved from the brink of bankruptcy by huge government loans, the industry did survive but it has yet to regain its earlier pre-eminence in the Ontario economy and just recently, executives from the Big Three came back to Ottawa to plea for more financial and political support, including retaining the import tax on South Korean automobiles.

Compared to the other regions of Canada, *Québec* stands out because of its historic connections with France and its amazing resilience and capacity to build on this cultural base. As the only region with a French-speaking majority, this precious cultural aspect of Québec represents a key defining element of Canada's identity. As the Auto Pact Agreement set the tone for Ontario's economy, so did the Quiet Revolution for Québec. Under this peaceful revolution, Québec took control of its destiny and, in doing so, ensured the well-being of its culture and language. The role of Hydro-Québec was an essential element in the transformation Québec's economy and society. Hydro-Québec harnessed the vast water resources found in its Canadian Shield, and by doing so not only became a world class hydro-electric company, but also opened opportunities for francophone in the public and private business worlds. In this process, the northern half of Québec was transformed into an industrial landscape where huge dams, diverted rivers and high voltage power lines allowed Québec to provide low cost energy to its businesses and residents as well as to sell 'clean' power to New England. At the same time, its manufacturing sector, like its counterpart in Ontario, was forced to deal with stiff global competition. Its textile manufacturing, for instance, suffered due to stiff competition from countries like China and Mexico where labour costs are much lower than in Québec. On the other hand, by shifting to the high end manufacturing where skilled labour dominates and where the products, such as high speed trains and aircraft, command a place in the global economy, Québec firms found space in the global economy. Culturally, Québec is unique among Canadian regions – its language is French; its politics includes separatist parties; and its daily life exhibits a *joie de vivre* not found in other regions.

Like other regions, *British Columbia* depends heavily on exports of its natural resources to the United States. Like other Canadian regions, the attraction of the US market is understandable – easy access, large market, and high prices. The downside for some natural resources, such as softwood lumber, is that demand evaporates and prices drop sharply when the United States slides into a recession. BC's forest industry provides a perfect example of this dependency. Unlike oil, forest products are not in high demand in the United States. Now that the US financial crisis has left its housing market in shambles, BC forest exports to the United States has sunk from 75% in 2001 to 50% in 2010 (Schrier, 2011). With little prospects for a rebound in the US house construction, Ottawa and Victoria scrambled to find alternative markets and they met with modest success in China due to active promotion of the Chinese market. For example BC forest companies (with federal and provincial funding) provided demonstration projects and training programs for constructing various types of buildings in China. By 2010, these efforts had a pay-off with China importing 22% of BC's compared to almost zero ten year previous (Schrier, 2011). While China is unlikely to replace the US as the major importer of BC forest products, sales to

China, if they keep growing, will provide a modicum of diversification and market security.

By the end of the first decade of the 21st Century, *Western Canada* had become the economic powerhouse of Canada. One measure of its economic success is extraordinarily low unemployment rates. Western Canada's turnaround was largely due to two factors. First, increasing global demand for its natural products, including coal, oil, and potash, caused an expansion in production. Second, prices for these natural resources reached new heights, thereby increasing their profitability and flow of royalties to provinces. Even its agricultural products, such as grain, may have latched on to this rising price platform. Demand for agricultural commodities reached new levels in 2010 causing prices to rise; and prospects for 2011 appear even brighter for Prairie farmers. Within this wide mix of resources, the Oil Sands of Alberta stand out in two ways: as a world-class reserve of oil and as a threat to the environment. This dilemma is exemplified by the Keystone pipeline which currently stretches from the Oil Sands to refineries in the Midwest of the United States. In order to ship bitumen (the oil product from the Oil Sands), this pipeline delivers diluted bitumen to heavy oil refineries in the United States and reduces its energy dependency on Middle East oil. Plans to extend this pipeline to Gulf Coast refineries would solve the American desire for a secure energy supply. Yet, the threat of spills that could pollute valuable ground water sources has complicated matters and delayed the environmental assessment process. While this debate swirls around the corridors of Washington, a decision is expected by 2012. In the meantime, Enbridge, a major pipeline company, is proposing to build the Northern Gateway pipeline from the Oil Sands to the Pacific Coast. In a recent development, the Northern Gateway pipeline project now has a Chinese partner.

Atlantic Canada remains a slow growing area of Canada. Much of its raison d'être was based on its vast fishing stocks, especially the cod fishery. In the 1990s, the collapse of the cod fishery dealt a devastating blow to the backbone of the region's economy. With little prospects of a recovery of cod stocks, many fishing communities have since disappeared and their residents moved to larger towns and cities. About the same time, development of off-shore oil reserves began to provide an alternative economic base for Newfoundland and Labrador. Since 1997, the Hibernia project has transformed Newfoundland and Labrador into a prosperous province, leaving the other three provinces stuck in the dependency category of *have-not* provinces. Even with off-shore oil and gas, Atlantic Canada's economy exhibits restrained advances and, as a result of double digit unemployment figures, remains the best recruitment area for Oil Sands workers. Some moved with their families to

Alberta while other opted to commute to the Oil Sands of Alberta, thus allowing them to remain based in their home region. Known as the Big Commute, perhaps as many as 5% of Atlantic Canada's workforce takes advantages of the much higher wages and guarantee of steady work in Alberta but still keeping their roots and their families in Atlantic Canada (Bone, 2011: 395).

The North, but particularly the Arctic, is feeling the effects of Global Warming. What impacts a warmer Arctic will have on Arctic sovereignty, resource development, and marine transportation and, most importantly, the well-being of the Inuit remains unclear, but these issues are receiving more attention from Ottawa than ever before. The frozen Arctic Ocean has already seen its ice cover diminish in the late summer months and, for a few days, leaving the fabled Northwest Passage ice-free and making travel on ice more hazardous for Inuit hunters. Arctic sovereignty is one consequence of global warming and the questions of ownership of the Northwest Passage and the seabed of the Arctic Ocean has caught the attention of Canada and other Circumpolar Nations. Canada claims that the Northwest Passage is part of its internal waters while the United States and other sea faring nations argue that this is an international waterway connecting two oceans. Negotiations between Canada and the United States over the ownership of the Northwest Passage have begun and urgency to find a solution is pushed by the melting of the Arctic Ice Pack. As well the race for ownership of the Arctic seabed with its promise of vast petroleum reserves is heating up. Canada is well on track to submit its claim based on scientific research with the United Nations in 2013.

Canada, the United States and the Global Community

As Canada enters the second decade of the 21st century, the centre of world geopolitical power has shifted toward the Pacific Rim countries at the expense of Europe and the United States. While Canada escaped remarkably well from the 2008/2009 global crisis, its major trading partner, the United States, did not. In fact, Canada is the only G7 nation to see its economy and employment levels return to pre-recession levels by 2010 (Cross, 2011). Canada's success was due largely to finding new markets for its natural resources and for keeping its financial matters in order. Most new markets were in the Pacific Rim countries led by China. Trade with our continental partner, United States, remains strong, but not as strong as before the 2008/2009 global collapse. Canada/US trade has shrunk from 71% of Canada's total trade in 2005 to 63% in 2009 (Petrova, Rowat and Rapnik, 2010). The reason is simple: the American economy, hobbled by high unemployment, a housing crisis and an enormous deficit, cannot easily dig itself out of its economic hole and, until it does, Canadian exports – save for energy – will be limited.

Forced by circumstances, Canada's economy had witnessed a redirection and this shift toward resources and away from manufactured goods has greatly affected each region of Canada. The western half of the country has gained while the traditional industrial heartland of Canada, Ontario and Québec, has lost. This west/east shift of economic power is based on two factors. First, the seemingly insatiable appetite from countries around the world for the natural resource so plentiful in the western reaches of Canada led by Alberta with its vast oil sands. Second, the downturn in demand for manufactured products especially in Ontario with its heavy emphasis on automobile manufacturing has put a damper on the economies of Central Canada. This sudden and dramatic shift in regional power is reflected in the drop in automobile exports from 25% of total exports in the mid-1990s to 14% in 2010; while oil exports doubled from 10% to 20% (McKenna, 2011). When the United States recovers from its economic malaise, the strong pull of geography with its north/south orientation continues to favour a continental-based economy for North America. Under those circumstance, Ontario and Québec's manufacturing sectors could regain their previous dominate positions among Canada's regions – unless, of course, the price of oil exceeds US $100/barrel, the US economic recovery stalls, and, worst of all, a thicker border win the day.

References

Anonymous (2011), "Global Index", *The Globe and Mail*, Feb. 7: A1.

Bone, Robert (2011), *The Regional Geography of Canada*, Fifth Edition, Toronto: Oxford University Press.

Bone, Robert (2012), *The Canadian North: Issues and Challenges*, Toronto: Oxford University Press.

Cross, Philip (2011), "How did the 2008-2010 recession and recovery compare with previous cycles?", *Canadian Economic Observer*, Statistics Canada, #11-010-X, 13 January. At http://www.statcan.gc.ca/pub/11-010-x/2011001/part-partie3-eng.htm.

Furlong, John and Gary, Mason (2011), *Patriot Hearts: Inside the Olympics that Changed a Country*, Vancouver: Douglas and McIntyre.

McKenna, Barrie (2011), "A country built on crude", *The Globe and Mail*, Feb. 7: B1&4.

Simpson, Jeffrey (2011), "Wake up, Americans, Your economic dream is a nightmare", *The Globe and Mail*, 19 Feb.: A13.

Petrova, Evgeniya, Miles, Rowat and Carlo, Rapnik (2010), *International Merchandise Trade Annual Review, 2009*, Statistics Canada, Report #65-208-X, 6 March, At: http://www.statcan.gc.ca/pub/65-208-x/65-208-x2009000-eng.htm.

Schrier, Dan (2011), "*Exports (BC Origin): 2001-2010*", BC Stats, Feb. At: http://www.bcstats.gov.bc.ca/pubs/exp/exp_ann.pdf.

Contributors

Rémy Tremblay (ed.)

Rémy Tremblay was born in Québec City and graduated from Université Laval with a B.A. and M.A. in Geography. In 2000, he received a PhD in Geography from the University of Ottawa. He was also a postdoctoral Fellow at *Institut National de la Recherche Scientifique-urbanisation, culture et société* in Montréal (INRS-UCS). His work investigates the many sociospatial dimensions of cities, as well as the tourism and migration patterns of Québec residents to Florida. Since 2005, Rémy Tremblay is professor at Télé-université (Université du Québec) in Montréal.

Rémy Tremblay, Télé-université, Montréal
email: tremblay.remy@teluq.ca

Hugues Chicoine (ed., translation Chapter 8)

Hugues Chicoine (M.A., Télé-université) is a research assistant (educational design) and further specialises in IT and learning environments.

Hugues Chicoine, Télé-université, Québec City
email: chicoine.hugues@teluq.ca

Susan Lucas (co-author Introduction)

Susan Lucas is lecturer of Geography at Temple University University of Pennsylvania. She was educated at Nene College (United Kingdom), Indiana University and Wilfrid Laurier University (Canada). She has articles published in The Professional Geographer, Urban Geography and Geographica Helvetica. Dr Lucas' teaching interests are urban geography, urban social geography, population geography and the regional geography of the US and Canada. She currently serves as chair of the Canadian Studies specialty group of the Association of American Geographers.

Susan Lucas, College of Liberal Arts, Temple University
email: susan.lucas@temple.edu

Stefan Freelan (Custom Cartography, chapters 10 and 12)

MS in Geography, BA in Liberal Arts (Environmental Ethics). GIS Specialist and Instructor, Institute for Spatial Information and Analysis, Huxley College of the Environment, Western Washington University. Teaching Introductory GIS and Cartography. Recently published a Map of the Salish Sea Basin as part of the Salish Sea naming effort: http://staff.wwu.edu/stefan/SalishSea.

Stefan Freelan, Huxley College of the Environment
email: stefan@wwu.edu

Peter Meserve (Chapter 1. Canada: Physical Geography, Landscapes and Lithology)

Peter Meserve has been studying Canada since his dissertation on Canadian-American water resource disputes. Over the past ten years he has served as Chair of the Canadian Studies Specialty Group of the AAG and has taught at the U. of Saskatchewan. Currently at Fresno City College, Dr. Meserve wrote the physical geography chapters of The National Geographic's Desk Reference (1999), recently returned from a Fulbright Teacher's Exchange in Poland, and is working on iconic locations of Canadian culture.

Peter H. Meserve, Fresno City College
email: peter.meserve@fresnocitycollege.edu

Chris Mayda (Chapter 2. Rural Canadian Renewable Energy)

A 600-mile pre-doctoral walk across the 49th parallel in 1995 transformed Chris Mayda from an American to a lover of Canada. Her research since that time has focused on the evolution of rural US and Canada, with a special interest in food production. She has maintained contact with the many Canadian and American friends she made on that walk and has learned about the region from the ground up. Her interest in sustainability and the environment inspired her to write a sustainable geography of US and Canada (Rowman and Littlefield), from the ground up, of course. She travelled (aimlessly wandered as her husband says) every state and province in the process talking and learning from those who work in the primary economy. Today Chris is an associate professor of geography and sustainability at Eastern Michigan University.

Chris Mayda, Eastern Michigan University
email: cmayda@emich.edu

Hélène Bélanger (Chapter 3. Population: The Canadian Mosaic. Co-author Chapter 6. Canada: An Urban Nation)

Hélène Bélanger is professor in urban planning/urban studies at the Département d'études urbaines et touristiques of the Université du Québec à Montréal. Her research interests include the socioresidential dynamics of households; the gentrification processes in cities of the North as well as of the South; the perceptions, meanings and appropriation processes of residential spaces. She is currently studying the meanings of public spaces in gentrifying neighbourhoods.

Hélène Bélanger, Université du Québec à Montréal
email: belanger.helene@uqam.ca

Alfred Hecht (Chapter 4. The Changing Canadian Economy: From Resources to Knowledge)

Alfred Hecht is Professor Emeritus and Associate Director of the Viessmann Research Centre on Modern Europe at Laurier. In addition he was the Director of Laurier International and International Relations from 1999 to 2005. He has published more than 85 articles/chapters and books in various local, national and international journals. His graduate degrees are from Clark University (PhD) and the University of Manitoba (MA).

In addition to his above-mentioned administrative positions, he has also been the chair of the Geography and Environmental Department at Wilfrid Laurier University for some 12 years. Furthermore, he also served for a three-year term as the Director for the joint graduate program in Geography between Laurier and the University of Waterloo. Since joining Laurier in 1972 he has taught in the areas of Economic and Urban, Geography and Regional Development. In addition he has spent nearly five years in Europe as either a visiting professor at the universities in Marburg, Berlin, Giessen, Kiel and Hof or as an Alexander von Humboldt Fellow in Germany (28 months) first focusing on regional development in the EU and later on the process on deindustrialization in former East Germany. He has devoted much time to create the VGT (Virtual Geography Texts) on Canada and Germany (http://www.v-g-t.de/). He is a committed internationalist!

Alfred Hecht, Wilfrid Laurier University
email: ahecht@wlu.ca; ahecht@rogers.blackberry.net

Heather Nicol (Chapter 5. Canadian Political Geography for the 21st Century: Navigating the New Landscapes of Continental Division and Integration)

Heather Nicol has been an Associate Professor in the Department of Geography, Trent University, Peterborough, Ontario since 2007, where she teaches courses in Human Geography. Prior to that, she taught at the University of West Georgia in the USA She is the author of several books and articles on Canada in the North, circumpolar geopolitics, the Canada-US border, and Canada-Cuba-US relations.

Heather Nicol, Trent University
email: heathernicol@trentu.ca

Yona Jébrak (co-author Chapter 6. Canada: An Urban Nation)

Yona Jébrak is professor in urban planning/urban studies at the Département d'études urbaines et touristiques, Université du Québec à Montréal. Her research interests include post-disaster urban reconstruction, built heritage, city image and representations. She currently studies urban resilience and the creation of narratives in post-natural disaster cities.

Yona Jébrak, Université du Québec à Montréal
email: jebrak.yona@uqam.ca

Norm Catto (Chapter 7. Atlantic Canada)

Norm Catto is a Professor at the Department of Geography, Memorial University of Newfoundland. His research interests include coastal landforms, natural hazards, sea level change; the impacts of climate and weather events to transportation, fisheries, and communities; fluvial landforms and flood risk assessment; aeolian geomorphology and anthropogenic stresses on dune systems; glacial landforms; loess and palaeosols; and Quaternary history. His research and related teaching has included projects and investigations in environments in Scandinavia, Estonia, Russia, Germany, Serbia, Iceland, Dominican Republic, Argentina, South Korea, USA, and throughout Canada.

Norm Catto, Memorial University
Editor-in-Chief, Quaternary International
email: ncatto@mun.ca

Sheridan Thompson (Chapter 7. Atlantic Canada)

Sheridan Thompson is a final-year student within the Geography Department at Memorial University of Newfoundland. She is currently working on a coastal erosion project at Mistaken Point Ecological Reserve, south eastern shores of the Avalon Peninsula, Newfoundland. Aside from coastal geomorphology Sheridan's research interests include paleoenvironments, and biogeography.

Sheridan Thompson, Memorial University
Masters Candidate, Geography Department
email: r.thompson-graham@mun.ca

Kelly Vodden (Chapter 7. Atlantic Canada)

Kelly Vodden is an Associate Professor (Research) in Environmental Studies at the Grenfell Campus, Memorial University. She is a former Assistant Professor in the Department of Geography at Memorial University and Research Coordinator and Instructor with the Centre for Sustainable Community Development at Simon Fraser University. She has also acted as a consultant to all levels of government, non-government and private sector organizations. Kelly completed her PhD and Master of Arts degrees at Simon Fraser University, British Columbia in the Department of Geography and received her undergraduate degree in Honours Business Administration from the University of Western Ontario. Dr Vodden's research focuses on collaborative governance, sustainable community and regional development, and community involvement in natural resources management, particularly in Canadian rural and small town communities and coastal regions. She is a Principal Investigator in the multi-year research project Canadian regional development: a critical review of theory, practice, and potentials. Her publications include the co-authored book Second Growth: Community Economic Development in Rural and Small Town British Columbia, along with numerous book chapters, journal articles, and reports.

Kelly Vodden, Grenfell Campus, Memorial University
email: kvodden@grenfell.mun.ca
website: http://ruralresilience.ca/?page_id=185

Isabelle Maret (Chapter 8. The Territories of Québec: Geography of a Territory Undergoing Transformation)

Isabelle Maret is Associate Professor at the University of Montreal. Her research focuses on urban vulnerability, community resiliency as well as urban sustainable development. She uses GIS (geographic information systems) to build geodatabases and conduct analysis essential for strategic regional planning. She has focused lately on comparing risk prevention, planning and mitigation in communities located in Quebec and in Louisiana, dealing more specifically with Montreal and New Orleans.

Isabelle Maret, Université de Montréal
email: isabelle.thomas.maret@umontreal.ca

Sandra Breux (Chapter 8. The Territories of Québec: Geography of a Territory Undergoing Transformation)

Sandra Breux is Research Professor at CNRS-USC. Her primary research interests include representations and perceptions of urban territories and their effects on the actions of individuals in society. She has also contributed to research studies on participatory democracy, municipal participation and reorganisation.

Sandra Breux, Institut national de la recherche scientifique, INRS – Urbanisation Culture Société (UCS)
email: sandra.breux@ucs.inrs.ca

Paul Lewis (Chapter 8. The Territories of Québec: Geography of a Territory Undergoing Transformation)

Paul Lewis is professor with the Urban Planning Institute of University of Montréal since August 1993. He is Director of the *Observatoire SITQ du développement urbain et immobilier*. He worked as a planner for the cities of Hull (1977-1979) and Gatineau (1979-1980) as an urban planner, and for Québec's Department of Municipal Affairs (1981-1985) and Québec Council of Universities (1985-1993), as a researcher.

Paul Lewis holds a PhD in Aménagement (Montréal). His thesis (1994) discussed the controls exercised on retail. His work is mainly concerned with transportation planning and retail. He has also worked on the spatial impacts of information technologies. In the last few years, he has completed with colleagues three research projects, on the impact of telework on mobility (for Montréal and Québec), on the possibility of improving mobility with land use planning measures in Montréal, and on the potential for transit and TDM in the Montréal region; he has recently completed a research on active transportation as related to elementary schools in Montréal and Trois-Rivières.

Paul Lewis, Université de Montréal
email: Paul.Lewis@umontreal.ca

Nairne Cameron (Chapter 9. Ontario: The Shield, Suburbs, and Restructuring)

Nairne Cameron is an Assistant Professor of Geography at Algoma University and an Adjunct Assistant Professor in the Centre for Health Promotion Studies and School of Public Health at the University of Alberta. Her teaching and research interests include urban and regional development, health geography, spatial analysis and geographic information systems (GIS), comparative studies, and transportation geography. Recently, her research has focused on spatial patterns of food access and legal restrictions blocking supermarket operations in urban neighbourhoods. Nairne has served as the Food and Health Day Coordinator for the Canadian Association of Geographers Geography Awareness Week. She is presently the Chair of the Association of American Geographers Applied Geography Specialty Group.

Nairne Cameron, Algoma University
email: nairne.cameron@algomau.ca

Douglas C. Munski (Chapter 10. The Prairie Provinces: A "Drive-thru" Region in Another Phase of Geographical Transition)

Douglas C. Munski is a professor and the graduate director in the Department of Geography at the University of North Dakota at Grand Forks. He joined this institution of higher education in the American portion of the Red River Valley of the North as a faculty member in 1978. Since coming to what was part of Rupert's Land, he has been an active member of the Prairie Division of the Canadian Association of Geographers. His participation has including a number of terms on the executive board of that regional division as a member-at-large as well as chairing the annual meeting on five different occasions over the past 30 years. Dr. Munski's duties are principally in teaching, service, and administration coupled with limited responsibilities in the scholarship of teaching and learning. His courses are mainly in cultural/human geography, regional geography (Canada as the specialty), historical geography, and geographic education. He is one of the three faculty members at the University of North Dakota who started in 2007 the Canadian Area Studies undergraduate minor, the first such curriculum within the North Dakota University System.

Douglas C. Munski, University of North Dakota
email: douglas_munski@und.nodak.edu

Gita J. Laidler (Chapter 11. The North: Balancing Tradition and Change)

Gita has been working in northern Canada since 1998 since she was an undergraduate Research Assistant at York University. She earned her Master's of Science at Queen's University in 2002, after conducting research on Boothia Peninsula, in Nunavut. She completed her PhD at the University of Toronto in 2006, focusing on Inuit knowledge of marine environments. Her current research focuses on the local importance of sea ice processes, use, and change based on Inuit expertise. She has been working collaboratively with the Nunavut communities of Pangnirtung, Cape Dorset, and Igloolik since 2003. This work has developed into ongoing, long-term partnerships to explore human-environment interactions and understandings at local and regional scales, facilitated by International Polar Year funding. She has taught several northern-focused courses at the University of Toronto, and Carleton University, and has recently contributed a number of articles to northern and geographic academic journals, as well as chapters in a several different book projects.

Gita J. Laidler, Carleton University
email: gita_laidler@carleton.ca

Andrey N. Petrov (Chapter 11. The North: Balancing Tradition and Change)

Andrey Petrov is Assistant Professor of Geography and Assistant Director of the GeoTREE Center at the University of Northern Iowa, USA. He earned a PhD in Economic Geography and Geographic Information Science from the University of Toronto and PhD in Social Geography from the Herzen University in Russia. Andrey's research interests include economic development strategies and regional differentiation in the Canadian and Russian North, knowledge-based and resource-based economies in remote regions, Indigenous demographics and labour migration in the North, and geospatial techniques in regional analysis.

Dr. Petrov is the PI on the National Science Foundation "Creative Arctic" project – a panarctic study focused on spatial analysis of the creative capital and its economic impact in remote regions of the Circumpolar North. He is also a PI and Co-PI on several other NSF and institutional grants. Dr. Petrov is a collaborator in the Arctic Social Indicators project, an International Polar Year initiative under the auspice of the Arctic Council. Dr. Petrov is a recipient of the 2010 Canadian Association of Geographers Starkey-Robinson Award for his graduate research on the geography of Canada.

Andrey N. Petrov, University of Northern Iowa
Andrey.Petrov@uni.edu

David Rossiter (Chapter 12. British Columbia: Geographies of a Province on the Edge)

David A. Rossiter is Associate Professor of Geography at Huxley College of the Environment, Western Washington University. With research interests in the historical geographies and political ecologies of British Columbia, he has published articles on the environmental discourses of logging protests, the cultural politics of native land claims, and the colonial geographies of industrial forestry. His current research investigates the political ecologies of Canada-US border space and the historical geographies of Vancouver's North Shore.

David Rossiter, Western Washington University
email: david.rossiter@wwu.edu

Robert M. Bone (Conclusion. Canada in the 21st Century)

Dr. Robert M. Bone is a Professor Emeritus of the Department of Geography at the University of Saskatchewan. As a human geographer, his areas of expertise focused on the Canadian North, Circumpolar World and Canada.

When Dr. Bone joined the University of Saskatchewan in 1963, he focused his attention on development issues affecting the Canadian North and its predominately Aboriginal population. He has supervised a number of Master and Doctorate students who have examined various aspect of northern development. Since his retirement, Dr. Bone has continued to serve on M.A. and Ph.D. thesis committees. Three recent examples are: In 2009, he served as the External Examiner on the Ph.D. Examining Committee of Brenda McLeod (University of Manitoba); in 2010, as a member of Andre Legare's PhD oral defense, and, in the same year, as an external examiner of Nicole Hamm, M.A. in Political Science.

His experience and understanding of the geography of northern lands and peoples came from his teaching and research at the University of Saskatchewan from 1963 to 2000; and from serving as Director of the Institute for Northern Studies at the University of Saskatchewan from 1972 to 1982. More recently, he served as Research Advisor to the Inuit Relations Secretariat of Indian and Northern Affairs Canada from 2007 to 2009; in 2008-2009, as an evaluator of 29 essays from graduate students across Canada on the topic, Canada's Role in the Arctic, and then as Chairman of the follow-up March 2009 symposium of nine graduate students where they presented their findings.

Dr. Bone holds a membership in a number of professional institutions, including the Canadian Association of Geographers and the Association of American Geographers. He is a Fellow of the Royal Canadian Geographical Society and the Arctic Institute of North America; and Policy Research Associate of the Aboriginal Policy Research Consortium International based at the University of Western Ontario.

Robert Bone has written over a hundred papers, reports and books on the peoples, resources and demography of the Canadian North. The 5th edition of *The Regional Geography of Canada* was published in 2010 and the 4th edition of *The Canadian North*: *Issues and Challenges* in 2012. In this 4th edition, Dr Bone has prepared a new chapter entitled the *Geopolitics of the Arctic*.

Robert M. Bone, University of Saskatchewan

"Canadian Studies": Series Titles

No.24 – Rémy TREMBLAY & Hugues CHICOINE (eds.), *The Geographies of Canada*, 2013, ISBN 978-2-87574-017-5

N° 23 – Jean-Michel LACROIX et Gordon MACE (dir.), *Politique étrangère comparée : Canada - États-Unis*, 2012, ISBN 978-90-5201-783-9

No.22 – Éric TABUTEAU & Sandrine TOLAZZI (eds./dir.), *"A Safe and Secure Canada". Politique et enjeux sécuritaires au Canada depuis le 11 septembre 2001*, 2011, ISBN 978-90-5201-715-0

No.21 – Héliane DAZIRON-VENTURA & Marta DVOŘÁK (eds.), *Resurgence in Jane Urquhart's Œuvre*, 2010, ISBN 978-90-5201-634-4

No.20 – Pierre ANCTIL, André LOISELLE & Christopher ROLFE (eds./dir.), *Canada Exposed / Le Canada à découvert*, 2009, ISBN 978-90-5201-548-4

No.19 – Gunilla FLORBY, Mark SHACKLETON & Katri SUHONEN (eds.), *Canada: Images of a Post/National Society*, 2009, ISBN 978-90-5201-485-2

No.18 – André MAGORD, *The Quest for Autonomy in Acadia*, 2009, ISBN 978-90-5201-476-0

No.17 – Carmen CONCILIO & Richard J. LANE (eds.), *Image Technologies in Canadian Literature. Narrative, Film, and Photography*, 2009, ISBN 978-90-5201-474-6

No.16 – John-Erik FOSSUM, Johanne POIRIER & Paul MAGNETTE (eds.), *The Ties that Bind. Accommodating Diversity in Canada and the European Union*, 2009, ISBN 978-90-5201-475-3

N° 15 – Pascale VISART DE BOCARMÉ & Pierre PETIT (dir./eds.), *Le « Canada inuit ». Pour une approche réflexive de la recherche anthropologique autochtone / "Inuit Canada". Reflexive Approaches to Native Anthropological Research*, 2008, ISBN 978-90-5201-427-2

N° 14 – Serge JAUMAIN et Nathalie LEMARCHAND (dir.), *Vivre en banlieue. Une comparaison France/Canada*, 2008, ISBN 978-90-5201-415-9

N° 13 – Nick NOVAKOWSKI & Rémy TREMBLAY (eds.), *Perspectives on Ottowa's High-tech Sector*, 2007, ISBN 978-90-5201-370-1

N° 12 – Jean-François DE RAYMOND, *Diplomates écrivains du Canada. Des voix nouvelles*, 2007, ISBN 978-90-5201-346-6

N° 11 – Claire OMHOVÈRE, *Sensing Space. The Poetics of Geography in Contemporary English-Canadian Writing*, 2007, ISBN 978-90-5201-053-3

N° 10 – Caroline ZÉAU, *L'Office national du film et le cinéma canadien (1939-2003). Éloge de la frugalité*, 2006, ISBN 978-90-5201-338-1

N° 9 – Serge JAUMAIN et Paul-André LINTEAU (dir.), *Vivre en ville. Bruxelles et Montréal (XIXe-XXe siècles)*, 2006, ISBN 978-90-5201-334-3

N° 8 – Madeleine FRÉDÉRIC et Serge JAUMAIN (dir.), *Regards croisés sur l'histoire et la littérature acadiennes*, 2006, ISBN 978-90-5201-333-6

N° 7 – Pierre ANCTIL & Zilá BERND (eds./dir.), *Canada from the Outside In. New Trends in Canadian Studies / Le Canada vu d'ailleurs. Nouvelles tendances en études canadiennes*, 2006, ISBN 978-90-5201-041-0

N° 6 – Anne MORELLI et José GOTOVITCH (dir.), *Contester dans un pays prospère. L'extrême gauche en Belgique et au Canada*, 2007, ISBN 978-90-5201-309-1

N° 5 – Britta OLINDER (ed.), *Literary Environments. Canada and the Old World*, 2006, ISBN 978-90-5201-296-4

N° 4 – Madeleine FRÉDÉRIC, *Polyptyque québécois. Découvrir le roman contemporain (1945-2001)*, 2005, ISBN 978-90-5201-096-0

N° 3 – André MAGORD (dir.), *Adaptation et innovation. Expériences acadiennes contemporaines*, 2006, ISBN 978-90-5201-072-4

N° 2 – Robert C. THOMSEN and Nanette L. HALE (eds.), *Canadian Environments. Essays in Culture, Politics and History*, 2005, ISBN 978-90-5201-295-7

N° 1 – Serge JAUMAIN & Éric REMACLE (dir.), *Mémoire de guerre et construction de la paix. Mentalités et choix politiques. Belgique – Europe – Canada*, 2006, ISBN 978-90-5201-266-7
